People & Ideas in Theoretical Computer Science

Springer
Singapore
Berlin
Heidelberg
New York
Barcelona
Budapest
Hong Kong
London
Milan
Paris
Tokyo

People & Ideas in Theoretical Computer Science

edited by

C. S. Calude

Springer

CENTRE FOR DISCRETE MATHEMATICS
AUCKLAND AOTEAROA
& THEORETICAL COMPUTER SCIENCE

Professor C.S. Calude
Department of Computer Science
University of Auckland
Private Bag 92019
Auckland
New Zealand

Library of Congress Cataloging-in-Publication Data

People and Ideas in Theoretical Computer Science / edited by C.S. Calude
p. cm. -- (Springer series in discrete mathematics and theoretical computer science)
Includes bibliographical references
1. Computer science. I. Calude, Cristian, 1952- . II. Series.
QA76.T426 1998
004--dc21 98-28655
CIP

ISBN 981-4021-13-X

Printed in Singapore

Typesetting: Camera-ready by editor
SPIN 10693708 5 4 3 2 1 0

Preface

Two scientific events made computers an engineering reality: the discovery of electrons by J.J. Thomson in 1895 and the craft of mathematical models of computation by A. Turing, E. Post, A. Church and K. Gödel, in the 1930s and 1940s. A. Turing and J. von Neumann, pioneers of theoretical computer science, are widely recognized as founders (among others) of computer science.

Theory and theoreticians have played a major role in computer science. Many insights into the nature of efficient computations were gained and theory was crucial for some of the most celebrated engineering triumphs of computer science (e.g., in compiler design, databases, multitask operating systems, to name just a few). Theoretical computer science functioned also as a communication bridge between computer science and other subjects, notably, mathematics, linguistics, biology. Recently theoretical computer science has been a champion in developing unconventional models of computation (DNA, quantum, reversible).

Theoretical computer science has gained both maturity and prestige, and time has come to have a more "social" picture of its history, state-of-the-art and future. This is the aim of this book: it collects personal accounts and reflections of fourteen scientists who have dedicated themselves to theoretical computer science. Naturally, the contributors have focussed on their specific interests, experiences, and reminiscences. The reader will notice a diversity of styles and approaches. Some papers are technical; other contributions concentrate on events and people, and others on personal reflections. Comments on research strategies, teaching matters, and the future of theoretical computer science alternate with more intimate stories: in fact, nothing related to the craft of the subject (even disputes, controversies, etc.) was excluded.

This collection includes contributions from some of the main architects of theoretical computer science and the picture presented by it is neither complete nor balanced—it is just one among other possible ones. Consequently, the book should not be regarded as the only view or the true story, but as a catalyst for further developments and continuations.

The idea of this book was supported by eminent people. Most scientists invited to contribute to this volume willingly agreed to write an article. Unfortunately, not all were able to commit themselves or, ultimately, to submit articles. Juris Hartmanis, whose involvement was instrumental in the early stages of this project (one of his contributions is the title of the book), was not able to write his paper due to heavy commitments with the NSF. With a sad voice I mention Ron Book, an enthusiastic supporter of the book, who died in 1997 before completing his contribution.

The support, comments and suggestions of contributors, Laci Babai, Greg Chaitin, Martin Davis, Edsger Dijkstra, Joseph Goguen, Helmut Jürgensen, Richard Karp, Solomon Marcus, Yuri Matiyasevich, Hermann Maurer, Grzgorz Rozenberg, Arto Salomaa, Anatol Slissenko, Boris Trakhtenbrot, were essential: once again, a warm thank you. Many other people have been extremely helpful in supporting this book. Special thanks go to Juris Hartmanis, Don Knuth, Anil Nerode, Ludwig Staiger (for advice and suggestions), Noam Chomsky, Tony Hoare, Robin Milner, Maurice Nivat (for encouragement), Mike Dinneen (for being my LaTeX sorcerous adviser), David Congerton and Peter Shields (for computer assistance), Penny Barry, Anita Lai, Elisa Mikkola, and Neena Raniga (for secretarial work).

The book edited by Joseph Gani, *The Craft of Probabilistic Modelling* (Springer-Verlag, New York, 1986), was a great source of inspiration.

I am grateful to Wilfried de Beauclair for granting permission to reproduce photos of early computers (Z22 and X1) from his book *Rechnen mit Maschinen* (Friedr. Vieweg & Sohn, Braunschweig, 1968). The additional pictures included in this book have been kindly sent by Laci Babai, Helmut Jürgensen, Solomon Marcus, Arto Salomaa, and Boris Trakhtenbrot: may I offer them my thanks. E.W. Dijkstra's picture was taken by Studio BM, Arch.nr. 4663-5, Nuenen.

It is a great pleasure to acknowledge the perfect cooperation with Springer-Verlag Singapore team, especially with Rebecca Ali, Gillian Chee and Ian Shelley.

I warmly thank Andreea and Elena for supporting me to cope with difficult circumstances, especially when dealing with recalcitrant software and machines.

C.S. Calude
Auckland
May 1998

Table of Contents

László Babai

László Babai is a professor in the Departments of Computer Science and Mathematics at the University of Chicago. Born in Budapest in 1950, Professor Babai received his degrees in mathematics from Eötvös University, Budapest, and the Hungarian Academy of Science, and he taught at Eötvos University for nearly two decades. He joined the University of Chicago in 1984 but maintained his affiliation with Eötvös University until 1994, splitting his time between the two continents. Professor Babai has served as an editor of 11 periodicals, evenly split between mathematics and computer science. He has written over 150 research papers in combinatorics, group theory, algorithms, and complexity theory, forging many links between these areas. As a "mathematical grandchild of Paul Erdős," he has used probabilistic considerations in all these contexts. He was one of the recipients of the first ACM SIGACT – EATCS Gödel Prize (1993). We quote the citation:

"The 1993 Gödel Prize is shared by two papers (Babai and Moran [BM], Goldwasser, Micali and Rackoff [GMR]). These papers introduced the concept of interactive proof systems, which provides a rich new framework for addressing the question of what constitutes a mathematical proof. The invention of this framework has already led to some of the most exciting developments in complexity theory in recent years, including the discovery of close connections between interactive proof systems and classical complexity classes, and the resolution of several major open problems in combinatorial optimization."

Professor Babai received the Erdős Prize of the Hungarian Academy of Science in 1983. His numerous plenary lectures include the European Congress of Mathematics (Paris, 1992) and the International Congress of Mathematics (Zürich, 1994). In 1990, he was elected a member of the Hungarian Academy of Science. In the past five years he gave lecture series in Canberra, Paris, Rome, Montreal, Bath (U. K.), Jerusalem, and at Penn State on a diverse set of topics, including algorithms in finite groups, interactive proofs, communication complexity, vertex-transitive graphs, and the life and work of Paul Erdős. Babai held the André-Aisenstadt Chair at the Université de Montréal in Autumn 1996. He is the 1996-98 Pólya lecturer of the Mathematical Association of America. He is a cofounder of the highly acclaimed study-abroad program "Budapest Semesters in Mathematics."

The Forbidden Sidetrip

A mathematician by background, I became interested in the theory of computing in the late 70s, several years after completing my doctorate. A key episode of the eventful period of transition, against the background of a bygone bipolar world, is described in this story.

Leningradskiĭ Vokzal[0]

I entered the huge, dimly lit hall with a definite sense of adventure. It was 9:40 p.m. on Friday, November 3, 1978. The place was Moscow, capital of the Soviet empire. I had the feeling that the events of the next day might accelerate a change in the direction of my career, a change that in turn would profoundly affect my future.

After a brisk walk in the autumn chill and drizzle, the dreary hall offered a welcome refuge. I joined one of the three unpromising lines leading to small windows at the distant end of the hall, gateways to joy and fulfillment. My supreme desire, a trip to Leningrad that very night, was apparently shared by hundreds of aspiring souls neatly lined up before me.

My hopes were pinned on the slim chance that I would ever advance to the coveted goal line, and that, once there, I would not be *exposed.*

I was not sure how well I would pass for a Russian. While merely standing in the line, I thought I should be all right. My shoes were arguably more comfortable than what the Brezhnev bureaucracy thought fit for the feet of their faithful citizens, but who would notice? My cap was actually inferior to that of the average local, at least in its capacity to protect against the elements that had beaten Napoleon. I admit my overcoat wasn't exactly from GUM[1], but it was still the product of a "socialist" economy. It was made in my native Hungary, then a "fraternal country" in the Soviet bloc, whose central planners were hardly fettered by an overwhelming sense of duty to heed the consumer's whimsical concerns such as fashion, comfort, or utility. I did not wear my jeans[2], or anything stylish. So I felt quite safe.

My case would become more tenuous, however, right at the *okoshko*, the legendary little window where fates are decided.

My Russian was good enough to conduct a conversation and even to give lectures, thanks to ten years of Russian language classes I had to take at school

[0] The Leningrad Railway Station: located in Moscow, this was the station of departure of trains to Leningrad. After the fall of Communism, that destination regained its old name, Saint Petersburg.

[1] Central shopping mall at the Red Square in Moscow

[2] I bought my first pair of jeans in Vienna the previous summer. Jeans, the products of Western decadence, were not available in the Soviet bloc, except on the black market.

(a splendid demonstration of how effectively guns can spread culture[3]) followed by an adventure-filled semester of study in mathematics in Leningrad, complete with visits to ten out of the fifteen Soviet republics.

Years of Russian studies notwithstanding, my Hungarian accent would give me away the instant I uttered a word. But this, I was hoping, would not necessarily be fatal. Among the many nations that had merrily joined the Soviet family under Stalin's guns (or the Russian empire under the Czar's), some, especially in the Baltic, so detested the Russification of their land and culture that they refused to learn Russian to any decent degree. I knew from experience that my Russian, however poor, was still competitive with that of the average Estonian. And all I needed was to pass for a Soviet citizen, not necessarily a Russian.

But being a Soviet citizen was a necessary pretense. Whether from fraternal Hungary or from the imperialist West, as a foreigner I was not supposed to stand in that line, to mix with the natives. Not in the ticket lines, and not on the train. The attention of the Soviet government to foreigners was painstaking. The offices of the Intourist were there specifically to serve our needs. Hotel? Airline or train ticket? Just go to the Intourist. No lines, no waiting, you are comfortably seated while an agent talks to you in any of several foreign languages, and you get tickets even while Soviet citizens are turned away under the pretext that the flight or the hotel has been fully booked. You will not need to share the crowded railway cars with the locals; this in turn also serves the noble purpose of protecting the virtuous workers of the worker-state from possibly harmful encounters with Western decadence.

There was one little thing, though, that you needed to show at the Intourist office: your passport. I had one, of course, and I also had a permit to be in Moscow. What I did not have, however, was a permit to visit Leningrad, my unforeseen destination. Alas, I was not able to take advantage of this superb, if compulsory, convenience.

Planned research in a planned economy

While waiting in the dull grey line at the "Leningrad" railway station in Moscow, I had ample time to reflect on how easy it would have been, with some foresight, to have a permit for Leningrad in my wallet.

Employed as an assistant professor at Eötvös University, Budapest, I was in Moscow on official scientific exchange pursuant to the scientific cooperation treaty between the Soviet and the Hungarian Ministries of Education. Notably, this fact alone restricted the choice of my host institutions to university departments and excluded the important mathematical research institutes of the USSR because those did not fall under the jurisdiction of the Ministry of Education.

I had applied for the one-month visit 18 months earlier, specifying the institutions to be visited and the names of the host scientists, as well as the scientific subject matters to be explored. Obviously, a responsible scientist would plan

[3] The same teaching tool may have played some role along the way as Shakespeare's tongue gained its universal status...

such things well in advance, rather than haphazardly following a random, if brilliant, new idea that might call for a collaboration with a different group of people at entirely different locations and possibly lead to dramatic breakthroughs extraneous to the meticulously designed multi-year plans.

I thought that this kind of clay-brained bureaucratism was a trademark of the Communist system. How enormous was my astonishment when I learned in recent years, that, following the strict guidelines of the European Union[4], mathematicians all over free Europe, on both sides of the former Iron Curtain, were fervently writing proposals for long-term collaboration between small groups of institutions! If person A has a mathematical problem to discuss with person B, then the colleague next door to A should have a project with the colleague next door to B, right? Post-cold-war Brussels has surely caught up with cold-war-time Moscow's oxymoron of "planned mathematical research[5]."

Having by nature been endowed with the requisite foresight, in May 1977 I set forth a plan calling for spending two weeks at Minsk State University in Belorussia, as Belarus used to be called, and another two at Moscow State University (MGU). At the university in Minsk I had a friend working in graph theory, and more importantly, this plan would give me the opportunity to meet several colleagues working on the interfaces of graph theory and algebra at the Mathematical Institute of the Belorussian Academy of Science, although I could not name that institution among my hosts for the aforementioned reason. As to Moscow, the city used to be a mathematician's Mecca. I had many colleagues at various institutions with whom I had desired to consult; none of them, unfortunately, affiliated with MGU and some of them at institutions altogether forbidden to foreigners.

For the proposal, however, I needed a host at MGU. Through a colleague in Budapest working in universal algebra I contacted professor Skornyakov of MGU, and he generously agreed to be my nominal host. For his kindness he would later be subjected to some unpleasantness. At this point, however, his consent cleared the way for the proposal which was subsequently approved, and

---

[4] Seated in Brussels, the European Union (European super-government) has recently emerged as a major sponsor of European research cooperation. If you want EU money for a "European research conference," EU bureacrats will tell you where you can hold it, they will grade your speakers according to their "degree of Europeanness" (if you invite too many Russians, Israelis, or Americans, your conference is canceled – even Poles, Czechs, Hungarians are not counted favorably, they are on the wrong side of the *EU curtain*), you even have to keep a balance between various West European countries (too many Britons may be a problem), and pay attention to "less developed regions" of Western Europe (Mainz is apparently less developed than Frankfurt). That you would prefer having the 12 best speakers you could get? Sorry, if their geographic distribution is inappropriate, please go elsewhere for funding.

[5] For an insider's view on this subject, I highly recommend "Lemma One," the captivating and hilarious story of a brilliant female math graduate student in Socialist East Germany, embroiled in a controversy over having dangerously *exceeded* the plan of mathematical production at her department with the proof of a famous conjecture ... [Kö]. The author is Helga Königsdorf, herself a bearer of a degree in mathematics.

I was instructed to wait until the host institutions would report their readiness to receive me. How this readiness was being determined remained a mystery; my friends at those places were certainly ready to see me any time.

While waiting for the green light which was supposed to arrive through the Ministry of Education, I received a letter from the New World; professor Bjarne Jónsson, one of the great innovators of universal algebra, invited me to spend a term at Vanderbilt University (Nashville). Anxious to make my debut in the U. S., I accepted, although universal algebra was not my primary interest, and the salary, to be covered by professor Jónsson's grant, was quite meager, considerably less than what I had been offered two years earlier, in 1976, at Syracuse (N. Y.).

The Syracuse visit fell through because, at the last minute, the Hungarian Ministry of Labor vetoed my employment abroad, and, as a consequence, my exit visa was denied[6] by the Ministry of Interior. I never got an official explanation, but it has been suggested that, at 26, I was not deemed mature enough to navigate the ideologically treacherous American terrain.

Thus denied a new learning experience, I nevertheless did have the pleasure of enjoying a decisively new, albeit somewhat different, and ideologically certainly less onerous, course of studies that same autumn, courtesy of another department of the government: in November 1976 I was drafted to the army as a "student" toward the degree of "reserve officer, rank sub-lieutenant." The memorable curriculum included everything from 2 a. m. alarm drills to spending a week inside a cozy A. P. C.[7] I took several unfinished mathematical manuscripts to the barracks in the expectation that there would be ample leisure time in which to complete the papers. My peers would have tolerated such diligence, everyone in our "class" had a university degree. And there was indeed no shortage of leisure time. My morale as a captive, however, was so low I was unable to make better use of the time than to play chess in the beginning and cards near the end. Mercifully, this interlude in my life lasted only five months. I celebrated my successful "graduation" from that institution of higher learning by completing five research papers in a single week, a feat never again to be repeated.

My newly acquired "degree" did little to quell my desire for an extended visit overseas. Professor Jónsson's plan called for a start in Nashville in January 1979. Graph theorist Pavol Hell, then at Rutgers, helped preface the journey by a month in New Jersey, so I had to leave Budapest on November 30, 1978. I thought nothing would ever come of the Minsk–Moscow visit and was busy with the arrangements for my long awaited first trip to the United States when I got the urgent message from the Ministry of Education that Minsk and Moscow

---

[6] Exit visa? Those who grew up in a free country might wonder what on earth that might be. Yes, without an exit visa, we could not *leave* our country. So, taking up U. S. employment *without* Hungarian authorization was not an option, I could not even cross the Hungarian border. Found that border on the map? The country within, indeed a wonderful country with rich history and vibrant culture, occupies an area of the *size of Indiana*. Just a bit claustrophobic for a scientist without an exit visa.

[7] Armored Personnel Carrier.

were ready to receive me. This happened at the end of September, and two weeks later my plane landed at Moscow's Sheremetyevo International Airport.

After my return to Budapest from the USSR, I would have three weeks to ready myself for what turned out to be a 16-month journey which included visits to about two dozen universities in the U. S. and Canada. I was aware that any irregularity during my USSR trip, such as an unauthorized side-trip, if noticed by the Soviet authorities, would cancel my exit visa to the West.

Strangely enough, my visa to the USSR contained very little information other than that I was on scientific exchange. The passport control officer at Sheremetyevo Airport completed my permit to stay in the USSR; he asked me the cities to be visited and the names of the host institutions, and did not require any official document to support my statements. I honestly gave him Minsk and Moscow as my destinations, which he dutifully entered on my permit, certifying the new entries with his rubber stamp.

Little did I know that the most important target of this trip should have been Leningrad. Had I had the presence of mind, I could have gained the option to visit Leningrad by simply naming that city, too, to the Sheremetyevo passport officer. As far as I can tell, there was no risk involved. However, what I had told him was now carved in stone, to change the permissions was virtually impossible.

I had seven hours before I would need to board my flight to Minsk[8] at the domestic Domodedovo airport. I used the time to pay a brief visit to my friend Imre Bárány, a discrete geometer from Budapest, who was working in Moscow at the time. This small detour would later acquire vital significance.

Minsk: a place for joy and sorrow

In Minsk I felt like I was coming home. I made new friends and renewed friendships dating back to my undergraduate semester spent in the Soviet Union. I gave a series of lectures (in Russian) to an interested audience; this was the place where my early work, on automorphisms of graphs, was most appreciated. A combination of algebra and combinatorics was a hallmark of a research group striving under the motherly direction of Regina Iosifovna Tyshkevich, a leading mathematician, the wife of algebraist Dmitri Suprunenko, member of the Belorussian Academy of Science, and mother of group theorist Ira Suprunenko.

I learned a thing or two about freedom of speech, too. Friends would take me to the middle of a great park in town, and when at safe distance from everyone else, they would pour out their souls. The talk was about life in the Soviet Union, in general as well as in mathematics, under the oppressive climate of the progressively rigid Brezhnev regime.

---

[8] Minsk lies slightly off the straight line connecting Budapest to Moscow; the distance of Budapest to Minsk is less than a third of the combined distances of Budapest–Moscow and Moscow–Minsk. Yet, although Minsk was my first destination, I had to enter, as well as leave, the USSR via Moscow. This instance of wastefulness turned out to be unexpectedly helpful a little later.

A few months earlier I had attended the International Congress of Mathematicians in Helsinki. Half of the 30 Soviet invited speakers were "unable to attend." (Eminent Soviet mathematicians apparently tended to have family problems, especially in the month of August. High teaching loads during the summer break may also have been a factor...). Congress participants used a session fully composed by absentee Soviet speakers as a forum of discussion and protest. "Free Sharanski and Massera" buttons were distributed to bring attention to the plight of two scientists who were prisoners of conscience, Sharanski (now a member of the Israeli government) in the Soviet Union, and Massera, a Communist, in South America.

In the hierarchically organized Communist society, every discipline or activity had to have a supreme leader, and the branches of mathematics were no exception. Since combinatorics was a relatively new discipline, not represented by any member of the Academy of Sciences, it was extremely difficult to organize conferences in the area.

L. S. Pontryagin, I. M. Vinogradov, and S. V. Yablonskiĭ had gained supreme power in Soviet mathematical life. Led by the virulently anti-semitic Pontryagin, the troika introduced a crippling bias across the entire spectrum of their activities, ranging from admissions exams at Moscow State University, the leading university of the country, to the awarding of higher degrees in mathematics throughout the USSR, publishing in prestigious journals, and the hiring policy at the Steklov Institute, the elite research institute of mathematics in Moscow then headed by Vinogradov. Emigration to Israel was virtually halted by the government, removing that glimmer of hope for escape.

Pontryagin was instrumental in preventing half the Soviet invited speakers from attending the Helsinki congress. Pontryagin's attention extended to the "struggle against Jewish influence" in secondary school curricula. It is tragic that Lev Semyonovich Pontryagin (1908–1988), the blind genius of 20th century mathematics, should have harbored such abysmal bigotry.

On my first weekend in Minsk, my hosts took me to a gripping World War II memorial. It had been the tactic of the Nazis to retaliate for a guerilla attack in the occupied parts of the USSR by murdering an entire village. They would round up all residents, lock them up in a barn, and then blow up the barn or set it on fire. The symbolic cemetery, set up in 1969 in the village Khatyn'[9] (near Minsk) had hundreds of burial marks, each representing not an individual but a village.

But Minsk was not all sorrow. Curiously, it was Minsk where I first heard "The Köln Concert" by Keith Jarrett and a record by Chico Hamilton. The records that somehow made it across the border were played at the home of a mathematician, jazz fan, and amateur acoustical engineer, played on his homemade stereo.

---

[9] Not to be confused with Katyn (Katin), a small town near Smolensk, the location of the massacre of thousands of Polish army officers at the hands of Stalin's NKVD (Soviet secret police) in 1939–1940. This event is not remembered in Khatyn'; until Gorbachev, the Russian public was not allowed to know about it.

One evening we went to a baroque concert. Another night, Regina Iosifovna took me to a performance of Macbeth which remains memorable for more than the merits of the play. We agreed to meet in the foyer of the theater. While I considered my jeans to be my regular academic attire, I thought it would not be appropriate for the evening, especially in the company of a lady I so respected, so I put on my slacks, a white shirt, and reasonable sweaters. Fortuitously it was my view that wearing a necktie would have been going too far. Mighty was my surprise when I saw Regina under the chandeliers – all dressed in blue denim[10]! Until then, she had not seen me wear anything but jeans, so she thought she would match my dress code....

The performance was splendid, and my unpredictability remained an anecdote Regina would retell even years later.

The mystery theorem

Amidst all the fun and camaraderie, mathematics was the main subject nevertheless. I learned about counting subgroups of the free group, association schemes, ultrametric spaces, automorphisms of trees, and approaches to the traveling salesman problem, to mention a few of the subjects. The Graph Isomorphism Problem, the focus of my interest at the time, was attractive to several colleagues whose work combined algebra, combinatorics, and algorithms.

I reported my unpublished result that the isomorphism problem can be solved in polynomial time for (directed) graphs with simple eigenvalues, linking linear algebra to the problem. The next question would be to extend the result to graphs with bounded multiplicity of eigenvalues, a problem I had no success with as yet. A colleague, who had attended a recent meeting in Odessa, replied that he had heard that a colleague from Leningrad by the name of Dmitriĭ Grigoryev had solved just that problem, at least for *vertex-symmetrical graphs.*

I found the claim somehow incongruous. Symmetry makes isomorphism testing only more difficult, so if someone could solve the problem under such a symmetry assumption, they should also be able to solve it without that assumption. I expressed my doubts about the accuracy of the report and my informer backed down, not being certain of the precise side condition. But he was adamant that the result pertained to the isomorphism of graphs with bounded multiplicity of eigenvalues.

It was clear that in spite of the perplexing side condition, this was an announcement I could not ignore. However, no one in Minsk was sure about the exact claim, not to mention how to prove it. I was beginning to ponder how I could get in touch with Grigoryev. My hosts knew that he worked at LOMI, the Leningrad branch of the Mathematical Institute of the Soviet Academy of Science. To this date, I cannot tell why did we not consider trying to make a telephone call. Such a call would not have enabled us to discuss the result in detail, but it would at least have confirmed what the exact result was.

[10] Denim: the fabric from which jeans are made.

A telegram from Moscow

One evening in Minsk, a young male colleague took me to the restaurant "Zhuravinka" which had a dance floor. The place was crowded; the waiter ordered us to join two young women already seated at a table for four. None of the four of us seemed to mind; that's how I met Lida.

On Sunday, Lida and I took the train and visited nearby Vilnius, the beautiful capital of neighboring Lithuania (then a Soviet republic), where I had acquaintances: students I had met a few months earlier in Krakow, Poland, at an international student camp where I had accompanied a group of students from Budapest. (Not having a permit for Vilnius, I broke the law of the land with this side-trip. Lida bought the tickets, and she did most of the talking on the train.)

As the last day of my visit in Minsk approached, my friends arranged for a farewell party in a restaurant. I asked if I could take Lida along. They agreed, somewhat reluctantly. They asked if I was sure she wasn't a KGB agent. I was pretty sure; even if all her shining sincerity and blushes had been fake (which I could hardly imagine), why was she not allowed to enter my hotel? She told me why I could not visit her family or even see the apartment building where she lived: she felt embarrassed to show me their piece of "Soviet reality."

The atmosphere at the farewell party was generally good, although my colleagues could not entirely put their suspicion about Lida to rest. And at a certain point they called me to a sideroom, out of Lida's hearing range. The news they wanted to tell me, and especially my reaction to it, was not for the ears of the KGB.

The news was that they had just received a cable from Moscow. The International Office of MGU informed them that my visit to Moscow had been canceled; I should spend the remaining two weeks of my Soviet exchange in Minsk. My friends assured me that they would be more than happy to continue to be my hosts.

After a bewildered moment, I made up my mind. Minsk was friendly, but I was ready to explore my segment of the vast Moscow mathematical scene. I had myriad business to conduct there. In my mind, Moscow had been my primary destination all along. I thanked my hosts for their kindness but told them that I would ignore the telegram and would return to Moscow as planned. They sighed and wished me luck.

Lida's feelings were bruised by this secret conference, an obvious expression of mistrust. I could not tell her what it was about.

The ninety-minute flight that took three days

Having possessed a permit to enter Moscow, I had no difficulty boarding Aeroflot flight number 1992 Friday afternoon. Ninety minutes later I landed at the familiar Domodedovo Airport in Moscow.

I had no hotel reservation, however, nor could I make one. The reason was simple; in the hotel, they would ask my passport. Since I was on an official exchange, my host institution was supposed to provide a stamped form requesting the hotel room for me. As I had learned the preceding day, such a stamp was not forthcoming from my purported host institution.

I was in no danger of having to sleep under the bridge, though. My friend Imre Bárány, whom I had visited for a few hours on my first arrival in Moscow, had at that time invited me to stay at his cozy apartment in Moscow, instead of a dull hotel room. I had gladly accepted, not knowing that in the end there would be no other option.

A few days later the Báránys would leave for a vacation, so I had both comfort and privacy.

All this, however, did not resolve a major headache. Although I had the right document to *enter* Moscow, I had no permit to actually *stay* there, even for a single night. That permit was only offered by police through the hotels, upon appropriate documentation of my status. I was not *supposed* to stay with friends; this was an *official visit*, after all.

While unlikely to be caught while in town, my undocumented status would inevitably get me, as well as my hosts, in serious trouble at the time of departure when my papers would be checked carefully. Not only would my transgression likely prevent me from traveling overseas a few weeks later, setting back my career by years, but worst of all, it could get my friends in Minsk in trouble for not reporting my defiance of the order to stay there.

I was hoping that with the help of Professor Skornyakov, I could persuade the International Office of MGU to change their minds; recognizing that I was there anyway, what else could they do, send me outside to freeze?

But it was Friday evening, there was nothing I could do about this during the weekend.

Saturday afternoon I met several colleagues led by Misha Klin, my old friend who was working on permutation groups, association schemes, and related subjects in algebraic combinatorics. The basic framework of my 1981 paper on primitive permutation groups [Ba81] owes some of its roots to Misha's work on maximal subgroups of the symmetric group [Kl].

Misha had largely been responsible for my Minsk connection and thereby for the feasibility of my entire trip. At this time (1978) Misha was at a relatively happier stage of his life; after a decade of isolation in Nikolayev, a military–industrial town near the Black Sea, he now lived in Kaluga, a town 200 kilometers south of Moscow. The "elektrichka" train would take him to Moscow in 3 or 4 hours, and he would make the trip rather frequently to maintain his close collaboration with Igor Faradzhev's group working in constructive and algebraic combinatorics.

While in Nikolayev, a town closed to foreigners, he obtained his Ph. D. under the long distance advisorship of group theorist L. A. Kaluzhnin in Kiev. Subsequently he worked at a university for ship-building engineers in Nikolayev; several of his important papers are buried among engineering papers in the inaccessible publications of that institute ([Kl]). The move to Kaluga, where he taught at the Pedagogical Institute, was just the beginning of his long odyssey. After six years of commute between Moscow and Kaluga, he was at last able to

Fig. 1. With algebraic combinatorist Misha Klin in front of the Opera House in Odessa, U. S. S. R., 1971.

procure a "propiska" (residence permit) for Moscow. It would take another two years in a "communal apartment" before Misha and his wife Inna could move into a tiny Moscow apartment in a concrete-block complex and yet another year before Misha got a decent job at the Moscow Institute for Organic Chemistry. And in 1988, at the age of 42, he was able to make his first trip abroad (to a meeting in Calcutta).

Throughout the years of hardship, Misha remained the long-distance mentor of generations of algebraic combinatorists. Even while he was in Nikolayev, he already had students and close associates in Kiev, Minsk, and Moscow. His energy and determination seemed inexhaustible.

It would take another several years before he could emigrate to Israel, only to replace hardship with insecurity; at 46, he had to face stiff competition in a small country with a great surplus of superb mathematicians. Misha was sustained for years by various grants. But, having overcome his initial shock over the realization that the struggle would never seem to end, he appears at last to have arrived at the happiest stage of his life; his daughter Hana was born in 1997 in Suva, capital of the Fiji Islands, while Misha taught there at the University of the South Pacific. And Misha now has a "secure" temporary position at Ben-Gurion University. "As the old Soviet saying goes, 'Nothing is as permanent as a temporary arrangement'[11]," he reported cheerfully in a 1998 e-mail message.

[11] Hungarian newpapers always referred to the Soviet occupation forces as the "Soviet army units temporarily stationed in our country." – In Hungarian, an apartment

On that late October Saturday in Moscow two decades ago, we had some discussion of algebraic combinatorics and computational group theory (Igor Faradzhev was a leader in that area), then the talk turned to politics, and finally we spent a spirited evening during which my friends would sing an interminable series of songs from the operettas of Hungarian composer Imre Kálmán, whom they hailed as one of the greatest musical geniuses of all time. I stood there in embarrassment, not even recognizing most of the melodies. I was dumbstruck by the recognition that a genre for which I had little appreciation had brought my tiny country fame and honor across the vast land of Russia.

My urgent job Monday morning was to legalize my status in Moscow. I met professor Lev Anatolyevich Skornyakov at the Department of Algebra on the 13th floor of the massive "Stalin baroque" building of MGU. Together, then, we went to the *Inotdyel*, the International Office. The spacious office was dominated by the towering figure of the head of the office, a lady about six inches taller than professor Skornyakov. Hardly had I stuck my head in the door when she greeted me, yelling across the room, "What are you doing here? Did they not tell you to stay in Minsk?" How did she know who I was? I had never seen her.

Poor Lev Anatolyevich never had a chance to utter a word. "Goodbye, both of you," we were told. Now this was rude even by the standards of Soviet officialdom. Outside in the hall professor Skornyakov, intimidated and resigned, advised me to seek help from the Hungarian Embassy. I was grateful for his word of wisdom.

I followed up on his advice and managed to get in touch with Mrs. Kohánka, a consular officer. She understood and was prepared to help. The next day I had my paper with the all-important rubber stamp, and was directed to report at the desk of Hotel "Universitetskaya."

So I did. A young woman, slender and fair, greeted me with a heartwarming smile at the reception. She considered my documents carefully. Very carefully. The rubber-stamp

"Residence permit from (date) until (date)"

lay next to her on the desk. But she would not lift it. Instead, she looked up, deep into my eyes, and asked in a soft voice: "It has taken you *three days* to get here from Minsk?"

I was mortified. The residence permit stamped in my visa at the hotel in Minsk expired on Friday. Again, my stupidity. On arrival in Minsk, I could have asked them to enter the full length of my stay in the Soviet Union. But now, that entry gave me away, a burning testimony to my misdeed.

---

you own (as opposed to rent) is called an "eternal apartment." During the decades of Communism, most city-dwellers resided in apartments rented from the state. In an incident in the 70s that caused many tongue-in-cheek comments in private, a government newspaper reported that the keys to a block of "eternal apartments" were handed to the happy new owners, "officers of the Soviet army units temporarily stationed in Hungary." – Well, the occupation did in fact not last forever. After a mere 47 years, our temporary masters were out.

A kind lady next to my tormentor observed this exchange and took my papers from her junior colleague. Within a minute I was on my way to room No. 1002, key in hand. I was deciphering the new rubber-stamp on my visa:

"Residence permit extended until (date)."

So I will not need to account for the three days. How clever. How thoughtful. Would such an act of mercy be possible in the age of computers, interlinked databases, bar-coded documents, and magnetic strips? I have seen numerous manifestations of a peculiar "civil liberty" in the Soviet Union which did not exist in the West: the "right" to lie to authority. (Privacy–rights warriors seem to be working hard on upholding the Western version of this waning liberty.)

I spent the next couple of days attending lectures and seminars, as well as meeting mathematicians from all over Moscow, including even the "Institute of Steel and Alloys." Among the interesting results I learned was Margulis's lovely explicit construction of small 4-valent graphs without short cycles, addressing a problem of Erdős. Back in Budapest, my friends László Lovász, Vera Sós, András Hajnal, and I were in the process of creating a new journal called *Combinatorica.* I solicited Margulis's paper for our journal. He sent me a greatly expanded version three years later, which I translated into English [Ma82]. This construction was one of Margulis's starting points for his celebrated construction [Ma88] of what have become known as "Ramanujan graphs," a term introduced by Lubotzky, Phillips, and Sarnak who simultaneously discovered the same remarkable class of objects. (Incidentally, the LPS paper was also printed in *Combinatorica* [LPS].)

I had known Margulis from his several earlier visits to Budapest. In Moscow, Margulis worked at the Institute for Problems of Information Transmission (IPPI), one of the bastions of free inquiry in Moscow, not affected by the madness of Pontryagin and company.

Misha Lomonosov, working in combinatorial optimization, was another friend from IPPI. A few years later Misha would join the ranks of "refusniks," those applying for emigration to Israel but whom the Brezhnev government would not allow to leave (which was the case for almost all). The typical consequence of application for emigration was that the applicants would lose their jobs and livelihoods and could only find menial labor. Not quite so at IPPI; when I saw Misha again in Moscow in 1981, he was barred from entering the IPPI building, but he still received his monthly paychecks.

Misha was a different person in 1981. The die having been cast, he was now a free individual. Presumably always a family man, he was enjoying his life at home, in a concrete-block medium-high-rise apartment in one of Moscow's crowded residential districts, yet outside of Soviet society. He and his family were in the process of converting from a previously secular life to orthodox Judaism in preparation for his aliyah in the indefinite future. Misha's mathematical work flourished. He was planning to submit a paper to *Combinatorica;* I would have to smuggle the paper out of the Soviet Union. A false name would be printed as author on the typewritten manuscript in case my suitcase should be examined on exit. (This was my advice, I had done similar service for other friends.)

I felt I had to discuss the manuscript in detail with Misha since communication would be very difficult after I left. So I decided to return to Moscow a day earlier than planned from a conference I attended in Leningrad (legally!). I arrived at Misha's Friday morning; we discussed his paper until sundown, at which time he put down his pen and retreated to prepare for the Sabbath. Shortly afterwards a boy from the neighbors' arrived, and we had a jolly Sabbath dinner, the first I had ever attended. Misha's paper appeared in *Combinatorica* in 1983 [Lo].

The Czar's pencil

Leningrad Railway Station, 11:30 p. m., November 3, 1978.

There were still several trains scheduled to leave before midnight so I hoped to be able to get on one of them. The last train would leave at 11:59 p. m. It is said that the reason for the midnight cutoff was in the law for reimbursement of travelers on official business. The *per diem* rate at the time was 2.60 Rubles (roughly, $ 2.60) if the train left before midnight, and considerably less (!) otherwise.

The timetable has remained stable for decades, even up to the present day. The trains bear odd numbers on their way from Saint Petersburg to Moscow and even numbers on the way back. Train number 1/2, "Krasnaya Strela" (Red Arrow) offers the most comfort, it is the train foreigners would generally use. *Krasnaya Strela* leaves Moscow (now as then) at 11:55 p. m. and arrives in Saint Petersburg at 8:25 in the morning, stopping only once for ten minutes in the town of Bologoye, roughly halfway between the two metropolises.

The Saint Petersburg – Moscow railroad was built between 1846 and 1851 on orders of ironfisted Czar Nicholas I.[12] Called the "Nicholas Road," the route follows a single 400-mile straight line, with just a small jog at Bologoye. According to legend, the Czar himself drew the line on the map, and his otherwise firm hand jiggled a bit halfways. Would an engineer dare to disobey the mighty emperor?

Technological progress came at an enormous human price. The poet Nikolai Alekseevich Nekrasov (1821–1877) described it, ostensibly for the edification of children, in his poem "The Railroad" (1864):

> 'There is a Tsar in the world – without mercy:
> Hunger his name is, my dear.
> . . .
> 'He, too, it was, who drove multitudes hither,
> Crushing the task that he gave:
> In the grim fight to bring life to these deserts
> Many have found here a grave.
>
> 'Straight is the road, and the track is but narrow,

[12] Nicholas I: Czar of the Russian Empire from 1825 until 1855. His reign started with the bloody suppression of the Decembrist revolt.

Poles, rails, and bridges abound on the way,
Everywhere, too, Russian bones lie beside them,
Vanka, how many, my dear, canst thou say?'[13]

On arrival at the Moscow Railway Station in Leningrad, *Krasnaya Strela* used to be greeted by grandiose music thundering from loudspeakers, the kind of music that is supposed to elate your spirits and make you feel that you are part of a great human enterprise, be it the breaking of the sound barrier in aviation or entering the City of the Revolution. It may be no coincidence that A. Glier, the composer of this "Hymn to the Great City," is also known for his movie scores.

An unrecorded contribution to queuing theory

If you stand in a queue, you may progress once in a while for one of two reasons: the person at the head of the queue was served; or someone ahead of you was fed up with waiting and left the queue. You may also regress as people join the queue ahead of you for various reasons (they spot a friend already waiting there, or press their way in front of an easily intimidated person, etc.).

Assuming a patient crowd (few leaving as long as there is a glimmer of hope for progress) and a long enough queue, there will be a point in the queue where progress is no longer expected because the rate at which people join the queue ahead of that point is no less than the rate at which people in the front are served. This point is called the *Dobrányi threshold*, a fundamental invariant of any given queue, whose discovery is attributed to unaccomplished electrical engineer and tavern philosopher Elemér Dobrányi [pron. dob-chun-yee]. I heard about this magnificent concept from my friend József Pelikán[14] on an earlier trip to Moscow; the USSR seemed to provide unlimited inspiration and an inexhaustible source of working models for queuing theorists[15].

---

[13] Translation by J. M. Soskice[Ne]. The actual title in [Ne] is "The Railway." Other translations give the title as "The Railroad."

[14] József Pelikán, two years my senior, was my first teacher of group theory, an area that became one of my main research interests, especially in its combinatorial and algorithmic aspects. Pelikán, then a 12th grade student at Fazekas high school (the same school I attended), a formidable problem solver and a compulsive instructor, invited me to his home for a weekly "group theory class." I treasure the memory of these afternoons which later decisively influenced my choice of research direction.

[15] On my earlier trip to Leningrad I witnessed on several occasions how, in a matter of seconds, a line spanning several blocks would form when a pushcart carrying a big keg of cold *kvass* (Russian beer made from barley, malt, and rye) would drop anchor at a street corner. The sudden emergence of a supply of any other kind of goods, from shoes to groceries, would trigger the formation of similar instant lines. In fact, people would get in the line before even knowing what the line was for. – A highly visible change brought about by the new era of Capitalism is the disappearance of such lines from the post-Soviet scene.

For the purposes of this accounting, the place of the person right in front of the *"okoshko"* counts as position zero. It is to be noted that the Dobrányi threshold increases as service at the head of the queue speeds up; on the other hand, the threshold goes down at slower service rates.

The queue I joined (at its tail, of course) was quite long, and there was lively activity in the middle, so I was curious to assess whether my position exceeded the Dobrányi threshold. I soon decided that it did, but it took me a while to establish the precise value of the Dobrányi threshold, for this determination required keen observation of the events far ahead of me in faint light and through a fog of smoke. Lacking other optical devices, I had to strain my eyes somewhat, but my then-impeccable vision eventually helped to unlock this tantalizing scientific mystery. The exact answer turned out to be zero. No tickets were handed out. None, zilch, period. The *okoshko* in the obscure distance was shut.

I cannot say, though, that no entertainment was provided. Although, as usual in Moscow, the lines were generally quite disciplined, many patrons were well stocked with spirits and as time passed, more and more of them dug into their supplies to get a lift of their spirits. Considerable loud interaction occurred between the lines, mostly friendly, but some rough, too, with occasional scuffles. I hoped to stay out of trouble's way by avoiding eye contact, while trying to observe the performance.

I am found out

The Soviet nation was preparing for the celebration of the 61st anniversary of Lenin's "Great October Revolution" which took place on November 7, 1917 (October 25, 1917, according to the Orthodox calendar). November 7, 1978 was a Tuesday, and Sunday through Wednesday became a four-day national holiday. (November 8 was always added.) The long holiday would be the time for many Muscovites to visit relatives in other towns; the traffic between Moscow and Leningrad was at a peak. I was not particularly surprised that getting on a train should be difficult, and I had anticipated that returning from Leningrad would be no easier.

The trouble with the return trip was that, for a reason not clear to me, round-trip seat reservation was not possible. However, getting stuck in Leningrad was a risk I could not afford. I had to return to Moscow the next night. Lida took a few days off from work in Minsk around the November 7 holidays; she would visit a friend in Moscow. I had to meet her at 8:38 Sunday morning at the Belorussian Railway Station in Moscow. There was no way to change this appointment, no telephone contact, and no alternative dates to choose from.

I had procrastinated the decision about the Leningrad trip to the point that it became even more difficult than it should have been. Finally sometime Friday afternoon I decided that I could not afford to pass up the opportunity to learn first-hand about the "mystery theorem," even though the risks included jeopardizing my visit to the U. S.

Once resolved, I set about securing the return trip. Around 5 p. m. I called my friend Ruvim (Reuven) Gurevich in Leningrad and explained the situation to him. I asked him to reserve a seat for me on a return train for Saturday night.

Then I went about other business. I was interested in two seminars at Moscow State University late in the evening: Professor Rybnikov's on graph theory on the 16th floor at 6 p. m., and Professor Skornyakov's on abelian groups on the 13th floor at 7 p. m. After the seminars I called Ruvim again; he confirmed that he indeed managed to get a return reservation for me. So I rushed back to my hotel, packed up for the trip, and took off for the Leningrad Railway Station.

The very efficient subway system got me rather quickly to the destination. By 9:40 I was standing in the line, entertaining the hope that I might be able to get to the *okoshko* before the last train of the night would depart.

Alas, it was not to be. After I had spent nearly two hours in the line without any progress, the *okoshko* finally did open. "Grazhdanye[16], there are no more seats available on tonight's trains," I seemed to be hearing a strong voice through noise and distance.

My interpretation of the announcement was immediately confirmed by what I saw, the crowd began to disperse. Some cursed, but nobody protested the fact that the authorities failed to make the announcement an hour or two earlier. Within minutes, the hall was virtually deserted.

I was one of a handful of slow-wits who failed to recognize the single prudent course of action. Rather than moving vigorously to catch the last subway, I just hung on, alternately resting my weight on one foot and then on the other, looking to the right, looking to the left, as if waiting for some miracle to happen.

I noticed a man in his 30s, about fifty feet away, doing much the same: looking to the right, looking to the left. Our eyes locked. I was no longer concerned about eye contact.

The man walked up to me.

"You want to go to Leningrad?" he asked (in Russian), revealing remarkable insight at the Leningrad railway station. (Well, my destination could also have been Murmansk, within the Arctic circle, but no other place from this station.)

"Da," I agreed obligingly.

"You a foreigner?" How did he guess? Was my meek "da" enough, or even that wasn't necessary.

"Da," I acknowledged humbly.

Chervonyets sverkhu!

Then came the surprise.

"Chervonyets sverkhu!", he pronounced with a wide gesture.

I was clueless. I knew that something like "chervonnyĭ" had something to do with the color "red," and "sverkhu" meant "up." But how does "red up" fit in this discussion, I could not guess. The man seemed rather encouraging,

[16] Fellow citizens.

so I asked him to explain perhaps in some detail what exactly he had in mind. "Chervonyets" turned out to be the nickname of the (red) ten-ruble bill. Now the ticket, he explained, was 13 rubles. And he would get me a ticket if I paid him a chervonyets on top of the 13.

This sounded like a terrific bargain. The ruble was then officially on par with the dollar (and much less on the black market); imagine an overnight trip of 650 kilometers (400 miles) at 23 dollars. My new acquaintance also volunteered to let me in on his secret; his aunt was working at the ticket booth and they would split the profits.

I agreed to the deal and was wondering how we should go about executing it.

"Just give me the 13 rubles now, I take it to my aunt and I'll return with the ticket. Then you give me the extra ten."

"Sounds great, but what happens if you take the money and I don't see you again," I asked, emboldened by my partner's friendly business style.

In lieu of a reply, he unfastened his wristwatch and handed it over to me. The watch worked, and it seemed to be in rather good condition. I estimated that it might be worth twenty rubles. So I handed over the price of the ticket, and saw the man disappear behind a door at the far end of the hall.

He returned a short while later, ticket in his hand. I returned his watch and gave him the *chervonyets.* He took me by the arm, guided me to the platform and to the designated rail car, and wished me a good trip. It was shortly before midnight.

I boarded my "platskartnyĭ vagon" ("reserved seat rail car"). The car had 54 sleepers (the lower ones were used to serve as 72 seats during the day). The car had no walls to break it into compartments, all passengers were in the same hall. There were stacks of sleepers perpendicular to the train on one side, and an aisle with single sleepers on the other side at the window, parallel to the train.

I found my place in the crowded car and assumed my least inviting expression so as to avoid being engaged into conversation. My place was in the aisle. I was gazing out the window, as if anything of interest could be seen there. It seemed like everybody was talking to everybody. People were eating, drinking, strangers became acquaintances. I could not make my bed until there seemed to be a consensus about the timeliness of that operation. It seemed like an eternity until finally the lights went off and the danger of being approached by a friendly fellow traveler waned.

Slowly the chatter subsided and all I could hear was the rhythmic clacking of the rails. I tried to imagine the straight line we were traveling. It took me a long while to fall asleep.

When I awoke, not well rested, seven hours later, people were getting ready for arrival. Arriving at nearly the same time as the *Krasnaya Strela,* our train, too, appeared to be greeted by Glier's *Hymn.* I did feel some elation, having survived the trip without having to utter a word. Here I was, illegally in the City of the Revolution, and more importantly, in the City of Grigoryev. The chilly breeze on the platform quickly restored me to my senses.

Ruvim, Dima

I had been in Leningrad before. As an undergraduate, I spent a semester at Leningrad State University. I learned the theory of simple algebras in the classes of Professor D. K. Faddeyev, and the rigors of winter in my dorm room where snow collected between the double window-panes.

Waiting in the corridor for Professor Yakovlev's course on Galois theory to begin, I met Ruvim, two years my junior, a short and slightly stooped boy with very thick spectacles over his lively eyes. Ruvim was an ardent problem solver and on our first encounter, he immediately proposed some questions and asked me to reciprocate. He easily engaged me in conversation and was ready to discuss any subject in mathematics. Logic, topology, algebra were our favorite topics of mutual interest. We became friends. A few years later he visited me in Hungary.

Another four years passed, and there I was in Leningrad. Ruvim was there, too, after a detour to a provincial town. After graduation he had been assigned a job as a programmer in the town of Syktyvkar in the Ural mountains. He reported from there that there was no milk and the only vegetables in the store were half rotten cabbage and sometimes potatoes. As for meat, an occasional dried fish was on the menu. He did not like the job and did not like the town, he feared for his fragile health. So he moved back to Leningrad, although he knew, that, having left the track of assigned jobs, he would have difficulty finding employment in Leningrad. He told me that he wanted to emigrate to Israel; he had even written a letter to Brezhnev. That, of course, was of little help.

I called Ruvim from the railway station and took the streetcar to his place. Ruvim lived in a very small apartment with his parents and his sister. I received a warm welcome, took a hot bath and had a fine breakfast (I recall that the menu included pickled mushrooms), and was treated to a number of quotations from Russian playwright Griboyedov (after whom one of the great canals of this "Venice of the North" is named) and other great literary characters. The combination of hygiene, cuisine, and poetry made me feel civilized again after the dirt and the sweat of the trip.

While I took a much-needed nap, Ruvim called the Grigoryevs. He learned that Dima was at work, at LOMI. This was somewhat surprising since it was Saturday, no seminars were scheduled.

It turned out that Dima had some urgent editorial work to do: proofreading a volume he was editing for the publisher "Nauka." The manuscripts had been typed on large special paper to make a camera-ready copy; formulas were inserted by hand, by calligraphers employed for this purpose (invariably women).

Early afternoon we went to LOMI which was located on the bank of the small and peaceful Fontanka canal. Dima was busy checking the beautiful formulas when Ruvim and I showed up in his office.

Dima was very friendly and forthright. I told him that I was interested in his work on the Graph Isomorphism Problem under a restriction on the spectrum of the graph. I told him about my partial results (I had solved the case of simple eigenvalues) and about the rumors I heard in Minsk that compelled me to take this trip. It turned out that my colleagues in Minsk had quoted Dima's result rather accurately; he had solved the problem under the more general assumption of bounded eigenvalue multiplicity, but with the side-condition of vertex-symmetry which I found so unnatural as to make me doubt the result. I expressed my conviction that less symmetry could make the problem only easier, so if his result was indeed correct, it should be possible to remove that extra condition, thereby obtaining a complete and appealing result.

After a couple of hours of discussion I was not only satisfied that Dima's result was indeed correct, but I managed to translate his ideas into "my world" in a way that Dima himself found enlightening. What seemed rather complicated at the beginning now appeared crystal clear. I felt that I had learned something; the *chervonyets* was worth its weight in gold.

It was clear to both of us that I had become the custodian of the problem. Dima's result seemed to be "halfway" to the real thing. I was grateful for the insight gained, and excited about the prospect of possibly finding the missing other "half," by eliminating the symmetry condition. Whether that "half" would be just a routine increment, or itself a new beginning, was yet to be seen. Of course, as with any unfinished work in mathematics, there lurked the logical possiblity that the residual problem was beyond my powers to tackle, but I was not (yet) willing to consider this danger as real.

The first attempt

I returned to Budapest on November 10. Two days later I found a way to overcome the obstacle.

I spent the next few weeks feverishly preparing for my planned year-long journey overseas. My plane landed at JFK on November 30. The afternoon before the transatlantic flight, I gave a seminar lecture in Budapest about the isomorphism test for graphs with bounded eigenvalue multiplicity. The details of the "second half" seemed a bit messy; indeed I was unable to fully reproduce them at the blackboard.

Most disturbing perhaps, I did not feel elated over the solution. It did not seem to make me any wiser.

A couple of months later, in the U.S., I returned to the problem. I realized that I needed to formulate a simplified situation in which to formalize the solution. To this end I considered the problem where the graph under consideration had colored vertices with bounded multiplicity of each color; isomorphisms would preserve colors by definition. The linear algebra was out, but the problem retained the combinatorial difficulty that separated Dima's result from the "real thing."

After a failed attempt to explain the idea to Gary Miller at M. I. T. in March 1979, I had to concede that my "solution" did not work even in this simplified model. The embarrassment was considerable since I had announced the result at several places.

For several weeks I was desperate. I felt that I had to solve a problem under the gun.

Las Vegas in Montreal

Finally, the the life-saving spark came in June 1979 during a conference in Montreal.

I clearly remember the moment of discovery. After a conversation with László Lovász, a friend from my high school and a combinatorist who has produced many of the most original thoughts in the field, I recalled what András Hajnal, the brilliant set theorist, combinatorist, and chess player used to say when confronted with an interesting chess position. "The question is, what would Alekhine[17] move in this situation," Hajnal would proclaim in deep thought. Paraphrasing Hajnal, I asked myself, how Lovász would approach my problem.

This thought was, of course, just as ridiculous as trying to guess Alekhine, yet it seemed to help.

Thanks to seminars conducted by Péter Gács[18] in Budapest in the preceding years, I was already familiar with the basics of the P/NP theory.

Rather than trying to find a polynomial time algorithm, it dawned on me that the right question to ask was to place the problem in $coNP$. Within hours, I saw the basic structure that would accomplish this: coset representatives for a certain chain of subgroups. From here, a *randomized* polynomial time algorithm was only a small step; the group on top of this chain was a direct product of groups of bounded size, therefore uniformly distributed random elements of the group were easy to obtain. A subsequent process of "sifting" down the chain[19] would fill the coset tables with large probability. Since the order of the group on the top was known, this algorithm had a deterministic verification and thus it would never give erroneous output (although with small probability it would report failure).

I found it important to distinguish this type of randomized algorithms from those which can produce (with small probability) an undetected error ("Monte Carlo algorithms") and a few months later (in Vancouver) I coined the term "Las Vegas algorithm" for the error-free variety. In spite of misgivings by American-born computer scientists ("why should Las Vegas be more respectable than Monte Carlo?" [Jo]), slowly the term caught on and is now generally accepted[20].

More importantly, the subgroup chain method fulfilled my expectation that the solution to the simplified case (bounded color-multiplicities) would extend to the problem of bounded eigenvalue multiplicity. The "unnatural condition" from Grigoryev's solution was, finally, removed.

---

[17] Alexander Alekhine (1892–1946): legendary Russian chess player, world champion for 17 years. Noted for the variety and elegance of his attacks.

[18] Gács was influenced by Leonid Levin with whom he worked in Moscow in those years. Later both of them emigrated to the U. S. and found a permanent home at Boston University.

[19] I called this the "tower of groups method." The term "sifting" was subsequently introduced by Furst, Hopcroft, and Luks [FHL].

[20] Some Canadian and European computer scientists did not seem to share Johnson's reservations and picked up the term almost immediately [Me,Br].

The most significant aspect of the "tower of groups" method was that it represented the first application of group theory, albeit on a very elementary level, to the design of a polynomial time algorithm for a problem which did not involve groups in its formulation.

Canada and the 72 questions

I love Canada. The towns in the East which are reminiscent of the Old World, the breathtaking scenery of the West from the Rockies to Vancouver Island, the mathematical culture with which I had so many points of affinity. Canada was the first country I ever visited on the other side of the Iron Curtain[21], by invitation of algebraic graph theorist Gert Sabidussi of the Université de Montréal, in summer 1972 while I was still an undergraduate.

My summer 1979 in Canada was special in many ways, and exceptionally productive. It should not be surprising that to be admitted to the realm which offers all these pleasures, first I had to demonstrate my valor to the Government's satisfaction. Jousting before the royals being out of fashion, the Canadian government kindly devised another type of obstacle course which only the most resolved could successfully complete.

While in Nashville in Spring 1979, I had to subject myself to a rather thorough physical exam, complete with a 16 × 24 inch chest X-ray which I had to mail to Ottawa. The form my examiner had to complete contained 72 questions, probing all my organs and bodily functions, a task of obvious relevance to an eight-month work permit. I learned the English word "embarrassing" from my examiner, a shy but meticulous young female nurse, when it came to checking for *inguinal hernia.*

I was planning to leave Nashville on May 5, briefly visit Atlanta, and then spend some time in New York and Boston before arriving in Montreal at the beginning of June.

Just as I was preparing to leave for Atlanta and was anxiously awaiting the message from Ottawa that I can pick up my visa and work permit in Atlanta, I received a message instead that my urine test had to be repeated. The doctor who had administered the test was on vacation but luckily happened to be in the hospital that day, so I retook the test, the doctor assured me that there was nothing to be found in it (the previous sample had apparently not been a "clean catch finding," a concept I will never forget given the role it played on my road to Computer Science). So I took the new test results along to Atlanta where in addition to visiting Emory University and Georgia Tech, I also had to appear at the Canadian Consulate.

I showed the new finding to the consular officer. He informed me that he could not judge the doctor's finding, the report had to be sent to Ottawa where it would be read. I told him about my schedule, that I had to move on, I could

[21] This was just another irony of the Iron Curtain; Vienna, also on the other side, is about 4 hours by train from my native Budapest. But as a high school kid I had to travel to East Germany to practice my German.

not wait for the papers to return. So the officer wanted to give me a temporary visa valid for 6 weeks. However, I was planning to stay in Canada for 8 weeks, then go to the U. S., and return to Canada at the end of September. "I cannot give you more than one entry because I am bound by law," he declared solemnly. I begged him that I absolutely needed two entries because I had to visit Stanford and Berkeley (I was hoping the names of these places might ring a bell) and then return to Canada with no chance of seeing consulates in the meantime. To which he replied, without blinking: "I cannot give you more than two entries because I am bound by law." I was grateful for this elasticity of the law but I had one more question. I respectfully inquired why the government of Canada was interested in my hemorrhoids, of all things. The answer was as prompt as it was undeniably to the point: "You will have to do research. You will need to sit. If you cannot sit, you cannot fulfill your work obligations."

The temporary permit I finally received was good enough to cover my summer and the reentry to Canada in September. But it fell short of securing further reentries (which I needed for brief visits to the U. S. such as attending FOCS'79 in Puerto Rico). I ended up spending several days at the Canadian Immigration offices in Montreal, Vancouver, and Toronto, but eventually I had to leave Canada, go to the Canadian Consulate in New York, and receive my "not-so-temporary" visa there. By the time I had the right visa, a mere six weeks were left of my tour of Canada.

Montreal to Vancouver

Peter Frankl, a combinatorist with multiple talents[22] and an old friend of mine from Eötvös University, Budapest, was one of the participants in the Montreal conference. Peter and I already had several joint papers in algebraic combinatorics and later we would write a book on "Linear Algebra Methods in Combinatorics" [BFr]. But in Montreal, Peter's principal contribution to my welfare was not strictly mathematical; he introduced me to Maria Klawe, then a junior member of the University of Toronto C. S. faculty.

Meeting Maria was easily the highlight of the Montreal conference; she was one of the most brilliant, and definitely the most energetic, woman I ever met[23].

[22] The author of a large number of path-breaking papers in extremal combinatorics, Frankl is conversant in ten languages (Swahili is his most recent addition), juggles 7 balls, and is a celebrity in his chosen country, Japan, where he regularly appears on TV shows, edits mathematical columns for several magazines, and helps train the Japanese high school team for the International Mathematical Olympiad. He is the author of 11 books in Japanese.

[23] I met Maria before her stellar career took off: founder and manager of the Discrete Mathematics group at IBM Almaden, Head of the Computer Science Department and subsequently Vice President of the University of British Columbia, Vice-Chair of the Board of the Computing Research Association, Chair of the Board of Trustees of the American Mathematical Society, creator of an award-winning educational CD-ROM game, ..., the list is mind-boggling, and just watch what is yet to come.

In no time she arranged that I should appear at the Theory of Computing seminar in Toronto the day after I would leave Montreal. I spent half the preceding night packing and squeezing my stuff (mostly papers) into the car I had bought from a colleague for $200 in Nashville. I hit the road at 2 a. m.

After my head nodded for the third time at the steering wheel, I decided that I had to pull over and take a nap if I am ever to arrive in Toronto. I parked the car facing a brick wall and fell asleep, leaning on the steering wheel. I had a nightmare; I dreamt that I was approaching a wall at full speed. I grabbed the steering wheel, woke up, bumped my head, and saw with terror that right there, two feet in front of me was – the wall! I screamed and braced myself for the imminent crash.

Of course, nothing happened. Now wide awake, I had no difficulty negotiating the rest of the trip. After another nap in the comfort of my hostess's home, I went to the seminar early in the afternoon. I gave a talk on the algorithm I had found two days earlier. Al Borodin and Steve Cook were present, and I think they liked what they heard. Five years later, the opinion from Toronto was critical in my bid for a visiting position at the newly established Computer Science department at the University of Chicago.

The Toronto seminar was the first time I gave a talk in one of the research centers of the Theory of Computing anywhere.

From Toronto I had to move on to Vancouver, for a workshop in "Algebraic Graph Theory" held in July 1979 at Simon Fraser University. After the workshop I had a week of complete solitude in a home with a scenic view of the bay. During those halcyon days I completed the "Las Vegas" paper, along with two papers in group theory. One of the latter two was on a problem of combinatorist Michel Deza on "sharply edge-transitive graphs" (two more coauthors joined in later on[BCDS]); the other was motivated by the complexity of graph isomorphism, but its real significance lay in group theory (it solved a century-old problem on primitive permutation groups), an interesting reversal of the direction of influence [Ba81].

On the road again

My next stop was Stanford (Péter Gács was there at the time). I received much-needed technical assistance from Professor Knuth's secretary in typing, according to prescribed format, my first FOCS manuscript which I had to prepare for an imminent deadline. The paper (joint work with L. Kučera [BK]) considered the average case complexity of graph isomorphism. It was Ron Graham who had pointed out to me half a year earlier the significance of the FOCS and STOC conferences. I had not been aware of these meetings before, and only later did I find out the feverish competitiveness of submissions.

I first met Ron Graham in Hungary in 1969 at a combinatorics conference. Ron, a close friend and collaborator of Paul Erdős in number theory, combinatorics, and discrete geometry, was a frequent visitor to Hungary, attending many of the numerous international conferences organized by Paul Erdős's Hungarian friends. The lucid and entertaining style of Ron's lectures was something I found worth emulating. I was even more affected by a message carried by Ron's lectures: the connection between Erdős's world and the new concepts of polynomial time and NP-completeness. A lecture by Ron on "Multiprocessor scheduling anomalies" [Gr], given in Hungary in the mid seventies, was among the first to make me think about problems of computational complexity.

Paul Erdős's enormous influence on the Theory of Computing is somewhat paradoxical because Erdős himself never showed any interest in the subject. While Erdős often jokingly discouraged his closest associates from "vasting their time" on computational problems, many of his disciples had become major players in complexity theory and algorithms, often using techniques they had developed in the course of work on Erdős's problems[Ba97].

I consider myself a mathematical grandchild of Paul Erdős; most of those who shaped my mathematical interests were Erdős's close collaborators, including my foremost mentor during the college years, number theorist and combinatorist Vera Sós, one of Erdős's closest friends. Over the years, I benefited tremendously from the warm welcome by the large circle of Erdős's friends, among whom Ron was the first non-Hungarian, and one whose continued guidance helped me a great deal in finding my way in the U. S.

My first joint paper with Uncle Paul was on the average case complexity of the graph isomorphism problem [BES]. The paper with Kučera strengthened these results, and, thanks to Ron's advice, accorded me a stage, which would, in the 20 minutes allotted, complete my initiation into the Theory of Computing community.

This ritual would take place in November 1979 in Puerto Rico. Meanwhile back in California that summer I did my best to make smaller-scale introductions. Double misfortune befell me the night before my much-awaited debut at Berkeley; I got a sore throat and locked myself out of my car. The locksmith cost $40, not a small sum while I was making less than $300 a month; the sore throat was left to take its own course. I hope Dick Karp did not catch it; if he did, I hereby offer my belated apologies.

A week later came a combinatorics conference in Arcata, CA, in the middle of the scenic forests of Northern California. I was all awe and admiration as I witnessed Phyllis Chinn, the organizer of the meeting, in her kitchen, with a big spoon over a large boiling pot in one hand, a telephone on her shoulder, and two sweet children occupying the full length of her other arm.

The meeting was attended among others by group theorist Bill Kantor, Ron Graham, and Michel Deza. It was reassuring to see that each of the three manuscripts I had written in Vancouver had found an illustrious reader. Ron acknowledged my terminological invention, "Las Vegas," with a smile of encouragement, and even suggested that I put it in the title of the paper. (The title was, and regrettably remained, "Monte Carlo algorithms....")

After California, I had to drive across the continent once again, the back seat of my increasingly ailing old Javelin filled to the brim with boxes of paper. My Gypsy lifestyle continued: October 1979 in Toronto (complete with a

Fig. 2. At a conference with Ron Graham, around 1980. *Photo: Adrian Bondy.*

heartbreak, nothing accomplished that month), November in Waterloo, and December in Montreal. Meanwhile the "Las Vegas" paper turned into an Université de Montréal tech report [Ba79].

It was at FOCS 1979 in Puerto Rico that I had the occasion to explain the Las Vegas algorithm to a wide audience, although the paper that allowed me to speak was the joint work with Kučera on a different aspect of the Graph Isomorphism problem. Bending the rules somewhat, I spent only half the time on the paper accepted by the program committee; I used the other half to communicate the more recent results. From then on there was no question that I was a member of the Theory of Computing community. John Hopcroft was among the audience; shortly after the meeting he described the "tower of groups" method and the Las Vegas algorithm in a survey paper on "Recent directions in algorithmic research" [Ho].

On New Years day I visited Péter Gács who had meanwhile moved to Rochester, N.Y. From Rochester I drove to my next destination, Columbus, Ohio. Finally, I had a real job with a decent salary; invited by professor Dijen Ray-Chaudhuri, I was teaching at Ohio State University for the Winter quarter 1980. It was there that I first taught a course in complexity theory (in addition to the calculus course I was paid for).

After a rewarding term in Ohio, already an extension of my planned trip to the New World, I returned to Budapest. I could have stayed longer at Ohio State, perhaps indefinitely, but I was determined to return to my native Hungary. Throughout the 16 months I spent in North America, there had never been a shade of doubt in my mind that I would return home, to help the wonderfully nurturing community which had lifted me up and paved my way. It was time for me to repay my debt, helping younger generations.

While in Ohio, I received a letter from Bucknell University in central Pennsylvania that would change my life forever.

The sender was Gene Luks, an algebraist specializing in nilpotent Lie algebras. Gene had decided that for a change, he would pursue his side-interest in computer science, so he spent the fall of 1979 at Cornell studying the theory of algorithms with John Hopcroft. Gene was a quick study indeed. Hopcroft called Gene's attention to the graph isomorphism problem, and on Hopcroft's return from Puerto Rico, he described my Las Vegas algorithm to Gene. Subsequently Gene discovered that my randomized procedure for finding coset representatives can be replaced by a deterministic algorithm; this was the content of his letter of introduction.

That discovery had far reaching consequences and laid the foundations of the polynomial time theory of permutation groups [FHL]. Shortly afterwards, Gene communicated to me his seminal discovery: a polynomial time isomorphism test for trivalent graphs, involving an ingenious application of the *divide-and-conquer* technique.

Before returning to Hungary, I visited Gene at Bucknell. I brought him a copy of a handwritten letter by Peter Cameron containing marvelous new results on permutation groups [Ca81] based on the (by then nearly complete) *classification of finite simple groups.* Within two months, Gene was able to make profound use of the techniques found in Cameron's notes, extending his polynomial time isomorphism test from trivalent graphs to graphs of bounded valence.

I returned to Bucknell for brief visits during the subsequent three years. This was the beginning of our long friendship and prolific collaboration with Gene in the algorithmic theory of finite groups.

Epilogoue

Permutation groups and graph isomorphism

While I procrastinated writing up our result with Dima, David Mount observed that Luks's divide-and-conquer can also be used to obtain a polynomial-time algorithm for isomorphism of graphs with bounded eigenvalue multiplicity. The two proofs finally appeared together in [BGM].

The "sifting" method turned out not to be as new as I had thought; in the context of the stabilizer chain of permutation groups, it had been widely used in computational group theory following the pioneering work of C. C. Sims more than a decade earlier [Si70,Si71]. Three aspects of the "tower of groups" method [Ba79], however, remained novel: first, the use of a different chain of subgroups; second, the application to subcases of the graph isomorphism problem; and third, the complexity analysis.

The complexity aspect was extended by [FHL] to Sims's context (managing permutation groups given by a list of generators), ushering in the era of complexity theory in algorithms for finite groups. This area has combined the paradigms of algorithm design and analysis with a great variety of methods ranging from elementary combinatorial ideas to the deepest results of group theory (cf. [Ba91a]). Much of the fundamental work in this area was done by Gene Luks; I feel fortunate to have had the privilege of a long period of collaboration with him. Some of the highlights are included in the bibliography. Ákos Seress, one of the chief architects of the theory, has also taken made invaluable contributions in turning the most efficient theoretical algorithms into code (using suitable heuristic shortcuts, of course); his programs are now part of the group theory system GAP [Sch+].

Graph Isomorphism is particularly intriguing for its unsettled complexity status. In spite of the considerable success of the group-theoretic approach, Graph Isomorphism is still not known to be solvable in polynomial time. The group-theoretic methods alone do not even yield a better than exponential (c^n) upper bound, where n is the number of vertices. Another input from Leningrad was needed to bring the exponent down to a fractional power of n (ultimately to $\sqrt{n \log n}$). On my 1981 visit to Leningrad I met a remarkable mathematician by the name of Victor Zemlyachenko (characteristically, Misha Klin made the introduction). Victor invented a strikingly elegant combinatorial trick which when combined with Luks's divide-and-conquer, allowed the reduction of the exponent (cf. [ZKT,BL]). The article [Ba95] contains an almost up-to-date survey on Graph Isomorphism.

Matrix groups and Interactive Proofs

Moving in a different direction, in a 1984 joint paper with Endre Szemerédi we considered the membership problem for matrix groups over finite fields, given by a list of generators [BSz].

The algorithmic obstacles to this problem are formidable, so the first goal was to settle the nondeterministic complexity. While a sweet combinatorial lemma puts the membership problem in NP, putting it in $coNP$ seems much harder; we seem to require a conjecture regarding finite simple groups (the "short presentation conjecture"). This conjecture has been verified for all but three of the infinite classes of finite simple groups ("rank-1 twisted groups") [BGKLP], so we can say we have an *almost theorem* that non-membership in matrix groups belongs to NP.

I then wished to complement this *almost theorem* with a genuine *theorem* that the non-membership problem belongs to *almost*-NP. The concept of *almost* NP would involve a marriage of randomization and nondeterminism; the attempt at defining the right combination of these two concepts naturally led to the "Arthur–Merlin hierarchy" of complexity classes [Ba85], the public coin variety of Interactive Proofs [GMR]. The connection of interactive proofs to many problems in algorithmic group theory is explained in [Ba92a].

The concept of interactive proofs, especially its generalization to multiple provers, has quickly led to a huge body of work with striking implications to the seemingly unrelated area of approximate optimization (cf. the survey [Ba92b]).

Curiously, interactive proofs made an important contribution to clarifying the complexity status of Graph Isomorphims (GI); it turns out that GI belongs to the class *coAM*, and therefore GI *cannot be NP-complete,* unless the polynomial time hierarchy collapses [GMW].

I should mention one last feedback loop involving interactive proofs: the results in [Ba92a] rest on a lemma on "local expansion" in groups. Subsequently this lemma, conceived in a nondeterministic context, found many algorithmic applications, to polynomial time and even nearly linear time algorithms [Ba91b,BCFS,BeB].

Ruvim Gurevich (1952–1989)

After living as a *refusnik* for many years, Ruvim Gurevich eventually managed to emigrate to the U. S. in 1985. Already an accomplished logician, he become a graduate student at the University of Illinois at Urbana. He obtained his Ph. D. in 1988 and subsequently joined the mathematics faculty of the University of Wisconsin at Madison.

Earlier that year Ruvim was diagnosed with cancer. The Soviet government allowed his mother to join him in December 1988. Ruvim passed away[24] on October 10, 1989, three days after his 37th birthday.

Maybe someday there will be a reunion of the old friends. If not at LOMI, or POMI, as it is now called, then over the Internet. To interact, we don't need passports, visas, and rubber stamps, we don't have to take the *platskartnyĭ vagon.* (We may need passwords, though.)

We shall miss Ruvim at that virtual reunion.

References

[Ba79] L. Babai: Monte Carlo algorithms in graph isomorphism testing, *Université de Montréal Tech. Rep.* DMS 79-10, 1979 (pp. 42). A transcript available on the web at `www.cs.uchicago.edu/~laci`.

[Ba81] L. Babai: On the order of uniprimitive permutation groups. *Annals of Math.* **113** (1981), 553–568.

[Ba85] L. Babai: Trading group theory for randomness. *17th ACM STOC*, 1985, pp. 421–429.

[Ba90] L. Babai: E-mail and the unexpected power of interaction. *5th IEEE Symp. on Structure in Complexity Theory*, Barcelona 1990, pp. 30–44.

[Ba91a] L. Babai: Computational complexity in finite groups. *In: Proc. Internat. Congress of Mathematicians,* Kyoto 1990, Springer-Verlag, Tokyo 1991, pp.1479–1489.

[Ba91b] L. Babai: Local expansion of vertex-transitive graphs and random generation in finite groups. *23rd ACM STOC,* 1991, pp. 164–174.

[Ba92a] L. Babai: Bounded round interactive proofs in finite groups. *SIAM J. Discr. Math.* **5** (1992), 88–111.

[Ba92b] L. Babai, Transparent proofs and limits to approximation. *In: Proc. First European Congress of Mathematics (1992)*, Vol. I, Birkhäuser Verlag, 1994, pp. 31–91.

[24] See [Gu].

[Ba95] L. Babai: Automorphism groups, isomorphism, reconstruction. Chapter 27 of the *Handbook of Combinatorics*, R. L. Graham, M. Grötschel, L. Lovász, eds., North-Holland - Elsevier, 1995, pp. 1447–1540.

[Ba97] L. Babai: Paul Erdős (1913–1996): His Influence on the Theory of Computing. *29th ACM STOC,* 1997, pp. 383–401.

[BCDS] L. Babai, P. J. Cameron, M. Deza, N. M. Singhi: On sharply edge-transitive permutation groups. *J. Algebra* **73** (1981), 573–585.

[BCFLS] L. Babai, G. Cooperman, L. Finkelstein, E. M. Luks, Á. Seress: Fast Monte-Carlo algorithms for permutation groups. *23rd ACM STOC,* 1991, pp. 90–100.

[BCFS] L. Babai, G. Cooperman, L. Finkelstein, Á. Seress: Nearly linear time algorithms for permutation groups with a small base. *In: Proc. ISSAC'91 (Internat. Symp. on Symbolic and Algebraic Computation)*, Bonn 1991, pp. 200–209.

[BES] L. Babai, P. Erdős, S. M. Selkow: Random graphs isomorphism. *SIAM J. Comp.* **9** (1980), 628-635.

[BFL] L. Babai, L. Fortnow, C. Lund: Nondeterministic exponential time has two-prover interactive protocols. *Comput. Complexity* **1** (1991), 3–40. (preliminary version: FOCS 1990)

[BFr] L. Babai, P. Frankl: *Linear Algebra Methods in Combinatorics, with Applications to Geometry and Computer Science*, book, Preliminary version 2, University of Chicago 1992, pp. 216.

[BGKLP] L. Babai, A. J. Goodman, W. M. Kantor, E. M. Luks, P. P. Pálfy: Short presentations for finite groups. *J. Algebra* **194** (1997), 79-112.

[BGM] L. Babai, D. Yu. Grigor'ev and D. M. Mount: Isomorphism of graphs with bounded eigenvalue multiplicity. *14th ACM STOC*, 1982, pp. 310–324.

[BK] L. Babai, L. Kučera: Canonical labelling of graphs in linear average time. *20th IEEE FOCS*, 1979, pp. 39–46.

[BL] L. Babai, E. M. Luks: Canonical labeling of graphs. *15th ACM STOC*, 1983, pp. 171–183.

[BLS] L. Babai, E.M. Luks, Á. Seress: Permutation groups in NC. *19th ACM STOC*, 1987, pp. 409–420.

[BM] L. Babai, S. Moran: Arthur-Merlin games: a randomized proof system, and a hierarchy of complexity classes, *J. Comp. Sys. Sci.* **36** (1988), 254–276.

[BNS] L. Babai, N. Nisan, M. Szegedy: Multiparty protocols, pseudorandom generators for Logspace, and time-space trade-offs. *J. Comp. Sys. Sci.* **45** (1992), 204–232 (preliminary version: STOC 1989)

[BSz] L. Babai, E. Szemerédi: On the complexity of matrix group problems I. *25th IEEE FOCS*, 1984, pp. 229–240.

[BeB] R. Beals, L. Babai: Las Vegas algorithms for matrix groups. *34th IEEE FOCS*, 1993, pp. 427–436.

[Br] G. Brassard: A time-luck tradeoff in relativized cryptography. *J. Computer and Syst. Sci.* **22** (1981), 280–311.

[Ca81] P. Cameron: Finite permutation groups and finite simple groups. *Bulletin of the London Math. Soc.* **13** (1981), 1–22.

[FHL] M. Furst, J. Hopcroft and E. M. Luks: Polynomial–time algorithms for permutation groups, *21st IEEE FOCS,* 1980, pp. 36–41.

[GMW] O. Goldreich, S. Micali, A. Wigderson: Proofs that yield nothing but their validity and and a methodology of cryptographic protocol design. *27th IEEE FOCS*, 1986, pp. 174–187.

[GMR] S. Goldwasser, S. Micali, and C. Rackoff: The knowledge complexity of interactive proof systems. *SIAM J. Comp.* **18** (1989), 186-208.

[Gr] R. L. Graham: Bounds on multiprocessor timing anomalies. *SIAM J. Appl. Math.* **17** (1969), 416–429.

[Gu] In memoriam - Reuven Gurevich. *ASL Newsletter,* Assoc. Symb. Logic, Nov. 1989.

[Ho] J. Hopcroft: Recent directions in algorithmic research. *In: Theoretical Computer Science*, Proc. Conf. Karlsruhe, Springer Lect. Notes in Comp. Sci. 104, 1981, pp. 123–134.

[Je] M. R. Jerrum: A compact representation for permutation groups. *J. Algorithms* **7** (1986), 60–78.

[Jo] D. S. Johnson: NP-completeness column. *J. Algorithms* **2/4** (1981), 393–405.

[Kl] M. H. Klin: On an infinite series of maximal subgroups of the symmetric groups. (In Russian.) *Memoirs of the Nikolayev Ship Building Institute* **41** (1970), 148–151.

[Kö] Helga Königsdorf: *Meine ungehörigen Träume* ("My unseemly dreams," short stories in German). Aufbau-Verlag, Berlin 1981. "Lemma 1," pp. 16–28.

[Kn] D. E. Knuth: Efficient representation of perm groups. *Combinatorica* **11** (1991) 33–43.

[Lo] M. V. Lomonosov: On the planar integer two-flow problem. *Combinatorica* **3** (1983), 207–218.

[LPS] A. Lubotzky, R. Phillips, P. Sarnak: Ramanujan graphs. *Combinatorica* **8** (1988), 261–278.

[Lu] E. M. Luks: Isomorphism of graphs of bounded valence can be tested in polynomial time. *J. Comp. Syst. Sci.* **25** (1982), 42–65 (preliminary version FOCS 1980)

[Ma82] G. A. Margulis: Explicit construction of graphs without short cycles and low density codes. *Combinatorica* **2** (1982), 71–78.

[Ma88] G. A. Margulis: Explicit group theoretic construction of combinatorial schemes and their application for the construction of expanders and concentrators. (In Russian; English translation exists.) *Problems of Info. Transmission* **24** (1988), 39–46.

[Me] K. Mehlhorn: Las Vegas is better than determinism in VLSI and distributed computing. *14th ACM STOC,* 1982, pp. 330–337.

[Ne] *Poems by Nicholas Nekrassov.* Translated by Juliet M. Soskice. Oxford University Press, 1929.

[Sch+] M. Schönert et. al.: *GAP: Groups, Algorithms, and Programming.* Lehrstuhl D für Mathematik, RWTH Aachen, 1994. `gap@dcs.st-and.ac.uk`

[Si70] C. C. Sims: Computational methods in the study of permutation groups. *In: Computational problems in abstract algebra* (J. Leech, ed.), Pergamon Press, 1970, pp. 169–183.

[Si71] C. C. Sims: Computation with permutation groups. *In: Proc. Symp. Symb. Alg. Manipulation* (S. R. Petrick, ed.), ACM 1971, pp. 23–28.

[ZKT] V. N. Zemlyachenko, N. M. Korneyenko, R. I. Tyshkevich: The graph isomorphism problem (in Russian). *In: Computational Complexity Theory* (D. Yu. Grigoryev, A. O. Slisenko, eds.), Nauka, Leningrad 1982, pp. 83–158.

Gregory C. Chaitin

Professor Gregory Chaitin is at the IBM Watson Research Center in New York. In the mid 1960s, when he was a teenager, he created algorithmic information theory, which combines, among other elements, Shannon's information theory and Turing's theory of computability. In the three decades since then he has been the principal architect of the theory. Among his contributions are the definition of a random sequence via algorithmic incompressibility, and his information-theoretic approach to Gödel's incompleteness theorem. His work on Hilbert's 10th problem has shown that in a sense there is randomness in arithmetic, in other words, that God not only plays dice in quantum mechanics and nonlinear dynamics, but even in elementary number theory. His latest book is *The Limits of Mathematics* (Springer-Verlag).

Elegant Lisp Programs[1]

Call a program "elegant" if no smaller program has the same output. I.e., a LISP S-expression is defined to be elegant if no smaller S-expression has the same value. For any computational task there is at least one elegant program, perhaps more. Nevertheless, we present a Berry paradox proof that it is impossible to prove that any particular large program is elegant. The proof is carried out using a version of LISP designed especially for this purpose. This establishes an extremely concrete and fundamental limitation on the power of formal mathematical reasoning.

[1] Lecture given at DMTCS'96 at 9 am, Thursday 12 December 1996 in Auckland, New Zealand. The lecture was videotaped; this is an edited transcript.

Introduction

Good morning, everyone! I'd like to talk about an old subject and give it a new twist. The subject I want to talk about is from the 1930's. It's Gödel and Turing's incompleteness results in their two famous papers from 1931 and 1936. I want to throw two new things into the stew. I'm going to use an approach more like Turing's than like Gödel's. So algorithm is very important the way I'll do it. But I'm going to throw in a new thing, which is program size—I'm going to look at the size of computer programs. And the other thing is that I'm not going to use Turing machines or lambda calculus or recursive function theory or fixed point theorems. I want to actually write out programs and run them on computers using current techniques that are used in the industry. You know, good software, 1996 vintage. So the idea is to look at some very old ideas from the 1930's revisiting them using the best software technology that we have available now, at this moment. So it's a mixture of extremely philosophical stuff that probably no mathematicians are interested in, because it deals with the limits of mathematics—and on the other hand I want to make it as practical as possible because I want to tell you about actually getting your hands dirty programming this, and getting it to run efficiently and fast on our current computers with current software.

To give you another hint of the difference in viewpoint, Gödel's approach to incompleteness is "This statement is false!" And instead I use an approach based on the Berry paradox, which is "the first positive integer that you can't name in a billion words". Or even better, "the first positive integer that this statement is too small to name". So there is no self-reference. Actually there is, but it's a much weaker kind of self-reference than Gödel needs to use. Also, I'll be using LISP as my programming language, versions of LISP that I have to invent. And a nice thing about LISP is that I can invent a LISP and program it on a new computer in a new programming language in about a week. So it's easy, it's small enough, it's less than a thousand lines of code to do a LISP. So I think that's a nice approach. So you won't see any fixed point theorems, you won't see any recursive function theory or lambda calculus—I want to actually run programs efficiently.

And I won't need to have a statement say of itself that it's false. For a statement to refer to itself you need to use some cleverness, right? I don't need to use cleverness. I only need to have a statement know how big it is. That's the only self-reference I need, in order for it to achieve something that it's too small to achieve. For a statement to know its own size is easy. To put a statement within itself is impossible, right? It doesn't fit! But just to put the size of a program or statement within itself is easy, because that's going to be about $\log N$, it's going to be very small compared to the object. So it's very easy to have an object know its own size and it takes much more cleverness to have an object know itself completely. So this is a rather different viewpoint, it's a much easier self-reference than Gödel's.

Why I Love (Pure) LISP

But let me start telling you why I think LISP should be loved by mathematicians. I think it's the only computer programming language that is mathematically respectable, because it's the only one that I can prove theorems about!

LISP

So why do I love LISP?! Well, the answer is, because it's really set theory, and all mathematicians love set theory!

Set Theory

LISP is just a set theory for computable mathematics rather than for abstract mathematics. Of course in set theory the basic object would be a list, say, of three objects

$$\{A, B, C\}$$

And as a joke one way to explain LISP is to say, well, take all the curly braces and make them into parentheses and take the commas out and make them blanks!

$$(A\ B\ C)$$

Syntactically that will show you what LISP is like. LISP objects are these parenthesized expressions which can be nested to arbitrary depth.

$$(A\ (B\ C)\ 123)$$

And objects are just separated by blanks. So this

$$(A\ (B\ C)\ 123)$$

is a list with three elements, first A, second $(B\ C)$, and third 123. The second element $(B\ C)$ is in turn a list with two elements. So these are just sets of sets. The only difference between a list and a set is that there is a first, a second, a third element, and elements can be repeated. But otherwise this is just a computerized version of set theory. Just like in set theory where you create everything out of sets—in fact, if you're an extremist, you create everything out of the empty set—in LISP this

$$(A\ (B\ C)\ 123)$$

is everything. This is your universal substance. This is the wood out of which you build the world! And it's simultaneously data and programs, they are these objects, which are called symbolic expressions (or S-expressions). And the things you can put inside S-expressions are words or numbers.

Also LISP is very mathematical in that you don't think about time, and you don't think about executing a program and that it does things that change the state of the world. What you think of in LISP is you think a program is an expression and you evaluate the expression giving you a value. But nothing happens! You don't think of time and you don't think of values being assigned to variables, you don't think of goto's. Instead in a LISP expression you define functions, you apply the functions to values, and the final thing you get is a final value. So it's very much a mathematical notion, of expressions giving values.

Now let me give you an example of a LISP program. I can't give you a complete course on LISP—if I had an hour I could! Of a reduced LISP that I've invented. So let me give you an example. Let's take factorial, which is a typical LISP function. By the way, for experts, my LISP is Scheme-ish, it's a LISP that looks a lot like Scheme, but I had to make some changes. So let's define factorial of N.

```
define (fact N)
```

And we're going to say that if N is equal to 1, then it's going to be 1.

```
if = N 1  1
```

Otherwise it's going to be N times factorial of N minus 1.

```
* N (fact - N 1)
```

So the final result, if we put in all the parentheses, is this.

```
(define (fact N)
(if (= N 1)  1
             (* N (fact (- N 1)))))
```

So we're just using Polish prefix notation. For example, this

```
(- N 1)
```

is N minus 1. Okay, so this program defines factorial. We're going to call it `fact` of variable N. If N is equal to 1, then factorial of N is 1. Otherwise it's the product of N times factorial of N minus 1. And then to use it to get 3 factorial, you write

```
(fact 3)
```

and this gives 6 after the definition of factorial has been processed.

```
(fact 3) ---> 6
```

Now actually I don't like to write all these parentheses. LISP programmers haven't heard about parenthesis-free Polish notation! So I just write

```
define (fact N)
if = N 1  1
        * N (fact - N 1)
```

The other parentheses are understood.

Also in theory what we just did was bad, because we defined a function in one S-expression and then we use it in another S-expression. These are two separate S-expressions. That means that the first S-expression is having an effect, it's leaving a definition. You don't want to do that in LISP. In theoretical LISP, an expression has to define within itself all the functions that it needs, and then use them. Because there is no lasting effect of evaluating a LISP expression. But you can define a function locally and then use it. In fact, the only way to get a value to be assigned to a variable is to have it be the argument of a function which binds the value to the variable within the appropriate scope. Anyway, I don't want to get into all the details. Let me just show you the correct one-expression version of our factorial example. It uses let-be-in, which is a three-argument function.

```
let (fact N) if = N 1 1 * N (fact - N 1)
(fact 3)
```

This expands to

```
('lambda (fact) (fact 3)
 'lambda (N) if = N 1 1 * N (fact - N 1)
)
```

or

```
((' (lambda (fact) (fact 3)) )
 (' (lambda (N) (if (= N 1) 1 (* N (fact (- N 1))))) )
)
```

whose value is 6. Here ' is the one-argument quote function, meaning no evaluation occurs, and triples of the form

```
(lambda (arguments) body)
```

are function definitions.

And if I add that `car` gives you the first element of a list, that `cdr` gives you the rest of the list, and that `cons` puts them back together, then you know essentially all of LISP! Oh, I forgot to say that `nil` is another name for the empty list `()`. And there's a way to test if something has elements or not, that's `atom`.

Okay, that's all the time I can devote to LISP. But I think that you can see from these examples that LISP is very pretty, it's very elegant, it's very mathematical, and it's not at all like a normal programming language.

Proving LISP Programs Are Elegant

Now let me give you an incompleteness result. I think it's a very dramatic incompleteness result, if you like LISP, that you can do with LISP. I emphasized that LISP programs are expressions. Now let's define an elegant LISP expression

Elegant LISP Expression

to be a LISP expression with the property that no smaller expression has the same value. So now we're looking at the size of LISP expressions. LISP expressions are written out in characters, and you just take some standard format for writing them out with blanks in the right places, and you ask, "How big is it?" You measure the size in characters. You count the blanks too, you have some standard format, and this gives a natural way to define the size of a LISP expression. And I'll say that a LISP expression is elegant if no smaller expression gives the same value that it does. Okay? By the way, the value of a LISP expression is also a LISP expression. Everything is an S-expression in this world.

So clearly for any LISP object there is a most elegant expression that gives it as its value, and there may even be several. But what if you want to **prove** that the LISP expression that you've got is elegant, that no smaller expression has the same value? Well, the surprising answer is that **you can't prove that!**

Now I'm going to prove this incompleteness result. I'll start with a hand-waving proof, and then I'll tell you the trouble you get into if you try to program out the proof.

The hand-waving proof goes like this. You start like the incompleteness result in Turing's original paper in 1936. You say, let's assume you have a set of axioms and a set of rules of inference which are so formal, so well specified, that there's an algorithm to check if a proof is valid. Then you run through all possible proofs in size order, check which ones are correct, and you get one by one all the theorems—they're in order of the size of the proofs. So you're given this formal axiomatic system

$$\text{FAS}$$

and you start running through all possible proofs and getting all the theorems.

$$\text{FAS} \longrightarrow \text{Theorems}$$

And I'll simplify the formal axiomatic system because I'm only interested in theorems which give elegant LISP expressions, where you prove that a particular S-expression is elegant. So I'll think of a formal axiomatic system as a computation which starts running and every now and then it throws out a LISP expression that it claims it's demonstrated is elegant. So it's just a black box that every now and then outputs an expression that's elegant, that it's demonstrated is elegant. Okay?

So you start doing this and you just keep going until you find a LISP expression that's elegant, that you've proved is elegant, but it's much more complicated than the formal axiomatic system. And then I'll show you that you get into trouble, you get a contradiction.

Oh, I forgot to say that at this point the formal axiomatic system is in the form of a LISP expression.

$$\text{FAS (S-expression)} \longrightarrow \text{Theorems}$$

I hadn't told you this before. LISP is a nice language for doing things like formal axiomatic systems, because it's a symbolic language, but I'll have to explain in more detail later how this works.

Now let me start over and be more precise. The way this proof goes, is you're going to have a large LISP expression which somewhere in it is going to have the formal axiomatic system, contained within it.

((formal axiomatic system))

The proof of this incompleteness result consists of exhibiting this large LISP expression which in fact is going to be exactly 410 characters of LISP bigger than the formal axiomatic system that it contains. So you put the formal axiomatic system that you want to show has limitations in the right place in this big expression. And there are 410 additional characters of LISP programming that I wrap around the formal axiomatic system. What are these 410 additional characters for?

(410 characters (formal axiomatic system))

What this large LISP expression does, is it starts running the formal axiomatic system, getting the theorems that it produces, which are elegant LISP expressions, until it finds an elegant LISP expression that is larger than it is. How does this large LISP expression know its own size? It gets its own size by adding 410 to the size of the formal axiomatic system that it was given; 410 is just a constant embedded in the large expression. So the large LISP expression

(410 characters (formal axiomatic system))

takes the formal axiomatic system and determines its size (we provide a built-in function for doing that), adds 410 characters to that, which happens to be the number of characters in the wrapping for the formal axiomatic system, and at that point this expression

(410 characters (formal axiomatic system))

knows its own size exactly. Then it starts running the formal axiomatic system looking for the first elegant LISP expression that is larger than it is. Once it finds this elegant LISP expression, it runs it to get the value of the elegant expression, and then it returns this value as its own final value. So the value of this big LISP expression

(410 characters (formal axiomatic system))

is the same as the value of an elegant LISP expression which is larger than it is. But that's impossible! This contradicts the definition of elegance, because this

(410 characters (formal axiomatic system))

large LISP expression is at least one character too small to produce that value.

In other words, we have a LISP expression which is 410 characters larger than the formal axiomatic system that it contains. It's given a formal axiomatic system, it measures its size and adds 410 to that, which happens to be the right way to calculate the exact size of the entire LISP expression. Then it starts running the formal axiomatic system searching for a proof that some LISP expression is elegant that's larger than this

(410 characters (formal axiomatic system))

whole thing is. And once it finds this elegant LISP expression, it runs it, and produces as value of this

(410 characters (formal axiomatic system))

expression the value of that elegant LISP expression. But this

(410 characters (formal axiomatic system))

is too small an expression to produce that value! That's the whole point! So either the formal axiomatic system was lying, and produced a false theorem, or in fact this

(410 characters (formal axiomatic system))

won't work, because it will never find the elegant LISP expression that it's searching for, it will never find an elegant LISP expression larger than it is, it will never find an elegant LISP expression that's more than 410 characters bigger than the formal axiomatic system that it's using.

So we've gotten an upper bound on the size of provably elegant LISP expressions. The upper bound is this: A formal axiomatic system whose LISP complexity is N cannot prove that a LISP expression is elegant if the expression's size is greater than $N + 410$. So at most finitely many LISP expressions can be shown to be elegant.

That's great, but I should emphasize that this overview of the proof sweeps a lot of programming problems under the rug! To get the 410 characters of LISP that I need to make the above proof work, I have to add some things to normal LISP. So now I'm going to tell you about these programming problems.

What We Have to Add to LISP

As I said, there are some problems programming all this. Normal LISP really isn't good enough. But any other programming language would be even worse! I had a version of this proof in 1970, and in words you can explain the idea; it's very simple. But let's say you want to actually program this out and run it on a computer, on an example, and check that it works. Well the answer is, no existing programming language is really adequate for the task. And I really want to run this on a computer. I'm a computer programmer, I earned a living as a computer programmer for many years! I think LISP is almost the right language. But it's still not quite right. So I had to take the heart of a normal LISP, pure LISP, LISP with no side-effects, and add a few things to it, to make things work.

The main thing that I added to LISP is this. Normal LISP is based on a function called `eval`.

```
eval
```

`Eval` is the LISP universal Turing machine, it's the LISP interpreter. LISP is not a compiled language, it's an interpreted language. So the LISP interpreter is always present while a LISP program is running. And since you have the interpreter there all the time, a LISP program can create a LISP program and then immediately run it. In a normal programming language, you have to compile a program before you can run it. But in LISP it works seamlessly, you just use `eval`.

So the LISP universal Turing machine is called `eval`, and it's built in, it's a primitive function that's provided for free. You could program it out in LISP, just like Turing programmed out his universal Turing machine. But in fact you're just given this

```
eval
```

as a built-in function. Unfortunately, this is not the right built-in function for my incompleteness proof. I need a time-limited `eval` that I call `try`.

```
try
```

Recall that in LISP notation

```
(f x y)
```

means just what

$$f(x, y)$$

means in normal mathematical notation, it's the function f applied to the arguments x and y. Now let me explain what `try` does. Here's how you use `try`. You give it a time limit and a LISP expression.

```
(try time-limit lisp-expression)
```

It's a way to try to evaluate the given expression for a limited amount of time, just in case the LISP expression goes on forever and never returns a final value.

Why do I need this? Well, in my proof I've got a formal axiomatic system, and it goes on forever producing theorems, it never stops. So `eval` would be no good. If somebody gives you a formal axiomatic system and you run it using `eval`, you never get anything back, it just goes on forever. So what I need is a time-limited `eval`, a way to run the formal axiomatic system for a certain amount of time and see which theorems show up before the time runs out. And then I'm going to loop and run the formal axiomatic system for more and more time, until I find the theorem that I'm looking for.

If you read the source code for a LISP interpreter, in a low-level language like C or, God forbid, machine language, well, it's just `eval`, that's all the interpreter is. `Eval` is constantly calling itself recursively. My interpreter isn't based on `eval`, it's based on `try` instead. `Try` plays the same role that `eval` does in a normal LISP interpreter.

So we have a formal axiomatic system, and we try running it like this

```
(try time-limit formal-axiomatic-system)
```

and then we gradually increase the time limit while we examine the theorems produced by the formal axiomatic system. How does `try` give us the information that we need to do this? More generally, what is the value of the following try?

```
(try time-limit lisp-expression)
```

`Try` always returns a value, it never gets stuck in an infinite loop. In fact, `try` always returns a triple of the following form

```
(success/failure value/out-of-time captured-intermediate-results)
```

If we're trying a formal axiomatic system, this triple will be

```
(failure out-of-time theorems)
```

`Success` means that the try was a success because the evaluation completed. `Failure` means that the evaluation did not complete. If the try was a success, then the second element will be the value of the LISP expression that was being evaluated. If not, it will indicate here that the evaluation ran out of time. And the third element will always be a list containing all the output, all the intermediate results, produced during the evaluation. In the case of a formal axiomatic system this will be a list of theorems. In fact, with the formal axiomatic systems that we considered before, it will be a list of elegant LISP expressions.

So `try` provides a way of handling infinite computations that output intermediate results instead of having a final value. It's the way that I deal with formal axiomatic systems in LISP. `Try` captures all the intermediate results, it gives us all the theorems. And the 410-character wrapping in my proof that you can't prove that large S-expressions are elegant uses `try` to run the formal axiomatic system for longer and longer amounts of time, until it finds an elegant LISP expression that's bigger than the formal axiomatic system and its wrapping. If such an elegant LISP expression is found, then it uses `eval`, which is just a `try` with no time bound, to get the value of the elegant LISP expression. That's the final value that

(410 characters (formal axiomatic system))

returns, and that's how we get the contradiction that proves my incompleteness theorem!

Okay, this is straightforward. It's a simple proof. The idea is simple, but it took me a quarter of a century to do the programming! It wasn't easy to come up with the LISP expression that proves this incompleteness result. But now that I've done the work, we're rewarded with a very sharp incompleteness result. To prove that an N-character LISP expression is elegant you need a formal axiomatic system whose LISP complexity is at least $N - 410$. Before, all we had here was $N - c$, not $N - 410$, and we had no idea how big c might be.

Discussion

So this is fairly straightforward, it's a very simple proof! And I have two claims about this piece of work that I want to discuss with you. First of all, I claim that **this is a very fundamental incompleteness result!** Secondly, I'm going to try to convince you that **LISP is beautiful!**

Why is this a very fundamental incompleteness result? The game in incompleteness results is to try to state the most natural problem, and then show that you can't do it, to shock people! You want to shock people as much as possible! Now there are many ways to shock people—this is the best that I can do. What's so shocking here? Well, the notion of an elegant LISP expression is very straightforward. There are lots of them out there! An infinity of them! But you can only prove that finitely many LISP expressions are elegant, unless you change the rules of the game by changing the formal axiomatic system. And you can't prove that a LISP expression is elegant if it's more than 410 characters bigger than the LISP implementation of the axioms and rules of inference that you're using to prove that LISP expressions are elegant.

I hope that computer scientists will find this shocking. After all, LISP expressions are very natural objects. Although the notion of an elegant LISP expression has no practical significance, it's not too farfetched, and it has a straightforward mathematical definition.

Of course, computer programmers don't usually want an elegant program, they want a program that works, that they can get running as fast as possible. Their boss wants the programs to be understandable in case a programmer quits and gets a better job elsewhere. And sometimes elegant programs are cryptic and hard to understand. Nevertheless, as a sport programmers sometimes try to outdo each other in the compact cleverness of their programs. The notion of elegance is not entirely foreign to the spirit, to the ethos of computer programming, to computer programming as a sport or as an art. Even though this may not be the way that a company that pays programmers wants them to do things!

And another good thing about this incompleteness result is that it's easy enough to understand what it is that you can't do, that it may be interesting that you can't do it!

Now another thing that's interesting about this is that in 1970 I had essentially this proof. In words you can explain it very easily. The novelty here is that I've taken the trouble to actually program this out on a real computer in a real computer programming language. And in spite of the fact that the ideas are simple, one could never really do this before. So the other message that I have for you theoreticians is that LISP, or some version of it, because I had to invent one, LISP really is beautiful from a mathematical point of view. I view LISP as the set theory of computational mathematics. And if I were in a university, which I'm not, and I wanted to give a first course in theoretical computer science, this theorem about elegant LISP expressions would be the very first thing that I would give the students. In fact, I would assume that the students knew no computer programming at all. I would give them LISP, a toy version, an elegant version, the heart of LISP, as their first programming language. And then I would hit them over the head with this incompleteness result! That would be my approach. Yes?

Question. Do you think you'd get tenure?!

Answer. No! You see why I'm not at a university!

Okay, unfortunately normal LISP wasn't quite good enough. I admit though that I made some changes just for the fun of it, because it's so easy to do a LISP, that every time anyone does a LISP, they always "roll their own" version. That's the problem with creativity! You can't stop it! That's why I've made many, many different LISP dialects. But in addition to the changes that I made mostly for the fun of it, I did have to invent a way for LISP expressions to produce an infinite amount of output. And I had to change `eval` into `try` in order to be able to carry out my incompleteness proof elegantly.

Now you may object that there was no reason to add `try`, and that I could have defined `try` in LISP, without adding any new primitive functions to LISP. Well, that's true, but it would be a gruesome piece of work. Though LISP programmers do love to show that you can do LISP in LISP, that you can program `eval` as a LISP function, that LISP is powerful enough to easily express its own semantics. But why should one program `eval` in LISP, when the interpreter itself is `eval`?! You're doing the same work twice! So I don't think that it is cheating to provide `eval/try` as a primitive function. It's not substantially more work to do my LISP built around `try` than to do a normal LISP built around `eval`, and this makes my proof run much, much faster than if `try` were programmed in LISP. So partly I do this for programming convenience, partly for execution speed, and partly because I'm trying to understand what are the right primitive functions, the right fundamental notions, for doing metamathematics.

By the way, let me mention that I originally wrote my LISP in Mathematica, because it's the most powerful programming language I know. The LISP interpreter is about three-hundred lines of Mathematica code. Then I redid it in C, and it's a thousand lines of C, and the program is incomprehensible, which means that I'm a good C programmer! The C version of the interpreter runs a hundred times faster than the Mathematica version, but the program is completely incomprehensible. You can find this software and my course on the limits of mathematics, from which my result on elegant LISP expressions is taken, you can find all this in my web site at

`http://www.cs.auckland.ac.nz/CDMTCS/chaitin`

In fact, it's a course that I gave at Rovaniemi, Finland, and the precise URL is

`http://www.cs.auckland.ac.nz/CDMTCS/chaitin/rov.html`

You can also find some of this material in my article in *J.UCS,* Vol. 2, No. 5.

Okay, so this is my fantasy for a first course on theoretical computer science, and I think that this could even work with bright high school students. There's very little in it that's technical, and the toy LISP is easier to learn than a real LISP. My goal has always been to teach quantum mechanics, general relativity and Gödel's incompleteness theorem to bright high school students! So obviously, the way I would teach such extremely bright youngsters programming would be with LISP—that's for theory. To actually get work done on the computer, I would teach them Mathematica, which is what high school students are learning in Rovaniemi!

Algorithmic Information Theory

Okay, I've used up half my time, but this discussion of elegant LISP expressions was actually just a warm-up exercise! This is not really the incompleteness result that I'm most proud of. Let me tell you what else you have to do starting in this spirit to get to my best incompleteness result. I'll outline the rest of my Rovaniemi course.

First I should tell you that I'm not going to use LISP expression size as my program-size complexity measure. That was simple and easy to understand, but it's not the right complexity measure for doing algorithmic information theory! So the first thing I want to do is define a new universal Turing machine.

UTM

Instead of taking in LISP expressions and putting out LISP expressions, which we did before, now the input program will be a bit string, and the output will be a LISP S-expression.

UTM: Bit String $\longrightarrow$ LISP Expression

So this is the computer that we're going to use to measure the size of programs. How does this computer work? Well, the program will be a long bit string, which the universal Turing machine will read from left to right.

Bit String—

The beginning of the program is going to be a LISP expression in binary, and I use eight bits for each character.

Bit String—LISP Expression (8 bits/char)

So every program for my universal Turing machine starts with a LISP expression telling you the other Turing machine to simulate, that's the idea. And LISP is a good language for expressing algorithms.

So to make a program, you take a LISP expression, and then you convert it into a list of 0's and 1's, and I have a primitive function in my LISP for doing that, and you end up with a very long bit string. It will have eight bits for each character. And I also provide a function to measure the size of a LISP expression, and another for determining the length of a list. In fact, I provide primitive functions for all the right things so that my proofs will be easy to do!

Okay, so the universal Turing machine starts off by reading in a LISP expression in binary, eight bits per character, and how does it know where the

LISP expression ends? Well, I'll just put a special character next to serve as endmarker, that'll be the next eight bits of the program. In UNIX there are a number of characters that people use to indicate the ends of things, and I've picked one of them, the newline character \n. In fact, my primitive function for converting a LISP expression into a bit string automatically supplies this special eight-bit pattern at the end.

Bit String—LISP Expression, NL

So there's this special character and when you get to it you know that you've finished reading the LISP expression.

After it finishes reading the LISP expression at the beginning of the bit string, the universal Turing machine starts running it, it starts evaluating the LISP expression. Think of this as the program, and as data we're going to give it the binary program for the Turing machine that's being simulated.

Bit String—LISP Expression, NL, Data

How does the LISP expression that's being evaluated get access to its binary data? Access is tightly controlled. Basically, the only way to access the data is by using a primitive function (with no arguments) that returns the next bit of the binary data. And it's very, very important that this primitive function can only return a 0 or a 1, but it cannot return an end of file indication. If you run out of data, the program aborts! This forces the program to be self-delimiting, which means that it has to indicate within itself how far out it goes. In other words, the initial LISP expression has to decide by itself how much data to read. For example, one simple convention is to double each bit of the binary data and then use a pair of unequal bits as an endmarker. But there are much more clever schemes for packaging binary data.

And if the LISP expression doesn't abort because it requested data that wasn't there, then the final value that it returns will be the final output produced by my new universal Turing machine. There may also be additional, intermediate output, which will also be LISP S-expressions. (The intermediate output is produced by a special primitive function that's an identity function, but which has the side-effect of outputting its argument.)

Why do we have to give binary data to the LISP expression? It's because the bits in a LISP S-expression are redundant due to LISP syntax restrictions. So you have to add to the LISP expression, which is a powerful way to express algorithms, raw binary data "on the side." There you have maximum flexibility, there you can really take advantage of each bit. And the initial LISP expression is going to determine the scheme that's used for reading in the raw binary data.

To convince you of the power of this self-delimiting scheme, let me show you how easy it is to use subroutines—you just concatenate them! The fancy way of saying this is that because programs are self-delimiting, algorithmic information content is subadditive. What does this mean? Well, let's say that you're given two programs to calculate two separate S-expressions. Then it's easy to combine these programs and get a program to calculate the pair of S-expressions. In fact, to do this you only need to prepend a particular 432-bit prefix.

Let me restate this. I like to use $H(.)$ for the size in bits of the smallest program to calculate something. $H(X)$ is the algorithmic information content or complexity of the S-expression X. Then the following basic inequality states that the complexity of a pair of S-expressions is bounded by a constant plus the sum of the individual complexities:

$$H((X\ Y)) \leq H(X) + H(Y) + 432.$$

This inequality states that algorithmic information is (sub)additive, that we can combine subroutines. How does it work? Well, there's a 432-bit prefix, which consists of a 53-character LISP S-expression and a \n. This 53-character expression reads in a LISP S-expression (there's a primitive function for doing that) and runs it (another primitive) to get X. Then it reads in a second S-expression and runs it to get Y. And then it returns the pair $(X\ Y)$. That's 53 characters of code in my LISP. That's the general idea; I don't want to go into the details.

By the way, this inequality

$$H((X\ Y)) \leq H(X) + H(Y) + c$$

has been around for a long time, in fact ever since I redid algorithmic information theory using self-delimiting programs in the mid 1970's. This inequality appears in a lot of papers (with $H(X,Y)$ instead of $H((X\ Y))$), but we never knew how big c could be. It depended on the choice of universal Turing machine. Well, now I've picked a particular universal Turing machine, and c is equal to 432! I think it's very interesting to get a specific value for c.

Algorithmic information theory is a theory of the size of computer programs, but up to now you've never been able to actually run these computer programs. I don't like that! So my new version of algorithmic information theory is very concrete and down to earth. You have to learn LISP programming, but you get a theory about the size of **real computer programs,** programs that you can actually run and use to get results. And after you put together a program and test it, you just look at its size and that gives you an upper bound on the program-size complexity of something! It's that easy! I think that this makes my theory much more concrete and much more understandable. And as a result of this, **I** certainly understand program-size complexity better. So hopefully this approach will also help other people!

Okay, in summary, we've got this universal machine. Its programs are like this:

Bit String—LISP Expression, NL, Data

There's a LISP expression in binary followed by a delimiter character followed by raw binary data. And you just measure the size of the whole program in bits, and that's our program-size complexity measure. The next thing you do with this theory, that I do in my course, is to show that information is additive.

$$H((X\ Y)) \leq H(X) + H(Y) + 432$$

In the course I construct the 432-bit prefix and you see it working. You actually run the programs.

The last time I gave this course was in Rovaniemi, Finland, this May, end of May, when it never got dark, and I had the pleasure of seeing a room full of people working on their computers, running programs on my universal Turing machine. They were able to do it and get results! That was really a thrill for me. Here's how you do it: First you write out a LISP expression—that's easy to do—then you use a primitive function to convert it to binary, and you append the binary data. Then you feed the result to my universal machine, which is just one line of code in my LISP. To define my universal Turing machine is very easy in this LISP.

So you see, I'm not using the size of LISP expressions as my program-size measure, but I am using LISP to define the universal machine, and I'm also using it to produce the programs that I feed to the universal machine.

And my universal machine really works! Not just for theory—we've always had universal Turing machines that work for theoretical purposes—right? since the 1930's—but you can actually run interesting programs on it. And the fact that I'm using a very high-level language, LISP, as my basis, means that set theory is essentially built into this universal machine. For example, it's very easy to program set union and intersection in LISP. And the kind of algorithms that I want to do, which are the ones that I use to prove my incompleteness results, are very easy to express in this language. That's the point: my universal machine is good not only for proving theorems, but also for writing programs and running them on examples in a finite amount of time.

I recently spoke to a professor at the University of New Mexico in Albuquerque who teaches recursive function theory. He said that his computer science students felt it was very strange that they were running programs on computers all day long, but never in their class on recursive function and computability theory! They felt a kind of cognitive dissonance, it just didn't make sense. Maybe, he said, my approach using LISP would work better, because then students could run the programs. LISP is a version of recursive function theory that really works! We've learnt much better how to write software since 1930!

But the version of `try` that I used to analyze whether one can prove that a LISP expression is elegant is not quite right. So let me show you the real `try`, which has an additional argument. Now it's like this:

```
(try time-limit lisp-expression binary-data)
```

There's a time limit, there's a LISP expression that we're going to try to run, to evaluate, and now there's also binary data "on the side," raw binary data. The binary data is just a list of 0's and 1's—that's the easiest way to represent a bit string in LISP—for example

```
(1 0 1 0 1 1 1)
```

So there are two parentheses and there are blanks between the bits, but otherwise it's just like the bit strings that you've always known and loved.

Just as before, we're going to try to evaluate the given LISP expression in the given amount of time. The new wrinkle is that while it's being evaluated this LISP expression has access to raw binary data on the side. So I'm using `try` to do a lot of things at the same time! It does a time-limited evaluation, and it also gives binary data to a LISP expression.

So while we're running the LISP expression, there's a time limit, and the LISP expression can use a zero-argument primitive function that says, give me the next bit of the binary data. And it'll get the next bit, if there is one. The LISP expression can also use another primitive function to read an entire LISP S-expression from the binary data. How does it do it? Well, it reads eight bits at a time, sees the characters that it gets, paying special attention to blanks and parentheses, until it gets to a newline character `\n`. Then it stops. So you don't always read individual bits from the binary data, you can also read big chunks in if you want to.

Now what value do you get back from this `try`? Remember, you always get a value back, even if the LISP expression that you're trying never terminates. That's why you use `try` instead of `eval`. Well, it'll still be a triple.

```
(success/failure value/out-of-time/out-of-data captured-displays)
```

It'll start as before with `success` or `failure`. If the try was a success, that means that evaluation of the LISP expression completed, and the second element of the triple will be the value that this LISP expression returned. On the other hand, if the try was a failure, then the second element of the triple is going to tell us **why** the evaluation failed. I already told you one way that the evaluation can fail: you can run out of time. An evaluation might also fail in many other ways, for example if you try to apply a primitive function to arguments which are not of a suitable type. Since I didn't want to be bothered with that kind of problem, I made the semantics of my LISP extremely permissive, so that I don't have to put a lot of error messages here telling you what went wrong. In fact, the only way that a try can fail is if it ran out of time or it ran out of binary data. You run out of binary data if you use it all up and then try to keep reading it. It's okay to read the last bit of the binary data, but then it's not okay to ask for another bit.

What about the third element of the triple returned by `try`, the captured intermediate results? Now it's "captured displays." Let me tell you why.

Intermediate results are important when you're debugging a big LISP expression. The final value may not be enough to tell what went wrong. The cute LISP solution to this problem is to add a one-argument primitive function that's an identity function. I call it `display`.

```
(display X)
```

So from the point of view of pure mathematics, `display` is useless, it just returns the value of its argument X. But it's extremely useful, because `display` prints the value of its argument on your screen. And to debug a large LISP expression, you just wrap `display`'s around interesting parts of the expression; that doesn't change the final value.

But in my LISP, in the game I'm playing, `display` is used for something much more important than debugging, it's used to put out theorems. `Display` enables a LISP expression to produce an infinite number of results. And that's very important because I want to model a formal axiomatic system, which is an unending computation that produces theorems, as a LISP S-expression. Normal LISP doesn't really have a way for a LISP expression to produce an infinite amount of output. So I had to create this mechanism in which `display` throws theorems and `try` catches them!

So in my LISP `display` has official status. It's not just a debugging mechanism. It's the way that a formal axiomatic system outputs each theorem, it's an important part of LISP.

So the last thing in the triple returned by a `try` is a list of captured displays. Every time that the LISP expression being tried tries to put something on your screen, it won't go there, it'll end up in this list instead. So if you're trying a formal axiomatic system, this will be a list of theorems.

`That's it!` This is all there is to it. It's the entire mechanism for programming algorithmic information theory in LISP.

The Halting Probability Ω

Well, I've shown you all the machinery, but what do I do with it! What's the next thing in the course on my web site? Well, there's not much time left! Just enough time for a quick summary!

I've shown you the basic tools that I need to program out my version of recursive function theory, my version of Turing and Gödel's incompleteness results. Basically, it's just `try`, that's all I have to add to LISP. It's what makes it possible for LISP to handle algorithmic information theory.

When I taught this course in Rovaniemi, I started with a historical introduction. Then I explained my LISP. Then I show off my universal Turing machine and run a bunch of simple programs on it. The next thing I do is I define the halting probability, it's capital omega.

$$\Omega$$

To define Ω you take my universal machine that has binary programs and produces LISP expressions, and you just feed it bits that you get by tossing a coin, that's independent tosses of a fair coin. And you ask, what is the halting probability? That's Ω. And I actually give a LISP program that calculates the halting probability in the limit from below. You'll find this program explained in my lecture transcript "An invitation to algorithmic information theory." It's in the DMTCS'96 *Proceedings.* When I gave that talk I was speaking three to five times faster than I am today—I don't know how I covered so much material! So if you want to see the LISP program to calculate Ω in the limit from below, it's in the DMTCS'96 *Proceedings,* or look at my Web site, where there are sample runs.

The next thing I do is I prove that Ω is irreducible algorithmic information. The precise result turns out to be this:

$$H(\Omega_N) > N - 8000.$$

What's Ω_N? Well, the halting probability is a real number, so write it in binary, and take the first N bits after the "decimal" point. So this inequality states that to get the first N bits of the halting probability, you need a program that's more than $N - 8000$ bits long. And the reason is that if you knew the first N bits of the halting probability, that would enable you to solve the halting problem for all programs up to N bits in size. That's how you prove that Ω is irreducible.

And then finally I get what I think is my most devastating incompleteness result, which says that a formal axiomatic system can't enable you to prove, to determine, more than

$$H(\text{FAS}) + 15328$$

bits of the halting probability. This follows from the previous inequality, the one that states that

$$H(\Omega_N) > N - 8000.$$

In other words, since a small program can't give you a lot of bits of Ω, a set of axioms can't enable enable you to determine substantially more bits of Ω than there are bits of axioms. That's what

$$H(\text{FAS})$$

is, it's the number of bits in the smallest program that makes my universal machine output all the theorems in the formal axiomatic system. In other words, it's the complexity of the formal axiomatic system, it's the number of bits in its axioms. So the more bits of Ω you want to determine, the more bits you have to add to your axioms!

My main theorem is a mathematical pun. Turing proved that the halting problem is undecidable. I prove that the halting probability is irreducible! Not only you can't compress bits of Ω into a program substantially smaller than the number of bits of Ω you calculate, you can't do it using reasoning either. Essentially the only way to get bits of Ω out of a formal axiomatic system, is if you just add those bits as axioms! So determining bits of Ω is a losing proposition! Mathematical reasoning doesn't help at all. Well, that's not quite right, because of the constant 15328.

$$H(\text{FAS}) + \mathbf{15328}$$

If it weren't for this, I'd say you get out exactly what you put in!

So if you want to prove more bits of Ω, essentially the only way to do it is to add them to your axioms. But you can prove **anything** by adding it to your axioms. And what Ω shows, is that sometimes—Ω's a fairly simple object from the point of view of non-constructive mathematics—in some relatively elementary branches of mathematics, the only way to get more out is to put it into the axioms. These are situations in which mathematical reasoning is really useless, really impotent! Because if you can get something out of a set of axioms only by putting it in as a new postulate, why bother! So Ω is really a worst case, it's our worst nightmare come true, because it's

Irreducible Mathematical Information

Conclusion

So to end, let me try to state in words what I find intriguing about Ω. As I said, Ω shows that in some cases you're in big trouble! But let me emphasize the philosophical discontinuity.

The normal view of mathematics is that if something is true, it's true for a reason, right? In mathematics the reason that something is true is called a proof. And the job of the mathematician is to find proofs. So normally you think if something is true it's true for a reason. Well, what Ω shows you, what I've discovered, is that some mathematical facts are **true for no reason!** They're true **by accident!** And consequently they forever escape the power of mathematical reasoning. Each bit of Ω has got to be a 0 or a 1, but it's so delicately balanced, that we're never going to know which it is.

I used to believe that all of mathematical truth, all the infinite variety of mathematical truth, **could be compressed** into a small set of axioms and methods of reasoning that we could all agree on, and that we learn as mathematics students. I felt this deep in my soul, it's part of what makes mathematics beautiful, the sharpness, the clarity—it seemed inhuman, even superhuman! Unfortunately the existence of irreducible mathematical facts shows that in some cases there is absolutely no compression, no structure or pattern at all in mathematical truth. I don't know why anyone would want to prove what bits of Ω are. But if you wanted to do that, it would be completely hopeless. Because, you see, the bits of Ω aren't 0 or 1 for any particular reason—they've got to be one or the other, it's a specific Ω, there's a LISP program for calculating it in the limit from below—but it doesn't happen for a reason, it happens by accident. If God were willing to answer yes/no questions, each bit of Ω would require a separate question, because there are no correlations, there is no redundancy!

But I'm being too pessimistic. After all, Fermat's last theorem was just demonstrated. And I wouldn't be too surprised if another determined, brilliant individual were to prove the Riemann hypothesis. In fact, clever mathematicians do succeed in settling famous conjectures—remember the announcement that "four colors suffice"?

In this direction

Irreducible Mathematical Information

you can't go any farther, right? But I think that it's an interesting question to understand how come in spite of these results it is in fact possible to do mathematics so well? I think that the interesting question now is not to prove incompleteness results, but to see how come mathematics is still so wonderful. It is! We can prove wonderful theorems, breathtaking theorems. And I think that it would be interesting now to try to understand better how this is possible.

I guess that's the story that I wanted to tell you. Thank you very much!

Martin Davis

Born in New York City in 1928, Martin Davis was a student of Emil L. Post at City College and his doctorate at Princeton in 1950 was under the supervision of Alonzo Church. Davis's book "Computability and Unsolvability" (1958) has been called "one of the few real classics in computer science." He is best known for his pioneering work in automated deduction and for his contributions to the solution of Hilbert's tenth problem for which latter he was awarded the Chauvenet and Lester R. Ford Prizes by the Mathematical Association of America and the Leroy P. Steele Prize by the American Mathematical Society. In 1983 he was a Guggenheim Foundation Fellow. His books have been translated into a number of languages including Russian and Japanese. Davis has been on the faculty of the Courant Institute of Mathematical Sciences of New York University since 1965, was one of the charter members of the Computer Science Department founded in 1969, and is now Professor Emeritus.

From Logic to Computer Science and Back

"My father and mother were honest, though poor –"
"Skip all that!" cried the Bellman in haste.
"If it once becomes dark, there's no chance of a snark–
We have hardly a minute to waste!"

"I skip forty years," said the Baker, in tears,
"And proceed without further remark ... "

–Lewis Carroll's "The Hunting of the Snark"

I was just over a year old when the great stock market crash occurred. My parents, Polish Jews, had immigrated to the United States after the First World War. My father's trade was machine embroidery of women's apparel and bedspreads. During the depression, embroidery was hardly in great demand, so we were dependent on home relief – what today would be called "welfare". Only with the upturn of the economy coming with the outbreak of war in 1939, was my father able to find steady work. In his spare time, he was a wonderful untaught painter. (One of his paintings is in the collection of the Jewish Museum in New York and two others are at the Judah Magnus Museum in Berkeley.) My mother, eager to contribute to the family income, taught herself the corsetiere's craft. Until I left New York for graduate school, the room where she conducted her business by day was my bedroom at night.

In the New York City public schools, I was an adequate, but not at all exceptional, student. I've always enjoyed writing, but an interest in numbers came early as well. I remember trying to find someone who could teach me long division before I encountered it in school. My parents, whose schooling was minimal, could not help. My first "theorem" was the explanation of a card trick. I learned the trick from a friend who had no idea why it worked; I was delighted to see that I could use the algebra I was being taught in junior high school to explain it.

It was at the Bronx High School of Science that I first found myself with young people who shared the interests I had been developing in mathematics and physics. My burning ambition was to really understand Einstein's theory of relativity. There were a number of books available in the local public library as well in the school library, but I couldn't understand many of the equations. Somehow I got the idea that it was calculus I needed to learn, so I got a textbook and taught myself. When I arrived at City College as a freshman, I was able to begin with advanced calculus. During those years, the math majors at City College were an enthusiastic talented group many of whom eventually became professional mathematicians. The faculty, on the other hand, was badly overworked, and, with a few notable exceptions, had long since lost their enthusiasm. Even by the standards of the time, teaching loads were excessive, and none of the usual amenities of academic life (such as offices and secretarial help) were available. Only very few of the most determined faculty members remained active researchers. In addition to these obstacles, Emil Post struggled against physical and psychological handicaps: his left arm had been amputated in childhood and he suffered from periodically disabling manic-depressive disease. Nevertheless, Post not only continued a program of important fundamental research, but also willingly accepted students for special advanced studies on top of his regular teaching load (16 contact hours). I absorbed his belief in the overriding importance of the computability concept and especially of Turing's formulation.

At City College my academic performance was hardly outstanding. I allowed myself the luxury of working hard only on what interested me. My A grades were in mathematics, German, history, and philosophy. My worst class was a required general biology course. I hated the amount of memorization of names of plant and animal parts I had no desire to know, and found genuinely difficult the "practicums", in which we were asked to identify specimens we viewed under the microscope. I actually failed the course, and even on the second try only managed a C.

During my Freshman and Sophomore years, my passionate interest was in the foundations of real analysis. I learned various alternate approaches and proofs of the main theorems. I spent weeks working out the convergence behavior of the sequence

$$s_0 = 1; \quad s_{n+1} = x^{s_n}$$

for $x > 0$. (It converges for $(1/e)^e < x < e^{1/e}$. The case $0 < x < 1$ is tricky because, although the even-numbered terms and the odd-numbered terms each converge, when $x < 1/e$ their limits are different.) I liked sequences and saw how to prove that every sequence of real numbers has a monotone subsequence as a way of obtaining the basic theorems. I even wrote quite a few chapters of a proposed textbook.

My fellow student John Stachel and I began to be interested in logic, and at his suggestion, we approached Post about a reading course in mathematical logic. Thus, in my junior year, we began studying an early version of Alonzo Church's textbook under Post's supervision. Unfortunately, it only lasted a few weeks: Post had made his discovery of the existence of incomparable degrees of unsolvability, the excitement precipitated a manic episode, and he was institutionalized. The following year Post was back and we spent a great deal of time talking about logic. He gave me a collection of his reprints and also referred me to Kleene's paper [30]. This was a paper Kleene had written in haste to get some results in publishable form before he was requisitioned for war work. For me this was a boon because it was written in a relatively informal style quite unlike Kleene's usual more opaque exposition. I spent a lot of time filling in the gaps, and in the process became enamored of the Herbrand-Gödel-Kleene equation formalism. In considerable part, my dissertation developed from that paper.

Kleene's paper showed that the sets definable in the language of arithmetic[1] formed a natural hierarchy in terms of alternating strings

$$\exists\forall\exists\ldots \quad \text{or} \quad \forall\exists\forall\ldots$$

of quantifiers applied to a computable relation: each additional quantifier makes it possible to define new sets. This result was applied to give short incisive proofs of Gödel's incompleteness theorem and the unsolvability results of Alonzo Church.

I would undoubtedly have remained at City College for my graduate studies to work with Post if that option had been available. But City College was strictly an undergraduate school, and I had to look elsewhere. I had offers of financial support from Princeton, where I could work with Church, and from the University of Wisonsin, where Kleene would have been my mentor. Post advised me to go to Princeton, and that is what I did. There was quite a culture

[1] that is, the language using the symbols $\neg \supset \vee \wedge \exists \forall =$ of elementary logic together with the symbols $0\ 1\ +\ \times$ of arithmetic.

clash between my New York Jewish working-class background and the genteel Princeton atmosphere, and at one point it seemed that my financial support would not be renewed for a second year for reasons having nothing to do with my academic performance. Although eventually I was given support for a second year, the unpleasantness made me eager to leave. Fortunately, the requirements at Princeton were sufficiently flexible that it was quite possible to obtain a Ph.D. in just two years, and that is what I did.

The problem that I knew would readily yield results was the extension of Kleene's arithmetic hierarchy into the constructive transfinite, what later became known as the *hyperarithmetic* sets. Post had shown that the successive layers of Kleene's hierarchy could also be generated using the jump operator,[2] and it was easy to see how to extend this method into the transfinite. But the problem that I found irresistably seductive was Hilbert's tenth problem, the problem of the existence of integer solutions to polynomial Diophantine equations. Post had declared that the problem "begs for an unsolvability proof" and I longed to find one. Not being at all expert in number theory, I thought that it was foolish to spend my time on Diophantine equations when I had a dissertation to write and a sure thing to work on. But I couldn't keep away from Hilbert's tenth problem.

Diophantine problems often occur with parameters. In general one can consider a polynomial equation

$$p(a_1, \ldots, a_m, x_1, \ldots, x_n) = 0$$

where p is a polynomial with integer coefficients, $a_1, \ldots, a_m$ are parameters whose range is the natural numbers, and $x_1, \ldots, x_n$ are unknowns. I began to study *Diophantine sets,* that is, sets that could be defined by such an equation as the set of m-tuples of values of the parameters for which the corresponding equation has a solution in natural numbers.[3] Another way to say this is that Diophantine sets are those definable by an expression of the form

$$(\exists x_1 \ldots x_n)[p(a_1, \ldots, a_m, x_1, \ldots, x_n) = 0].$$

It was not hard to see that the class of Diophantine sets is not only a sub-class of the class of recursively enumerable (r.e.) sets,[4] but also shares a number of important properties with that class. In particular, both classes are easily seen to be closed under union and intersection, and under existential quantification of the defining expressions. A crucial property of the class r.e. sets, a property that leads to unsolvability results, is that the class is *not* closed under taking complements. I was quite excited when I realized that the class of Diophantine

[2] The *jump* of a set A of natural numbers may be understood as the set of (numerical codes of) those Turing machines that will eventually halt when starting with a blank tape and able to obtain answers to any question of the form "$n \in A$?".

[3] For example, the "Pell" equation $(x+1)^2 - d(y+1)^2 = 1$ has natural number solutions in x, y just in case d belongs to the set consisting of 0 and all positive integers that are not perfect squares; hence that latter set is Diophantine.

[4] A set of natural numbers is *r.e.* if it is the set of inputs to some given Turing machine for which that machine eventually halts.

sets has the same property. This was because if the Diophantine sets were closed under complementation, then the de Morgan relation

$$\forall = \neg\exists\neg$$

would lead to the false conclusion that all of the sets in Kleene's hierarchy, all arithmetically definable sets, are Diophantine. (False because there are arithmetically definable sets that are not r.e. and hence certainly not Diophantine.) Although this proof is quite non-constructive,[5] the result certainly suggested that the classes of r.e. sets and of Diophantine sets might be one and the same. If every r.e. set were indeed Diophantine, there would be a Diophantine set that is not computable which would lead at once to the unsolvability of Hilbert's tenth problem in a particularly strong form. So, I began what turned into a twenty year quest, the attempt to prove that every r.e. set is Diophantine, what Yuri Matiyasevich much later called my "daring hypothesis".

During the summer between my two years at Princeton I was able to prove that every r.e. set is definable by an expression of the form

$$(\exists y)(\forall k)_{\leq y}(\exists x_1 \ldots x_n)[p(k, y, a_1, \ldots, a_m, x_1, \ldots, x_n) = 0]$$

where p is a polynomial with integer coefficients. From a purely formal point of view, this result (later known as "Davis normal form") seemed tantalizingly close to my conjecture; the only difference was the presence of the bounded universal quantifier $(\forall k)_{\leq y}$. However, there was no method in sight for getting rid of this quantifier, and I couldn't help agreeing with Church's assessment when he expressed disappointment that the result was not stronger.

Meanwhile, I had a dissertation to write. I didn't think at the time that my normal form by itself would suffice, although in retrospect I think it likely would have been accepted. In any case, I worked out an extension of Kleene's hierarchy into the constructive transfinite using Kleene's system of notations for ordinals.[6] Kleene had defined a set O of natural numbers and a partial well-ordering $<_O$ on this set. Each $m \in O$ represented an ordinal $|m|$, and

$$m <_O n \iff |m| < |n|.$$

With each $m \in O$ I associated a set L_m in such a way that $m <_O n$ implied that L_m is computable relative to L_n as oracle, but not the other way around. Then to extend Kleene's hierarchy, it was only necessary to consider the sets many-one reducible to the L_m. I studied their representation in terms of second order quantification and obtained the ridiculously weak result that up to ω^2 all of these sets were indeed so representable.[7] In addition I defined a constructive ordinal γ to be a *uniqueness ordinal* if whenever $|m| = |n| = \gamma$ the Turing degrees

[5] It furnishes no example of a Diophantine set whose complement is not Diophantine.

[6] Actually, Kleene's system S_3.

[7] Actually without any bound on the ordinal all the sets in the hierarchy are representable with only one second order function quantifer.

of L_m and L_n are the same. I proved that every $\gamma < \omega^2$ is indeed a uniqueness ordinal.[8]

I presented the results from my dissertation in brief talks at two professional meetings. The Diophantine result was given at a small meeting of the Association for Symbolic Logic in Worcester, Massachusetts in December 1949, which I attended with my first wife a few days after our marriage. Eight months later I attended the first post-war International Congress of Mathematicians at Harvard University, and spoke about my results on hyperarithmetic sets. This time I was alone – our marriage had proved short-lived; my wife had left me shortly before the Congress. At the Congress I met the great logician Alfred Tarski who showed considerable interest in my work, and, of particular significance, I also met Julia and Raphael Robinson. I had studied some of their published work, and was very pleased to meet them. I was surprised to find that Julia was presenting a short contributed paper on Diophantine sets. It turned out that we had approached the subject from opposite directions. While I had been trying to find a general representation for arbitrary r.e. sets, as close as possible to a Diophantine definition, she had been seeking such definitions for various particular sets. Her result that turned out to have the most important consequences was that from the existence of a single Diophantine equation with two parameters, one of which grows exponentially as a funtion of the other, she could obtain a Diophantine definition of $\{< a, b, c >| \ c = a^b\}$.

I would like to say that I expressed my pleasure at finding another Hilbert's tenth problem enthusiast. However, in Julia's sister Constance Reid's memoir, *The Autobiography of Julia Robinson*,[9] based on conversations with Julia shortly before her tragic death of leukemia, she quotes Julia as remembering me saying when we met that I couldn't see how her work "could help solve Hilbert's problem, since it was just a series of examples". I do not want to believe that I said anything so ungracious and so foolish. Julia is also quoted as remembering my "presenting a ten minute paper" at that Congress on my Diophantine results, and as that was not the case, I can comfort myself with the thought that her recollection of what I had said may also have been mistaken.

A few days after the Congress, I was on a plane from New York to Chicago, my first experience of air travel. After considerable difficulty in landing a job in a tight market, with my specialization in logic a definite disadvantage, I had had a stroke of luck. My former fellow student Richard Kadison having received a coveted National Research Fellowship, turned down the offer from the University of Illinois at Champaign-Urbana to be "Research Instructor". As their second choice, the position was offered to me, and I was delighted to accept. Research Instructors were expected to be recent Ph.D.'s and were required to teach only one course per semester at a time when the regular faculty taught three. In addition, we were given the opportunity to teach a second graduate course in our own specialty if we wished. I was very happy to take advantage of this

8 Clifford Spector showed that the result remains true for all constructive ordinals in his dissertation, written a few years later under Kleene's supervision.

9 in [32] p.61.

possibility: I taught mathematical logic in the fall and recursive function theory in the spring. In this second course, I decided to begin with Turing machines. Kleene had applied Gödel's methods of arithmetic coding to develop his results for the equation calculus. I saw that the same could be done for Turing machines and that this had certain technical advantages. However, in order to develop the necessary machinery, I had to design Turing machines to carry out various specific function; without realizing it, I was being a computer programmer!

Edward Moore (later known for his basic work on sequential machines), also a very recent Ph.D. in mathematics, was an auditor in my course. He came up to the front of the room after one of my classes and showed me how one of the Turing programs I had written on the blackboard could be improved. Then he said something very much like the following: "You should come across the street; we've got one of those machines there." In fact a superb engineering group were just finishing a computer called ORDVAC of the "johniac class" on the University campus. I had been paying no attention to computers, and up to that moment had not considered that Turing's abstract machines might have some relation to real world computing machines. It would make a better story if I said that the next day I took Ed up on his invitation. But the truth is that it was the Korean War and the hot breath of the draft that led me to take that walk "across the street" some weeks later. It was clear to me that if I remained in my faculty position, I would be inducted into the army, and it was equally clear to me that that was something I wanted to avoid.

A faculty group, led by the physicist Frederick Seitz, determined to contribute to the war effort and convinced of the military significance of automated systems, started a project within the university called the Control Systems Laboratory (C.S.L.). I was recruited for the project and, with the promise of a draft exemption, accepted. My boss was the mathematician Abe Taub, an expert in relativity theory and shock waves. It was a heady time. Norbert Wiener's *Cybernetics* heralding a new age of information and control had appeared a few years earlier, von Neumann had developed the basic computer architecture still used today and was investigating the use of redundancy to obtain reliable results from unreliable components, and the transistor had just been developed at Bell labs. There was much discussion of all this at the C.S.L., and after some vacillation, a report from the battlefield on the need for better fighter plane support for the front line troops decided the direction of the first major effort.

A working model was to be produced of an automated system for navigating airplanes in real time. The "brain" of the system was to be the newly constructed ORDVAC. And the job of writing the code fell to me. My instruction in the art of computer programming was delivered by Taub in less than five minutes of "This is how it is done". I also had as textbook the basic reports by von Neumann and Goldstine with many sample programs. Of course, the project was ludicrously over-ambitious given the technology available in 1951. The ORDVAC had 5 kilobytes of RAM; memory access required 24 microseconds. Addition time was 44 microseconds, and multiplication time a hefty kilosecond. From a programmer's point of view, interpreters, compilers, or even assembly language were all non-existent. There were no index registers. Inductive loops had to be coded by incrementing the address portion of the instructions themselves. And of course all the code had to be written in absolute binary. The RAM was implemented as static charge on the surface of cathode ray tubes, which tended to decay rapidly, and was continuously being refreshed. This worked so long as the programmer was careful not to write loops so tight that the same position on the CRT's was bombarded by electrons too rapidly for the refreshing cycle to prevent spillover of charge to neighboring positions. To a contemporary programmer, these conditions seem nightmarish, but in fact it was lots of fun (especially when I let myself forget what it was all supposed to be for).

My experience as an ORDVAC programmer led me to rethink what I had been doing with Turing machines in the course I had just finished teaching. I began to see that Turing machines provided an abstract mathematical model of real-world computers. (It wasn't until many years later that I came to realize that Alan Turing himself had made that connection long before I did.) I conceived the project of writing a book that would develop recursive function theory (or as I preferred to think of it: computability theory) in such a way as to bring out this connection. I hardly imagined that seven years would go by before I held in my hand a printed copy of *Computability & Unsolvability.* I enticed a group of my C.S.L. colleagues into providing an audience for a series of lectures on computability; the notes I provided for the lectures were a rough draft of the first part of the book.

Champaign-Urbana in the early 1950s was not an ideal locale for a young bachelor looking for a social life. In the university community, young men outnumbered young women by something like 10 to 1. (Even among undergraduates the ratio was 4 to 1.) But I was lucky enough to attract the interest of Virginia, a graduate student. By the spring of my first year there, she had moved into my apartment, an arrangement far more unusual in those days than it would be today. In fact, the university administration took an active interest in students' intimate lives. Female graduate students (and only female students) were subject to expulsion if they were found cohabiting with a male. So our menage was somewhat dangerous, especially as Virginia's parents didn't find me a particularly desirable suitor. We planned to marry on the earliest date after the legal formalities offically dissolving my first marriage were complete. That date turned out to be the first day of autumn just about a year after my arrival in Champaign-Ursula; we were married by the local Unitarian minister in a simple ceremony with only three friends present. My second marriage has proved somewhat more durable than the first; as I write this, our 46th anniversary is a month away.

Christmas week 1951 provided an occasion to drive East and introduce Virginia to my New York friends. It also enabled me to attend a mathematical meeting in Providence where I heard Kurt Gödel deliver his astonishing lecture in which he proposed that reflecting on his undecidability results would force one to adopt ontological assumptions characteristic of idealistic philosophy. The lecture was published only recently, after Gödel's death, but the audacious ideas he propounded have remained with me ever since I heard the lecture.

The spring of 1952 marked a major change for the ORDVAC. It had been built under contract with Army Ordnance, and it was time for its delivery to the Proving Grounds in Aberdeen, Maryland. The computer group had been busy working on a twin (not quite identical) to the ORDVAC dubbed the ILLIAC (later ILLIAC I). But here I was with my code and no computer to debug it on until the ILLIAC came on line. So I was sent to Aberdeen. Virginia came with me and we stayed in a motel in the nearby town Havre de Grace. It was in that motel that we conceived our first child.

The ORDVAC had been installed in the the building housing the Ballistics Research Laboratory along with two older, indeed historic, computers: the EDVAC and the ENIAC. The ENIAC consisted of racks of vacuum tube circuits and plugboards such were used by telephone switchboards, filling the four walls of a large room. The building was locked from the inside and one could only leave by first going to the ENIAC room and asking one of the people there to unlock the door. The ORDVAC was in use by Aberdeen people until 4PM, after which it was made available to me. Instead of the watchful crew in Urbana used to babying their creation, the computer operator was a sergeant whose main qualification was that in civilian life he had been an amateur radio operator. I was soon operating the machine myself, something I never would have been permitted to do in Urbana. One evening, I noticed that the machine seemed to be making many errors. I also noticed that I was getting very warm, but it didn't occur to me to connect these facts. Finally, when I saw a 0 change to 1 on a CRT at a time that the computer was not executing any instructions, I gave up and left. The folks back in Urbana were furious with me. The air conditioning had broken down, and there had been a very real danger that the ORDVAC could have been destroyed by the heat. It should have been powered down at once.

Back in Urbana, I found myself increasingly unhappy with what I was doing at the C.S.L. The Office of Naval Research came to my rescue with a small grant that enabled me to spend two years as a visiting member at the Institute for Advanced Study in Princeton. I thought that with that sponsorship, I would probably be safe from the draft. My proposal was to work on connections between logic and information theory. That was a really good idea: the great Russian mathematician Kolmogoroff and Gregory Chaitin showed what could be done with it quite a few years later. However, I found myself moving in other directions.

The Institute for Advanced Study in those years was directed by J. Robert Oppenheimer. On the faculty were Albert Einstein, Kurt Gödel, and John von Neumann. Einstein and Gödel, good friends, were often seen walking to or from the Institute buildings together. I well remember the first time we encountered them walking down the middle of Olden Lane together: Einstein dressed like a tramp accompanied by Gödel in a suit and tie carrying his briefcase. "Einstein and his lawyer" was Virginia's vivid characterization.

I had met Norman Shapiro as an undergraduate in Urbana..He had come to Princeton University as a graduate student and was writing a thesis on recursive functions. He and I organized a logic seminar. Among the regular attendees were Henry Hiz, John Shepherdson, and Hao Wang. Hilary Putnam, with whom I was later to do some of my best work, gave a philosophical talk which Norman and I mercilessly attacked. In my research, I was struck by the fact that the phenomenon of undecidability in logic could be understood abstractly in terms of the way each particular logical system provided a mapping from recursively enumerable sets[10] to subsets of their complements. I was particularly struck by the fact that Gödel's famous result about the unprovability of consistency could be expressed simply as the fact that the iteration of this map always produces the empty set. Some years later I told one of my first doctoral students, Robert Di Paola, about this, and he based his dissertation on studying that mapping. Gödel himself was uninterested when I summoned the courage to tell him about my ideas.

I occasionally thought about Hilbert's tenth problem, and I worked on my book. The chapter on applications of computability theory to logic gave me particular trouble. The problem I faced was giving a coherent exposition without writing a whole book on logic. I rewrote that chapter many times before I was satisfied. A problem of another kind was the difficulty I had in getting the Institute typists to produce a decent copy from my handwritten manuscripts. Our son was born in January 1953. After he was weaned, a year later, Virginia took a job at the Princeton Public Library. I imagined I could take care of the baby and work on my book at the same time. Of course this did not work out very well.

My arrangement with the Office of Naval Research left me free to seek employment during the summer months. We certainly needed the extra money. I was able to spend the summer of 1953 working at Bell Labs, a short commute from Princeton. My boss was Shannon, the inventor of information theory, and I was able to renew my aquaintance with Ed Moore. Shannon had recently constructed a universal Turing machine with only two states. He posed the question of giving a well defined criterion for specifying when a Turing machine could be said to be universal. I liked that question and wrote two short papers dealing with it.[11] The intellectual atmosphere at Bell Labs was stimulating and open to fundamental research. I could well understand how a fundamental breakthrough like the transistor could develop in such an environment. Shannon himself was treated like the star he was. He had a small shop with two technicians available to build any of his whimsical gadgets. His "mouse" that successfully solved mazes was already famous. Less well known was his desk calculator "Throwback I" that used Roman numerals. Shannon was also an expert unicycle rider. One day he brought his unicycle to the labs and created mass disruption by riding it down the long corridors and even into and out of elevators bringing swarms of Bell Labs employees streaming out of their offices to watch. Another thing

[10] Actually, indicies of r.e. sets.

[11] [1, 2].

I remember about that summer is the excitement of a real workers uprising in East Berlin against the Communist regime.

For the summer of 1954 I thought about applying the programming skills I had learned in Urbana to a logical decision procedure. My first choice was Tarski's quantifier elimination algorithm for the first order theory of real closed fields. But on second thought I saw that this was going to be too difficult for a first try, and instead I settled on Presburger's procedure for arithmetic without multiplication, since this was a much simpler quantifier elimination procedure. Had I known the Fischer-Rabin super-exponential lower bound for Presburger arithmetic (proved 20 years later), I would presumably have hesitated. But I went blithely ahead with the blessing of the Office of Ordnance Research of the U.S. Army which agreed to support the effort. I was able to do the work without leaving Princeton, using the original johniac at the Institute for Advanced Study. To my dismay the code used all of the 5 kilobytes of RAM available and was only able to deal with the simplest statements on the order of "The sum of two even numbers is even". My report on the program, duly delivered to the Army on its completion and included as well in the Proceedings of an important Summer Institute of Logic at Cornell in 1957 (about which, more later), ended with the understatement:[12]

> The writer's experience would indicate that with equipment presently available, it will prove impracticable to code any decision procedures considerably more complicated than Presburger's. Tarski's procedure for elementary algebra falls under this head.

An anthology [33] of "classical papers on computational logic 1957-1966" published in 1983 begins its preface with the sentence:

> In 1954 a computer program produced what appears to be the first computer generated mathematical proof: Written by M. Davis at the Institute of Advanced Studies (sic), USA, it proved a number theoretic theorem in Presburger Arithmetic.

In the spring of 1954 my two years at the Institute were drawing to a close, and I needed to find a job. Again the market was rather tight. We had a few possibilities, but opted for the one that took us furthest west: an Assistant Professorship at the University of California at Davis. For the first time we experienced what was to be repeated over a dozen times in our lives: the trip by automobile across the United States with a stopover in Kansas City to visit Virginia's parents. As we approached our new home, the road signs seemed to be directing us: "Davis use right lane".

The liberal arts program was newly instituted at Davis which had been exclusively (and is still to a considerable extent) an agricultural school. In 1954, the population of Davis was just about 5000 people. It was not a cultural center. Amusements were in such short supply that we would drive to the local soft ice-cream drive-in just to watch the customers come and go. It was not a year in

---

[12] The report was reprinted in [33], pp. 41-48.

which I accomplished much scientifically. The teaching load was quite modest: just two courses per semester. When I taught calculus (to students majoring in agricultural engineering), I had to speak in a loud voice to be heard above the clatter of the turkeys in the building next door.

Virginia was pregnant again and we needed to find an obstetrician. There were none in Davis itself, but there was a hospital at the county seat, Woodland, a few miles away. Virginia found the local obstetrician there quite unacceptable. Sacramento, the state capitol was perhaps 18 miles away, but we had heard too many obstetrical horror stories coming from that quarter. So we headed for progressive Berkeley 80 miles away, where Virginia found an excellent obstetrician. Today there is an excellent superhighway connecting Davis with Berkeley, but in 1954 the drive took at least two hours. The highway ended in Richmond with the rest of the route being through city streets. Virginia's first labor had been swift and uneventful, so we knew that we had to be prepared for the possibility of not making it to Berkeley in time. We obtained a government pamphlet for farmers on delivering babies, and bought a second-hand obstetrics textbook. We were in Berkeley a few days before our Nathan made his appearance, and Virginia was assured that all was well. Back in Davis, we were awakened at 2AM by a flood of amniotic fluid drenching the sheets. By 4AM the crying baby had arrived. Virginia tells people that I "delivered" him, although really I just watched. Except for one detail: Nathan was born with his umbilical cord wrapped around his neck. Before I had time to think about it, I had lifted the loop away.

People we hardly knew had strong opinions about what had happened. Those who thought we had done something praiseworthy in defense of the natural seem to have been outnumbered by those who thought we had behaved in an irresponsible manner. There was even the suggestion that we should be imprisoned. We were convinced that Davis was not for us, and were determined to leave. I found a position at Ohio State University in Columbus and quickly accepted. For the summer, I got a job at the Moore School of Electrical Engineering in Philadelphia where the ENIAC and the EDVAC had been built. So, we set out for Princeton in our 1951 Studebaker sedan with Harold and Nathan in the rear. Our plan was to spend the summer there, an easy commute to the Moore School.

The summer at the Moore School was a pretty complete disaster. They wanted me to prove a particular kind of theorem about certain numerical methods for solving ordinary differential equations. I knew very little about such things, but I saw no reason to believe that there was a theorem of the kind they wanted. I did not accomplish much for them. The best part of the summer was getting to know Hilary Putnam who was living in the same prefab housing complex for graduate student and junior faculty families where we had subleased an apartment for the summer. To my surprise, he was very interested in Hilbert's tenth problem and proposed that we collaborate. Nothing much came of this until a few years later.

A major plus of my new position at Ohio State University was that Kleene's student Clifford Spector was on the faculty. His brother was a close friend of a good friend of mine, and on his brother's advice, he had written me some years earlier about his interest in logic. Apparently, this interest had been actively discouraged at Columbia University where he had been informed that there are no interesting problems in logic. I had suggested a number of possibilities for graduate study in logic including Madison, Wisconsin with Kleene. Somewhat to my surprise, I detected something not entirely friendly in Clifford's welcome. It was several months before he became open. I learned that Kleene had been rather displeased with me. Kleene had gone to considerable trouble to get a fancy fellowship for me at the University of Wisconsin, and I had not only gone to Princeton instead, but had written a dissertation largely in areas where Kleene himself had been working. Kleene had given Spector the uniqueness ordinal problem left open in my thesis as an appropriate topic for his dissertation. Clifford reported that Kleene had whipped him on with the warning that "Davis is working on it" emphasizing the importance of reaching the goal first. In fact, I hadn't been thinking about uniqueness ordinals at all. In any case Spector was a more powerful mathematician than I. In his excellent dissertation, he not only proved that every constructive ordinal is a uniqueness ordinal (thus settling the question raised in my dissertation), but also proved a deep result in the theory of degrees of unsolvability.[13]

The one year we spent in Columbus was not a happy one. Among other difficulties, we were feeling financially pinched. I received my last paycheck from Davis at the beginning of June and the first from Columbus only in November. The money from the Moore School helped, but I had to return half of the money for moving expenses I had received from Davis because I had left after only one year, and Ohio State did not cover moving expenses. And apparently impossible to please, Virginia and I just didn't like life in Columbus very much. To save money, we moved into an apartment with just one bedroom that we give to our two babies, while we slept on a convertible couch in the living room. The Chair of the department helped by offering me the opportunity to teach an off-campus advanced calculus course to Air Force officers at the nearby Wright-Patterson base in the summer. In the hot Ohio summer, I often taught wearing short pants. I later found that a Colonel had complained about my attire to the department Chair.

One morning that summer, at the breakfast table, Virginia pointed to an advertisement in the New York Times. An anonymous "long established university in the northeast of the United States" was seeking teachers of engineering subjects including calculus and differential equations. Salaries, the ad said would be "comparable to industry". I sent off a letter at once, and I was interviewed and hired. My academic year salary increased from $5100 to $7900 and we felt rich. The "long established university" turned out to be Rensselaer Polytechnic Institute (R.P.I.). The position was not at the main campus in Troy, New York, but at the Hartford Graduate Division in Eastern Connecticut. In 1956 the nation was experiencing an acute shortage of engineers. In the Connecticut valley, the United Aircraft Company with its Pratt-Whitney subsidiary (a major manufac-

[13] the existence of "minimal" degrees. Only 31 years old, Clifford Spector died quite suddenly in 1961 of acute leukemia, a tragic loss.

turer of jet engines) had been finding it extremely difficult to hire the engineers it needed. To help to solve this problem, R.P.I. was asked to form the Hartford Graduate Division so United Aircraft engineering employees could take courses leading to a master's degree, with tuition to be paid by the company. This had helped, but not enough. So, liberal arts graduates who satisfied the minimum requirement of having completed a year of calculus and a year of physics were hired by United Aircraft and sent to the Hartford Graduate Division to study mechanical engineering. Those who completed the forty week program received a certificate and were put to work. They were also eligible to apply to R.P.I.'s master's program.

Faculty was needed to teach in this new program, and that was the reason for the ad I had answered. The Hartford Graduate Center was housed in a one-story, industrial-style building with a huge parking lot on the main highway between Hartford and Springfield, Massachusetts. Friends had predicted that moving to an environment with no research aspirations, to do elementary teaching would be the end of my research career. In fact it turned out to be an excellent move. From a personal point of view, Eastern Connecticut was beautiful and an easy drive to New York where there were friends, the amazing resources of that city, and my mother's apartment in the Bronx where she would cheerfully serve as baby sitter, and where we could spend the night. But it turned out very well professionally also. The student body were relatively mature interesting people of varied background who were fun to teach. And as the forty week program wound down, I moved into the master's level program, teaching a variety of courses far more interesting than what would have been available to a lowly assistant professor at Ohio State. Student notes for a course in functional analysis later became a short book.[14] The clerical staff turned out to be cheerful and competent and quite willing to turn my mangled and worked over manuscript for *Computability & Unsolvability* into a typescript I could send to publishers. There were mixed reviews, including one that derided the connection I was proposing with actual computers and included an invidious comparison with Kleene's recently published book (with which, by the way, the overlap was not extensive). It turned out that McGraw-Hill had chosen Hartley Rogers as their reviewer, and he not only wrote the kind of laudatory review that gladdens an author's heart, but also produced an astonishingly detailed helpful critique. The book was published in McGraw-Hill's series in "Information Processing and Computers" appearing in 1958. It was eventually translated into Japanese and Italian,[15] and, reprinted by Dover in 1982,[16] it remains in print today.[3]

The summer of 1957 was an exciting time for American logicians. A special "institute" on logic was held at Cornell University. For five weeks 85 logicians participated: established old-timers, those in mid-career, fresh Ph.D.'s, and graduate students. There was even Richard Friedberg, still an undergraduate, who

[14] [9].

[15] The translator for the Italian version called it a "classico".

[16] A review of the Dover edition by David Harel referred to the book as one of the few "classics" in computer science.

had just created a sensation by proving the existence of two r.e. sets neither of which is computable relative to the other thus solving Post's problem. There were seminars all day. The gorges of Ithaca were beautiful, and swimming under Buttermilk Falls was a summertime pleasure. Hilary Putnam and I seized the opportunity to work together. Our two families shared a house with an unusual distribution of the quarters: there were three small separate apartments; the adult couples each got one of them, and the third went to the three children, our two boys and Hilary's Erika who was two days younger than our Nathan. Hilary and I talked all day long about everything under the sun, including Hilbert's tenth problem. Hilary tended to generate ideas non-stop, and some of them were very good. I tended to be cool and critical and could be counted on to shoot down ideas that were pretty obviously bad. Hilary's idea that turned out to be very good indeed was to begin with the normal form from my dissertation and to try to get rid of that bounded universal quantifier that blocked the path to my "daring hypothesis" by using the Chinese Remainder Theorem to code the finite sequences of integers that the quantifier generates.[17] Using little more than the fact that congruences are preserved under addition and multiplication, we obtained two relations with rather simple definitions about which we were able to show that their being Diophantine would imply that every r.e. set is likewise Diophantine.[18]

We resolved to seek other opportunities to work together. Hilary suggested we try to get funding so we could spend summers together. He proposed investigations of possible computer implementations of proof procedures for first order logic.[19] I guess we thought we'd have more luck being funded with that than with Hilbert's tenth problem. We agreed to work through R.P.I. By the time we got our proposal together, it was too late to be funded for the summer of 1958 by any of the usual agencies. Someone suggested that we try the National Security Agency (NSA). I'd never heard of them, but sent the proposal along. Our idea was to define a procedure that would generate a proof of a sentence by seeking a counter-example to its negation in what later became known as its Herbrand universe. This involved generating ever longer Herbrand expansions, and testing periodically for a truth-functional inconsistency. When I was called to NSA headquarters, it turned out that it was this test for truth-functional inconsistency that interested them. They told me that this was a very hard problem, and seemed dubious of our ability to make serious inroads in just one summer, but, finally, they did agree to sponsor our work. We were to provide a report at the

---

17 $(\forall k)_{\leq y}(\exists u)$ ... is equivalent to saying that there exists a sequence $u_0, u_1, \ldots u_y$ of numbers satisfying ... The use of the Chinese Remainder theorem to code finite sequences of integers had been used by Gödel to show that any recursively defined relation could be defined in terms of addition, multiplication and purely logical operations. I had used the same device in obtaining my normal form.

18 [4].

19 Abraham Robinson had proposed similar investigations in a talk at the Cornell Institute. I attended that talk, but Hilary didn't. Somehow, I didn't connect the two ideas until years later when I noticed Robinson's paper in the proceedings of the Institute.

end of the summer. However, unlike typical funding agencies, they specifically asked that their support *not* be acknowledged in the report. Told that I'd never heard of the NSA, the reply was that their "publicity department" was doing a good job.

I found a summer cottage on Lake Coventry for Hilary and his family. As I said elsewhere about my summers with Hilary:

> We had a wonderful time. We talked constantly about everything under the sun. Hilary gave me a quick course in classical European philosophy, and I gave him one in functional analysis. We talked about Freudian psychology, about the current political situation, about the foundations of quantum mechanics, but mainly we talked mathematics.[20]

My first copy of *Computability & Unsolvability*, smelling of printer's ink arrived that summer. Elated, I showed it to Hilary. He smilingly offered to find a typographical error on any page I'd select. Determined to show him, I turned to the reverse side of the title page containg the copyright notice, only six lines. Giving the page a quick glance, Hilary noted that the word "permission" was missing its first "i".

Our report for the NSA, entitled *Feasible Computational Methods in the Propositional Calculus* is dated October 1958. It emphasizes the use of conjunctive normal form for satisfiability testing[21] (or, equivalently, the dual disjunctive normal form for tautology testing). The specific reduction methods whose use together have been linked to the names Davis-Putnam are all present in this report.[22]

After that first summer, our research was supported by the U.S. Airforce Office of Scientific Research. It was in the summer of 1959 that Hilary and I really hit the jackpot. We decided to see how far we could get with the approach we had used at Cornell if, following Julia Robinson's lead, we were willing to permit variable exponents in our Diophantine equations. That is, we tried to show that every r.e. set could be defined by such an exponential Diophantine equation. After some very hard work, using Julia Robinson's techniques as well

[20] [32] p.93.

[21] What has become known as *the satisfiability problem.*

[22] These are:

1. The *one literal rule* also known as the *unit rule.*
2. The *affirmative-negative rule* also known as the *pure literal rule.*
3. The *rule for eliminating atomic formulas.*
4. The *splitting rule,* called in the report, the *rule of case analysis.*

The procedure proposed in our later published paper used rules 1,2, and 3. The computer program written by Logemann and Loveland discussed below used 1,2, and 4. The first of these is the "Davis-Putnam procedure" generally considered for worst case analysis. The second choice is the one generally implemented, and still seems to be useful.

as a good deal of elementary analysis,[23] we had our result, but, alas, only by assuming as given, a fact about prime numbers that is certainly believed to be true, but which remains unproved to this day, namely: *there exist arbitrarily long arithmetic progressions consisting entirely of prime numbers.* As we wrote up our summer's work, we decided to include an account of a proof procedure for first order logic based on our work on the propositional calculus from the previous summer. Our report to the Air Force included the work on Hilbert's tenth problem, the proof procedure, and a separate paper on finite axiomatizability. Years later Julia Robinson brought a copy of this report[24] with her to Russia where the mathematicians to whom she showed it were astonished to learn that this work was supported by the U.S. Airforce. It was the proof procedure that brought some notoriety to the Davis-Putnam partnership. Published in the *Journal of the Association for Computing Machinery*,[25] it proposed to deal with problems in first order logic by beginning with a preprocessing step that became standard–Skolemization and reduction to conjunctive normal form–followed by a continuing Herbrand expansion interrupted by tests for satisfiability along the lines mentioned above.

We submitted our work on Hilbert's tenth problem for publication and at the same time sent a copy to Julia Robinson. Julia responded soon afterwards with an exciting letter:

> I am very pleased, surprised, and impressed with your results on Hilbert's tenth problem. Quite frankly, I did not think your methods could be pushed further ...
> I believe I have succeeded in eliminating the need for [the assumption about primes in arithmetic progression] by extending and modifying your proof.

She sent us her proof soon afterwards; it was a remarkable tour de force. She showed how to get all the primes we needed by using, instead of an unproved hypothesis about primes in arithmetic progression, the prime number theorem for arithmetic progressions which provided a measure of how frequently primes occurred "on average" in such progressions. We proposed that we withdraw our

---

[23] Among other matters, we needed to find an exponential Diophantine definition for the relation:

$$\frac{p}{q} = \sum_{k=1}^{n} \frac{1}{r + ks}.$$

We didn't go about it in the easiest way. We used the fact that

$$\sum_{k=1}^{n} \frac{1}{\alpha + k} = \frac{\Gamma'(\alpha + n + 1)}{\Gamma(\alpha + n + 1)} - \frac{\Gamma'(\alpha + 1)}{\Gamma(\alpha + 1)},$$

expanded Γ'/Γ by Taylor's theorem, and used an estimate for Γ'' to deal with the remainder.

[24] AFOSR TR59-124.

[25] Reprinted in [33], pp. 125-139.

paper in favor of a joint publication, and she graciously accepted. She undertook the task of writing up the work, and (another surprise), she succeeded in drastically simplifying the proof so only the simplest properties of prime numbers were used. Combined with Julia's earlier work, this new result showed that my "daring hypothesis" that all r.e. sets are Diophantine was equivalent to the existence of a single Diophantine equation whose solutions grow exponentially (in a suitable technical sense of the word).[26] The hypothesis that such an equation exists had been raised by Julia in her earlier work, and Hilary and I called it JR.

For years I thought of myself as an exile from New York. Now came an opportunity to move there. From the Courant Institute of Mathematical Sciences at New York University came an invitation to visit for a year. Although this was just a visiting appointment, I was confident that we would not be returning to Connecticut. Cutting our bridges behind us, we sold the house we had bought just a year before. Virginia was as enthusiastic as I about our new life. We moved into an apartment overlooking the Hudson River in the Upper West Side of Manhattan. At NYU, I was asked to teach a graduate course in mathematical logic which was a great pleasure. One of the students in that course, Donald Loveland, later became one of my first graduate students, and, still later, a colleague. One result of my association with the Courant Institute was access to their IBM 704 computer along with graduate student assistants to do the programming. I jumped at the chance to try out the proof procedure Hilary and I had proposed. Loveland and his friend George Logemann were assigned to me to do the programming. Donald was a particularly apt choice because he had been involved at IBM with Gelernter's "geometry machine", a program to prove theorems in high school geometry. It was found that the *rule for eliminating atomic formulas* (later called *ground resolution*) which replaced a formula

$$(p \vee A) \wedge (\neg p \vee B) \wedge C$$

by

$$(A \vee B) \wedge C$$

used too much RAM. So it was proposed to use instead the *splitting rule*[27] which generates the pair of formulas

$$A \wedge C \qquad B \wedge C$$

The idea was that a stack for formulas to be tested could be kept in external storage (in fact a tape drive) so that formulas in RAM never became too large. Although the "Davis-Putnam" rules proved very successful in testing ground formulas consisting of thousands of clauses, the program was overwhelmed by the explosive nature of the Herbrand expansion in all but the simplest examples.[28]

As I had expected and hoped, I was offered a regular faculty appointment at NYU. At that time, there were three more or less separate mathematics departments at NYU: the graduate department, the undergraduate department at

[26] Cf. [5].

[27] See footnote [22].

[28] Our paper was reprinted in [33] pp. 267-270.

the main campus in Greenwich Village, and another undergraduate department at the Bronx campus. The appointment I was offered was in the undergraduate department at the main campus. Although not what I had hoped for, I would certainly have accepted this offer, had not the first Sputnik gone aloft a few years earlier. The Soviet launching of a satellite in 1957 had provoked a furore in the United States. We were "falling behind" in science and technology. All at once, science became a growth industry. And that was why I received a very attractive offer from Yeshiva University.

Yeshiva College is housed in a building with a curiously Middle Eastern flavor in the part of Manhattan known as Washington Heights (because Washington fought a rearguard action against the British there as the revolutionary forces were retreating from New York). It takes its name from the traditional East European yeshivas, institutions of advanced religious training based on the Talmud with instruction mostly in Yiddish, training that could lead to rabbinical ordination. Yeshiva College adjoined to this traditional curriculum, a liberal arts program in the American mode. Various schools were added to the complex, most of them secular, leading to the "university" designation in 1945. The Mathematics Department at Yeshiva College was the home of the periodical *Scripta Mathematica*, specializing in mathematical oddities and issued regularly beginning in 1932. Abe Gelbart was a mathematician at Syracuse University who had become involved with the Scripta Mathematica effort. He began to imagine the possibility of building a first-rate graduate program in mathematics and physics in this milieu. He was able to convince the Yeshiva University administration that in the post-Sputnik atmosphere, external funding would be readily available, and he received a go-ahead to found a new Graduate School of Science (later the Belfer Graduate School of Science) with himself as dean.

My teaching load at Yeshiva was to be two graduate courses per semester with every encouragement to develop a program in logic as opposed to the NYU offer which would have required three *undergraduate* courses per semester with the option to conduct a logic seminar on my own time. In addition the Yeshiva offer came with a salary of $500 more than I would make at NYU. For various reasons I would have preferred to remain at NYU, but they were unwilling to respond to the Yeshiva offer, and so, I phoned Gelbart and accepted. Late that spring I was informed that the Courant Institute had reconsidered and were now willing to coming closer to the Yeshiva offer. However, I felt that I had made a commitment to Yeshiva that it would have been unethical to break. I was told to keep in touch and let them know if circumstances changed. It was five years before I took them up on this suggestion.

Gelbart found a home for the new school in a building not far from Yeshiva College. When I was taken to see it, I was quite startled. The building had previously been a catering palace, and I remembered it very well. It had been the scene of the celebration of my ill-fated marriage to my first wife a decade earlier. Gelbart turned out to be difficult and ill-tempered, but eager to please so long as his beneficence was duly acknowledged. The faculty, mathematicians and physicists all together on one floor, formed a very congenial group and a good deal of first-rate research was accomplished. I worked to develop a logic program and was successful in having an offer made to Raymond Smullyan which he accepted. Although Donald Loveland's degree was awarded by NYU, he was effectively part of our logic group at Yeshiva. From the beginning Robert Di Paola was at Yeshiva in order to work with me. Both Di Paola and Loveland received their doctorates in 1964.

I was able to publish several papers that were spin-offs of the work with Hilary and Julia (one of them joint with Hilary).[29] I also worked on proof procedures, a field that was beginning to be called automatic theorem proving (ATP). In fact my work with Logemann and Loveland was continuing after I had left NYU, the Courant Institute kindly continuing to make its IBM 704 available to us. It was after the program was running and its weaknesses were apparent to us that a reprint arrived in the mail that had a major influence on my thinking. It was a paper by Dag Prawitz[30] in which he showed how the kind of generation of spurious substitution instances that overwhelmed our procedure could be avoided. However, the procedure he proposed was subject to a combinatorial explosion from another direction. I set as my goal finding a procedure that combined the benefits of Prawitz's ideas with those of the Davis-Putnam procedure. I believed that we were on the right track in using as our basic data objects sets of disjunctive clauses (each consisting of *literals*) containing variables for which substitution instances could be sought. Prawitz had proposed to avoid spurious substitutions from the Herbrand universe by forming systems of equations the satisfaction of which would give the desired result. I came to realize that for problems expressed in our form, the required equalities always were such as to render literals complementary. That is given a pair of clauses one of which contains the literal $R(u_1, u_2, \ldots, u_n)$ while the other contains $\neg R(v_1, v_2, \ldots, v_n)$, what was needed was to find substitutions to satisfy the system of equations

$$u_1 = v_1 \quad u_2 = v_2 \quad \ldots \quad u_n = v_n.$$

I also saw that for any system of substitutions that was successful in producing an inconsistent set of clauses, there necessarily had to be a subset of that set which was *linked* in the following sense:

> A set of clauses is *linked* if for each literal ℓ in one of the clauses of the set, the complementary literal $\neg\ell$ occurs in one of the remaining clauses.[31]

I had the opportunity to explain these ideas and to place them in the context of existing research at a symposium organized by the American Mathematical Society on *Experimental Arithmetic* held in Chicago in April 1962 to which I was invited to participate. The ideas developed in the paper that was published in the proceedings of the conference[32] turned out to be very influential (although I

[29] [6, 7].
[30] Reprinted in [33] pp. 162-199.
[31] Here if $\ell = \neg R(c_1, c_2, \ldots, c_n)$ then it is understood that by $\neg\ell$, the literal $R(c_1, c_2, \ldots, c_n)$ is meant.
[32] Reprinted in [33], pp. 315-330.

believe that many of those whose work was ultimately based on this paper were unaware of the fact).

Around this time I had been invited to spend several hours weekly as a consultant at Bell Labs in Murray Hill, New Jersey. I was delighted to have the opportunity to see some of my ideas implemented. Doug McIlroy undertook to produce a working program for Bell Labs' IBM 7090, and did so in short order. The problem of finding solutions to the systems of equations needed to establish "links" was dealt with in McIlroy's program by using what was later called *unification.* Peter Hinman joined the effort as a summer Bell Labs employee and found and corrected some bugs in the McIlroy version. We wrote up our work and submitted it for publication. It was accepted with some rather minor changes. These changes were not made, and the paper never appeared.

The year 1963 brought great excitement to the world of logic. Paul Cohen invented a powerful new method he called forcing for constructing models of the axioms of set theory, and he had used this method to show that Cantor's continuum hypothesis could not be proved from the standard axioms for set theory, the Zermelo-Fraenkel axioms together with the axiom of choice. This settled a key question that had been tacitly posed by Gödel when more than two decades previously, he had shown that the continuum hypothesis couldn't be disproved from those same axioms. I was astonished to receive a letter from Paul Cohen dated November of that year reading in part:

> I really should thank you for the encouragement you gave me in Stockholm. You were directly responsible for my looking once more at set theory. ... Of course, the problem I solved had little to do with my original intent. In retrospect, though, the basic ideas I developed previously played a big role when I tried to think of throwing back a proof of the Axiom of Choice, as I had previously thought about throwing back a proof of contradiction.

In the summer of 1962 I had attended the International Congress of Mathematicians in Stockholm. These conferences are scheduled to occur every four years, but this was my first since the 1950 Congress at Harvard. At the Congress, I talked briefly with Paul Cohen. I knew that although he was not primarily a logician, he was a very powerful mathematician who had been attempting to find a consistency proof for the axioms of set theory. He indicated that some logicians he had talked to had been discouraging, and I urged him to pay no attention. That was really the total extent of my "encouragement". Of course, I was very pleased to receive the letter.

It was in 1963 that we realized that we were outgrowing our apartment overlooking the river. Our two sons had been sharing a bedroom. Now aged eight and ten, they had quite different temperaments. If we were to have any peace, they would have to have separate rooms. At this point Virginia found a "brownstone" town house a mile south of our apartment that we were able to buy . Although the price we paid was ridiculously low by later standards, the house and its renovation put an enormous strain on our budget. Of course, it turned out to be far and away the best investment we ever made. We lived there for 33 years. In order to make ends meet, I found myself becoming an electrician and a plumber. With the help of a friend (as much a novice as I), I even installed a new furnace.

A project that was absorbing a good deal of time and energy at this time was the preparation of my anthology of fundamental articles by Gödel, Church, Turing, Post, Kleene, and Rosser.[33] I wrote some of the commentaries for the book while attending a conference in the delightful town of Ravello south of Naples.

Meanwhile my relationship with Abe Gelbart was becoming more and more difficult. Things were brought to a head in the spring of 1965 when, interviewing a prospective faculty member, someone I had very much hoped would be a colleague, Gelbart behaved in an insulting manner. I decided that I had no choice but to resign my position. I called a friend at the Courant Institute and reminded him of the suggestion that I let them know if I were interested in a position at NYU. Soon enough an offer arrived. It was not quite what I had expected: the position was half at the Bronx campus and only half at the Institute. However, I was urged to regard the relative vacuum on the Bronx campus as an opportunity to develop a logic group there, and I was assured that I would be treated as a regular member of the graduate faculty. I accepted the posititionn and remained with the Courant Institute until my retirement in September 1996.

I took to the notion of developing a logic group at the Bronx campus with avidity. My old friend and student Donald Loveland was already there, and he was soon joined by the Yasuharas, Ann and Mitsuru. I was able to provide support in the form of released time for research from teaching from my continued funding by the Air Force. During these years Norman Shapiro, with whom I had organized a logic seminar in Princeton many years earlier, and my former student Bob Di Paola were both at the RAND Corporation. Norman arranged for me to be able to spend summers at RAND. Our family found housing in the hills above Topanga Canyon near the Malibu coast, and I enjoyed the daily drive along the beach to the RAND facility in Santa Monica. I was required to obtain security clearance at the "Secret" level, not because I did any secret work there, but because classified documents in the building were not necessarily under lock and key. I was tempted once to use my clearance. The Cultural Revolution was in full swing in China, and I was thoroughly mystified by it. Di Paola urged me to seek enlightenment by looking at the intelligence reports from the various agencies easily available to me. I did so, feeling that I was losing my innocence, and was greatly disappointed. Not only did these "secret" reports contain nothing that couldn't be found in newspapers and magazines, they turned out to be anything but unbiased, clearly reflecting the party line of the agency from which they emanated.

What I did at RAND was work on Hilbert's tenth problem, specifically I tried to prove JR. I used the computing facility at RAND to print tables of Fibonacci numbers and solutions of the Pell equation looking for patterns that would do

[33] [8].

the trick. I also found one interesting equation:

$$9(x^2 + 7y^2)^2 - 7(u^2 + 7v^2)^2 = 2.$$

I proved that JR would follow if it were the case that this equation had only the trivial solution $x = u = 1; y = v = 0$.[34] In fact the equation turns out to have many non-trivial solutions, but the reasoning actually shows that JR would follow if there are only finitely many of them, and this question remains open.

In the academic year 1968-1969 I finally had a sabbatical leave. I would have been due for one at Yeshiva, and as part of my negotiation with NYU, I secured this leave. I spent the year in London loosely attached to Westfield College of the University of London where I taught a "postgraduate" course on Hilbert's tenth problem. I continued efforts to prove JR (and thereby settle Hilbert's tenth problem). I found myself working on sums of squares in quadratic rings, but I didn't make much progress. Meanwhile "Swinging London" was in full bloom with the mood of the "sixties" very much in evidence. Although not quite swept away by the mood, I did not entirely escape its influence.

While I was in London my old friend Jack Schwartz, now a colleague at NYU, was working to found a computer science department in the Courant Institute. I was pressed by the Courant Institute to become part of the new department. I accepted but not with alacrity. Among other issues, it meant abandoning my logic group in the Bronx. In fact, Fred Ficken, the amiable Chair at the Bronx campus was about to retire and the applied mathematician Joe Keller had agreed to take on this role with the intention of making the Bronx campus a bastion of applied mathematics. So my group didn't have much future. What neither Joe nor I knew was that the entire Bronx campus would be shut down a few years later because NYU found itself a financial crunch.

I had been back in New York only a few months when I received an exciting phone call from Jack Schwartz. A young Russian had used the Fibonacci numbers to prove JR! Hilbert's tenth problem was finally settled! I had been half-jokingly predicting that JR would be proved by a clever young Russian, and, lo and behold, he had appeared. (I met the 22 year old Yuri Matiyasevich in person that summer at the International Congress in Nice.) After getting the news I quickly phoned Julia, and about a week later, I received from her John McCarthy's notes on a lecture he had just heard in Novosibirsk on the proof. It was great fun to work out the details of Yuri's lovely proof from the brief outline I had. I saw that the properties of the Fibonacci numbers that Yuri had used in his proof had analogues for the Pell equation solutions with which Julia had worked and I enjoyed recasting the proof in those terms. I also wrote a short paper in which I derived some consequences of the new result in connection with the number of solutions of a Diophantine equation.[35]

To make the proof widely accessible, I wrote a survey article for the *American Mathematical Monthly* which was later reprinted as an appendix to the Dover edition of *Computability & Unsolvability* [14]. In addition, I collaborated with

[34] [10].
[35] [11],[13].

Reuben Hersh on a popular article on the subject for the *Scientific American* [15]. Suddenly awards were showered on me. For the *Monthly* article I received the Leroy P. Steele prize from the American Mathematical Society and the Lester R. Ford prize from the Mathematical Association of America, and for the Scientific American article Reuben Hersh and I shared the Chauvenet prize, also from the Mathematical Association of America. I was also invited by the Association to give the Hedrick lectures for 1976.

In May 1974, the Society sponsored a symposium on mathematical problems arising from the Hilbert problems, and of course, Yuri was invited to speak on the tenth. But he was unable to get permission form the Soviet authorities to come. So, Julia was invited instead, and she agreed on condition that I be invited as well to introduce her. When it came to writing a paper for the proceedings of the symposium, we agreed that it should be by the three of us. Yuri's contribution faced the bureaucratic obstacle that any of his draft documents had to be approved before being sent abroad. But there was no such problem with letters. So he would send letters to Julia on his parts of the article generally beginning,

> Dear Julia,
>
> Today I would like to write about ....

One of his topics was "Famous Problems". The idea was that the same techniques that had been used to show that there is no algorithm for the tenth problem could also be used to show that various well-known problems were equivalent to the non-existence of solutions to certian Diophantine equations. One of Yuri's letters did this for the famous Riemann Hypothesis, an assertion about the complex zeros of the function $\zeta(s) = \sum_{n=1}^{\infty} \frac{1}{n^s}$ which remains unproved although it has important implications for the theory of prime numbers. The Hypothesis can be expressed in terms of the values of certain contour integrals, and Yuri's technique was to approximate these integrals by sums in a straightforward way. It was done in a very workman-like way, but it seemed very inelegant to me. I went to my colleague Harold Shapiro, an expert in analytic number theory, and he told me what to do. Julia was so pleased by the result that she sent me a note I kept on my bulletin board for years saying "I like your reduction of RH immensely". [36]

My joint appointment in mathematics and the new Computer Science Department defined my new situation at Courant when I returned from London. I had been flirting with computers and computer science for years. But now I had come out of the closet and identified myself as a computer scientist. The new Computer Science Department was developing its own culture and clashes with the mathematicians at Courant were not infrequent. It didn't help that among the mathematicians were some outstanding scientists that thoroughly outclassed our young department. To begin with our graduate students took the same exams as the mathematics students, and hiring was done by the same committee that hired mathematicians. The evolution towards autonomy was slow and painful. In the spring of 1973, two applied mathematicians and I constituted a

[36] [16].

hiring committee for computer science. The mathematicians were both heavily involved with scientific computing, but neither had any real appreciation or understanding of computer science as an autonomous discipline. I remember all too well a particular tense lunch meeting on a Friday in June of that year in which possible appointments were discussed in an atmosphere I did not find friendly. When I left the meeting I became aware of a sensation like a brick placed on my chest. I continued to experience this disagreeable sensation through the weekend and finally entered the hospital where a myocardial infarction was diagnosed. Although in retrospect I had done plenty to bring this about by poor diet and lack of exercise, I have always thought that that disagreeable meeting played a precipitating role. Before my heart attack, I had been an enthusiastic New Yorker even when in exile; but now I began to yearn to live someplace where I could have a rich professional life without the tension that I found in everyday life in New York.

Over the years I continued to have doctoral students from both the Mathematics Department and the Computer Science Department. I certainly taught hard-core computer science courses, beginning programming and data structures, and on the graduate level, theory of computation, logic programming, and artificial intelligence. In the early years I could also count on teaching mathematical logic, set theory, and even nonstandard analysis. Unfortunately for me this flexibility gradually vanished, a fact that contributed to my decison to retire from NYU in 1996. But this is getting ahead of the story.

In the 1970s the Italian universities were still not offering Ph.D. programs. Having received their bachelor degree, graduates were entitled to use the title "Dottore". The C.N.R., the agency of the Italian government involved with scientific research, concerned to do something about the inadequate training young Italian mathematicians received, set up summer programs in which they were exposed to graduate level courses. I was invited to give such a course on Computability in the lovely town of Perugia during the summer of 1975. This was the beginning of a connection between Italian computer science and the Courant Institute. A number of my students from that course became graduate students at NYU. Two in particular, Alfredo Ferro and Eugenio Omodeo, obtained Ph.D.'s in computer science from NYU. Ferro went back to his home town of Catania in Sicily where he started a computer science program at the university there, and sent his own students back to NYU. The relationship continues to be active.

As an undergraduate I had tried briefly to rehabilitate Leibniz's use of infinitesimal quantities as a foundation for calculus. It was easy enough to construct algebraic structures containing the real numbers as well as infinitesimals; the problem that baffled me was how to define the elementary functions such as sin and log on such structures. It was therefore with great excitement and pleasure that I heard Abraham Robinson's address before the Association for Symbolic Logic towards the end of 1961 in which he provided an elegant solution to this problem using techniques that he dubbed *nonstandard analysis.* Some years later together with Melvin Hausner, my roommate at Princeton and now a colleague, I started an informal seminar on the subject. We had available Robinson's treatise and some rather elegant lecture notes by Machover and Hirschfeld. Hausner was inspired to apply the technique to prove the existence of Haar measure. Reuben Hersh and I wrote a popular article on nonstandard analysis, also for the *Scientific American.*[37] Nonstandard analysis really tickled my fancy. As I wrote in the flush of enthusiasm:

> It is a great historical irony that the very methods of mathematical logic that developed (at least in part) out of the drive toward absolute rigor in analysis have provided what is necessary to justify the once disreputable method of infinitesimals. Perhaps indeed, enthusiasm for nonstandard methods is not unrelated to the well-known pleasures of the illicit. But far more, this enthusiasm is the result of the mathematical simplicity, elegance, and beauty of these methods and their far-reaching application.

I taught nonstandard analysis at Courant and benefited from class notes prepared by my student Barry Jacobs. In the summer of 1971, I taught it again at the University of British Columbia. Finally I wrote a book (the quotation above is from the preface).[38]

For the academic year 1976-77, I was able to go on sabbatical leave. I had spent two summers in Berkeley and was eager to try a whole year. John McCarthy (who had been a fellow student at Princeton) hired me to work for the month of July at his Artificial Intelligence Laboratory at Stanford University. I loved the atmosphere of play that John had fostered. The terminals that were everywhere proclaimed "Take me, I'm yours", when not in use. I was encouraged to work with the FOL proof checker recently developed by Richard Weyhrauch. Using this system, I developed a complete formal proof of the pigeon-hole principle from axioms for set theory. I found it neat to be able to sit at a keyboard and actually develop a complete formal proof, but I was irritated by the need to pass through many painstaking tiny steps to justify inferences that were quite obvious. FOL formalized a "natural deduction" version of **F**irst **O**rder **L**ogic. The standard paradigm for carrying out inferences was to strip quantifiers, apply propositional calculus, and replace quantifiers. I realized that from the viewpoint of Herbrand proofs, each of these mini-deductions could be carried out using no more than one substitution instance of each clause. I decided that this very possibility provided a reasonable characterization of what it means for an inference to be *obvious*. Using the LISP source code for the linked-conjunct theorem prover that had been developed at Bell Labs, a Stanford undergraduate successfully implemented an "obvious" facility as an add-on to FOL. I found that having this facility available cut the length of my proof of the pigeon-hole principle by a factor of 10. This work was described at the Seventh Joint International Congress on Artificial Intelligence held in Vancouver in 1981 and reported in the Proceedings of that conference [20].

---

[37] Actually, as I remember it, we worked on that article and the one one Hilbert's tenth problem for which we received the Chauvenet prize pretty much at the same time. The one on nonstandard analysis appeared in 1972 [12], a year before the prize-winning article.

[38] [17].

During the 1976-77 academic year, it was a great pleasure to be able to interact with the Berkeley logic group and especially with Julia Robinson. We worked on the analogue of Hilbert's tenth problem for strings under concatenation, but didn't make much progress. It had at one time been thought that proving this problem unsolvable would be the way to obtain the desired unsolvability result for the Diophantine problem. Julia guessed that the string problem was actually decidable, and she turned out to be right as we learned when we got word of Makanin's positive solution of the problem. At Berkeley that year, I taught two trimester courses, an undergraduate computability theory course for Computer Science and a graduate course in nonstandard analysis for Mathematics. For the nonstandard analysis course, I was able to use my newly published book as a text. It was a class of about thirty students, and a little intimidating. It was clear to me that among these Berkeley educated students were a number who were far better versed in model theory (the underlying basis for nonstandard analyis) than I.

Ever since my days with Hilary Putnam, I have had a continuing interest in the foundations of quantum mechanics. A preprint I received from the logician Gaisi Takeuti caught my attention as having important ramifications for quantum theory. This paper explored Boolean-valued models of set theory using algebras of projections on a Hilbert space. Boolean-valued models (in which the "truth value" of a sentence can be any element of a given complete Boolean algebra, rather than being restricted to the usual two element algebra consisting of {`true, false`}), had been studied as an alternative way to view Paul Cohen's forcing technique for obtaining independence results in set theory. What Takeuti found was that the real numbers of his models were in effect just the self-adjoint operators on the underlying Hilbert space. Since a key element in "quantizing" a classical theory is the representation of "observables" by such operators, I felt that the connection was surely no coincidence. I wrote a short paper about the application of Takeuti's mathematics to quantum mechanics, and I was very pleased when it was published in the *International Journal of Theoretical Physics* [18].

I worked at John McCarthy's AI lab again, and this time John asked me to think about some questions involving so-called non-monotonic reasoning. I wrote a pair of short notes which John later arranged to be combined for publication in *Artificial Intelligence* [19].

I spent the academic year 1978-79 as a Visiting Professor at the Santa Barbara campus of the University of California. There was some mutual interest in a permanent appointment, but it all faded away as a consequence of wrangling over the status of the campus's two computer science programs: the one in the Mathematics Department and the one in Electrical Engineering. On my return to New York, I met a new faculty member Elaine Weyuker with whom I was to find a number of shared interests. Although trained as a theoretical computer scientist, she had moved into the turbulent field of software testing. Of course all software must be tested before it is released. Often, in practice this testing phase is ended simply because some deadline is reached or because funding runs out. From an academic point of view, the field invites attention to the problem of finding a more rational basis for the testing process. Elaine and I wrote two papers attempting to provide an explication for the notion of test data adequacy.[39]

I had been teaching theory of computation for many years, and had developed lecture notes for some of the topics covered. For a long time I had wanted to produce a book based on my course, but had never found the time or energy to complete the task. Elaine came to the rescue adding the needed critical dose of energy. In addition, she produced lots of exercises, and tested some of the material with undergraduates. The book was published and was sufficiently successful that we were asked to update the book for a second edition. Neither of us being willing to undertake this, we coaxed Ron Sigal, who had written a doctoral dissertation under my supervision, to join the team as a third author largely in charge of the revision [23].

The CADE (Conference on Automated Deduction) meetings were occurring annually devoted to theoretical and practical aspects of logical deduction by computer. The organizers of the February 1979 CADE meeting in Austin, realizing that that year was the centennial of Frege's *Begriffsschrift* in which the rules of quantificational logic were first presented, thought that it would be appropriate to have a lecture that would place their field in a proper historical context. Their invitation to me to give such a lecture fundamentally changed the direction of my work. I found that trying to trace the path from ideas and concepts developed by logicians, sometimes centuries ago, to their embodiment in software and hardware was endlessly fascinating. Since then I have devoted a great deal of time and energy to these questions. I've published a number of articles and given many lectures with a historical flavor.[40] For 1983-84, when I was again on sabbatical leave, I received support from the Guggenheim Foundation for this work.

One key figure whose ideas I tried to elucidate was my old teacher Emil L. Post. I lectured on his work on a number of occasions including one talk at Erlangen in Germany. It was very much a labor of love to edit his collected works.[41] My current project is a semi-popular book (to be called *Leibniz's Dream*) tracing ideas that have eventually turned up in computational practice, from the seventeenth century to the present. I am particularly eager to emphasize the importance of ideas being pursued for their own sake without necessarily expecting the immediate practical payoff that nowadays is generally what is sought.

For the two academic years 1988-90, I was Chair of the Computer Science Department at NYU. I had always felt that I would not be happy in an administrative position, and this experience did nothing to change my mind. I would have been hopelessly swamped without the help of the department's capable and ultra-conscientious adminstrative assistant Rosemary Amico. The NYU central administration had been increasingly unhappy with the fact that the Courant

[39] [24, 26].
[40] [21, 22, 25, 27–29].
[41] [31].

Institute as a whole was running an increasing deficit each year. At the same time, the CS department was encouraged to improve its national standing among research-oriented CS departments. The administration was said to be surprised and pleased that our department was rated among the top twenty in the nation, and we were urged to produce a plan showing how we could move up to the top ten. Assuming that the central administration understood that this would require their providing additional resources, the department prepared an ambitious plan calling for expansion in a number of directions. The central administration did not deign to reply.

After my term of office was over, it was time for another sabbatical leave. The fall 1990 semester was spent in Europe. Our first stop was Heidelberg where I lectured at the local IBM facility and at a logic meeting at the university. Next, a series of lectures on Hilbert's tenth problem at a conference in Cortona in the north of Italy. Then a month visiting Alfredo Ferro at the University of Catania in Sicily. The fall semester was completed with a stay at the University of Patras in Greece sponsored by Paul Spirakis, and we were home in time for Christmas. I had completed an important article the day before our departure from Patras, had printed it, and left a copy on a secretary's desk with a note asking her to make copies. Our departure was to be by car ferry to Italy scheduled for the following midnight. The next morning I arrived on the campus to discover that students had occupied the building where I'd been working, and were permitting no one to enter. This was dismaying. I had no copy of my article; it was stored in a VAX that I couldn't access, and the only hard copy was on the secretary's desk. At this point a faculty member, who had become a friend, appeared and, ascertaining the problem, spoke briefly to one of the students. Evidently a deal was struck. I got out my key to the massive doors locking the computer science section, and the three of us entered. There was the hard copy of my article where I had left it and a copying machine, and I soon had several copies one of which my friend kept to send to the editor in Germany. Meanwhile the student helped himself to the copier to duplicate a handwritten document, doubtless a manifesto. We left and I was permitted to lock up.

Virginia and I took our friend and his wife to dinner that evening. Finally I asked him what he had said to the student that turned the trick. His reply was not at all what I had expected. "I reminded him that he was applying to Courant, and told him that you are the Chairman." Our ship due to sail at midnight didn't actually leave before 3AM. It turned out that the stabilizers were not functioning, and the voyage to Ancona took a day longer than scheduled with me being seasick most of the way. We drove to Paris in time for our flight back to New York. But our stay in New York was very short. Over the years Virginia had accompanied me on many trips. Now it was my turn to accompany her. Virginia had become a textile artist with an international reputation. She had become particularly adept at mastering folk techniques and using them to make works of art. For 1991 she had been awarded a three month Indo-American Fellowship[42] to study textiles in India. Of course I came along.

Our scheduled departure date was January 15. That was also the date on which President Bush's ultimatum to Saddam Hussein demanding that his forces leave Kuwait was expiring. Friends urged us to abandon our travel plans at such an uncertain time, but we decided to go ahead. After a delay caused by a bomb scare at Kennedy airport, we arrived in New Delhi to learn that bombs were dropping on Baghdad. Given the chaos just outside airports in India with throngs insistently offering their services we were delighted to be met by representatives of the American Institute for India Studies (AIIS) who took us to their guest house. The next morning we found other American fellowship recipients in a state of panic. The U.S. State Department had issued an advisory to the effect that non-essential American personnel leave India at once. Most of the others agreed to postpone their fellowship periods and left. We decided to remain. So we were in India for the entire duration of the Gulf War. In an odd way, the situation was advantageous for us. The lack of tourists meant that it was easy to get reservations and services, and the AIIS guest house was always available. The U.S. embassy, which had been transformed into a virtual fortress, was the target of virtually daily vituperative demonstrations by militant Muslim groups, but we ourselves had no problems.

The textiles Virginia was most eager to study were in the state of Orissa, one of the poorest states in India, just south of Calcutta, and we spent most of our time there. I had a new job: I was Virginia's camera man. My job was to use the video camera to record textile processes; we accumulated twelve hours of raw footage. There was a week-long tour of some of the the small villages of Orissa, where often, there were no hotels even minimally acceptable by U.S. standards. In one village, we were put up in the guest house of a cotton spinning factory.

In India the contrasts between the best and the worst is enormous. We saw people lying on the sidewalks of Calcutta waiting to die, and we had lunch with a matriarch whose huge family estate is guarded by a private police force and whose foot was kissed by her servants when she permitted them to take the lunch leftovers home to their families. The best educational and scientific research institutions are first-rate by any standard. On my previous sabbatical, I had spent a month as the guest of the Tata Institute of Fundamental Research (TIFR)in Bombay, and I was able to visit them again briefly this time. In addition I lectured at the Indian Institute of Technology (IIT) in New Delhi, an outstanding school whose entrance examinations in mathematics are quite formidable. (At IIT and TIFR I was able to collect my email.) But I also lectured at colleges, allegedly institutions of higher learning, that were sadly weak.

On our way back to New York from India, we stopped in Europe. I spent a week at the University of Udine as the guest of Professor Franco Parlemento who had been a student in my Computability course in Perugia two decades earlier. Then we went to the wonderful mathematical research institute at Oberwolfach

[42] These fellowships are administered by the CIES, the same office that manages Fullbright awards.

in Germany, an institute that started its successful life as an effort by German mathematicians to save their talented young people from becoming cannon fodder during the second world war. There are week-long conferences through the year on a great variety of mathematical subjects. On this occasion, it was on automatic theorem proving organized by Woody Bledsoe and Michael Richter, and a follow-up to a similar meeting fifteen years earlier.

Back in New York, and back to teaching, I was approaching that sixty-fourth birthday the Beatles had sung about, and beginning to wonder how I wanted to spend the rest of my life. The things that really interested me seem to be of less and less importance to my colleagues. I had my very own course called *Arithmetic Undecidability*; in a whirlwind semester I covered the elements of first order logic through the Gödel completeness theorem, Hilbert's tenth problem, and the essential undecidability of Peano arithmetic. I taught it for the last time in the spring 1993 semester, and was rebuffed in my request to teach it again. I taught the introductory programming course for computer science majors, and indeed supervised the sections taught by others, for three successive years. I love to program, and at first, I enjoyed these courses. But after a while, I did ask myself: do I really want to be teaching Pascal to classes of 60 students not all of whom are especially receptive, at this stage of my life? A triple coronary bypass operation in January 1994 brought matters to a head. The operation was very successful, but it certainly forced me to face my mortality. In short I decided to investigate retirement possibilities. May 17, 1996 was "Martin Davis Day" at the Courant Institute. Organized by my old friends Jack Schwartz and Ricky Pollack, there were eight speakers: two from Italy (my student Eugenio Omodeo, a Perugia veteran, and Mimmo Cantone, one of Alfredo Ferro's protégés), my first two students Bob Di Paola and Don Loveland, Hilary Putnam, Elaine Weyuker, Ron Sigal (another ex-student and Elaine and my third man), and my college chum John Stachel.

As I write this, I've been retired for a year. My study is in a house in the Berkeley hills, and I am enjoying the dazzling reflection of the late afternoon sun in San Francisco Bay. My older son, his wife and their very tiny baby (born a month ago, three months early, but doing very well) are here in Berkeley. The facilities of the University of California are available to me, and I have to pick and choose among the seminars given by members of the outstanding logic faculty. I have completed three chapters of *Leibniz's Dream*, and, other projects like this one permitting, I spend most of my time working on it. Life is good.

August 1997

References

1. Davis, Martin, "A Note on Universal Turing Machines," *Automata Studies*, C.E. Shannon and J. McCarthy, editors, Annals of Mathematics Studies, Princeton University Press, 1956.
2. Davis, Martin, "The Definition of Universal Turing Machine," *Proceedings of the American Mathematical Society,* vol.8(1957), pp. 1125-1126.

3. Davis, Martin, *Computability & Unsolvability,* McGraw-Hill, New York 1958; reprinted with an additional appendix, Dover 1983.
4. Davis, Martin and Hilary Putnam, "Reductions of Hilbert's Tenth Problem," *Journal of Symbolic Logic,* vol.23(1958), pp. 183-187.
5. Davis, Martin, Hilary Putnam and Julia Robinson, "The Decision Problem for Exponential Diophantine Equations," *Annals of Mathematics,* vol.74(1961), pp. 425-436.
6. Davis, Martin and Hilary Putnam, "Diophantine Sets over Polynomial Rings," *Illinois Journal of Mathematics,* vol.7(1963), pp. 251-255.
7. Davis, Martin, "Extensions and Corollaries of Recent Work on Hilbert's Tenth Problem," *Illinois Journal of Mathematics,* vol.7(1963), pp. 246-250.
8. Davis, Martin ed., *The Undecidable,* Raven Press 1965.
9. Davis, Martin, *First Course in Functional Analysis,* Gordon and Breach, 1967.
10. Davis, Martin,"One Equation to Rule Them All," *Transactions of the New York Academy of Sciences,* Sec. II, vol.30(1968), pp. 766-773.
11. Davis, Martin, "An Explicit Diophantine Definition of the Exponential Function," *Communications on Pure and Applied Mathematics,* vol.24(1971), pp. 137-145.
12. Davis, Martin and Reuben Hersh, "Nonstandard Analysis," *Scientific American,* vol.226(1972), pp. 78-86.
13. Davis, Martin, "On the Number of Solutions of Diophantine Equations," *Proceedings of the American Mathematical Society,* vol.35(1972), pp. 552-554.
14. Davis, Martin, "Hilbert's Tenth Problem is Unsolvable," *American Mathematical Monthly,* vol.80(1973), pp. 233-269; reprinted as an appendix to the Dover edition of [3].
15. Davis, Martin and Reuben Hersh, "Hilbert's Tenth Problem," *Scientific American,* vol.229(1973), pp. 84-91; reprinted in Abbott, J.C. (ed.) *The Chauvenet Papers,* vol. 2, pp. 555-571, Math. Assoc. America, 1978.
16. Davis, Martin, Yuri Matijasevic and Julia Robinson, "Hilbert's Tenth Problem: Diophantine Equations: Positive Aspects of a Negative Solution," *Proceedings of Symposia in Pure Mathematics,* vol.28(1976), pp. 323-378.
17. Davis, Martin, *Applied Nonstandard Analysis,* Interscience-Wiley, 1977.
18. Davis, Martin, "A Relativity Principle in Quantum Mechanics," *International Journal of Theoretical Physics,* vol.16(1977), pp. 867-874.
19. Davis, Martin, "Notes on the Mathematics of Non-Monotonic Reasoning," *Artificial Intelligence,* vol.13(1980), pp. 73-80.
20. Davis, Martin, "Obvious Logical Inferences," *Proceedings of the Seventh Joint International Congress on Artificial Intelligence,* 1981, pp. 530-531.
21. Davis, Martin, "Why Gödel Didn't Have Church's Thesis," *Information and Control,* vol.54 (1982), pp. 3-24.
22. Davis, Martin, "The Prehistory and Early History of Automated Deduction," Siekmann, Jörg and Graham Wrightson (eds), *Automation of Reasoning, vol. 1,* Springer Verlag, 1983, pp. 1-28.
23. Davis, Martin and Elaine J. Weyuker, *Computability, Complexity, and Languages,* Academic Press, 1983. Second edition with Ron Sigal 1994.
24. Davis, Martin and Elaine J. Weyuker "A Formal Notion of Program-Based Test Data Adequacy," *Information and Control,* vol.56(1983), pp. 52-71.
25. Davis, Martin, "Mathematical Logic and the Origin of Modern Computers," *Studies in the History of Mathematics,* pp. 137-165. Mathematical Association of America, 1987. Reprinted in *The Universal Turing Machine - A Half-Century Survey,* Rolf Herken, editor, pp. 149-174. Verlag Kemmerer & Unverzagt, Hamburg, Berlin 1988; Oxford University Press, 1988.

26. Davis, Martin and Elaine J. Weyuker., "Metric Space Based Test Data Adequacy Criteria," *The Computer Journal,* vol. 31(1988), pp. 17-24.
27. Davis, Martin, "Influences of Mathematical Logic on Computer Science," in *The Universal Turing Machine - A Half-Century Survey,* Rolf Herken, editor, pp. 315-326. Verlag Kemmerer & Unverzagt, Hamburg, Berlin 1988; Oxford University Press, 1988.
28. Davis, Martin, "Emil Post's Contributions to Computer Science," *Proceedings Fourth Annual Symposium on Logic in Computer Science,* pp. 134 - 136, IEEE Computer Society Press, Washington, D.C.,1989.
29. Davis, Martin, "American Logic in the 1920s," *Bulletin of Symbolic Logic,* vol. 1 (1995), pp. 273-278.
30. Kleene, S.C.,"Recursive Predicates and Quantifiers," *Trans. Amer. Math. Soc.,* vol. 53(1943), pp. 41-73. Reprinted in [10], pp. 254-287.
31. Post, Emil L., *Solvability, Provability, Definability: The Collected Works of Emil L. Post,* edited by Martin Davis with a biographical introduction. Boston, Basel, & Berlin. Birkhäuser 1994.
32. Reid, Constance, *Julia: A Life in Mathematics,* Mathematical Association of America, Washington, D.C. 1996.
33. Siekmann, Jörg and Graham Wrightson (eds), *Automation of Reasoning 1: Classical Papers on Computational Logic 1957-1966,* Springer-Verlag, Berlin 1983.

Edsger W. Dijkstra

Professor Edsger W. Dijkstra (1930) worked for 10 years as programmer at the Mathematical Centre in Amsterdam, for 11 years as Full Professor of Mathematics at the Eindhoven Institute of Technology, for 11 years as Research Fellow of Burroughs Corporation, and now—i.e., 1997—for 13 years as Professor of Computer Sciences at the University of Texas at Austin.
Professor Dijkstra has done pioneering work in graph algorithms, programming language implementation, operating systems, semantic theories and programming methodology in the broadest sense of the word, and has invented a number of seminal problems (mutual exclusion, dining philosophers, distributed termination detection, and self-stabilization).
His current interests focus on the formal derivation of proofs and programs, and the streamlining of the mathematical argument in general.

EWD1166: From my Life

(Written because people ask me for these data.)

I was the third of four children. My father was an excellent chemist, who worked in the town of Rotterdam, first as a teacher of chemistry at a secondary school, later as its superintendent. His knowledge was vast and he was perfectly able (and willing) to teach cosmography or physics when needed. He made a few inventions, one of which really sophisticated. Among chemists all over the country he was well-known—he had, for instance, been President of the Dutch Chemical Society—but steadfastly refused all forms of scientific promotion, explaining that he would rather be the first among the counts than the last among the dukes, but the net effect of adhering to this lofty principle was that he lost the colleagues that were intellectually his equals to the university campus. In the vague idealism after WWII, Dutch secondary education deteriorated to such an extent that he forbade me to become a schoolteacher. My mother was a brilliant mathematician who had no job—as was customary in those days, even for most women with a college degree. Her mind was much less encyclopaedic than that of my father but very quick; she had a great agility in manipulating formulae and a wonderful gift for finding very elegant solutions, and I learned a lot from her though she had not the patience to be a good teacher.

During the last class of the Gymnasium—that was 1947/48—I considered going to University to study law, as I hoped to represent my country in the United Nations. Then we did our final exams. Before we got our grades and were told whether we had passed, I was very depressed, for I reviewed all the mistakes and stupidities I had committed and I got convinced that I would fail. (My consolation was the conviction that I would get a second chance, always having been the best pupil in the class.) My grades turned out to be better than most teachers had ever seen at a final examination, and older and wiser people convinced me that it was safer to base my choice on ability than on idealism, and that it would be a pity if I did not devote myself to science. So I went to Leyden University to study mathematics and physics during the first years, and theoretical physics during the latter part.

On the whole I enjoyed the years a the university very much. We were very poor, worked very hard, never slept enough and often did not eat enough, but life was incredibly exciting. (Now, more than 40 years later, I still remember the first "international" piano recital I attended: Clara Haskil playing Mozart!) Another thing I remember—because it surprised me—was that a number of my professors—Kloosterman, Haantjes, Gorter, Kramers—took me seriously.

My father, who was a subscriber to "Nature", had seen the announcement of a three-week course in Cambridge on programming for an electronic computer (for the EDSAC, to be precise), and when I had passed the midway exam before the end of my third year—which was relatively early—he offered me by way of reward the opportunity to attend that course. I thought it a good idea, because I intended to become an excellent theoretical physicist and I felt that in my effort to reach that goal, the ability to use an electronic computer might come in handy. The course was in September 1951; by accident, A. van Wijngaarden, then Director of the Computation Department of the Mathematical Centre in Amsterdam, heard of my plans, invited me for a visit, and offered me a job. Due to obligations in Leyden I could accept that invitation only in March 1952, when I became the first Dutchman with on the payroll the qualification "programmer". Professor H.A. Kramers, who had accepted me as student of theoretical physics, died, I did not like his successor, and for a while I considered continuing my studies in Utrecht, but it turned out that the laboratory work I would have to do in Utrecht was incompatible with my half time job at the Mathematical Centre. That was the moment I had to choose between theoretical physics and computing; I chose the latter and my study of physics in Leyden became a formality. I got my degree in the Spring of 1956, moved immediately from Leyden to Amsterdam, and from that moment on I worked full time at the Mathematical Centre.

My closest collaborators there were Bram J. Loopstra and Carel S. Scholten, who had been hired (while they were still students of experimental physics at the Municipal University of Amsterdam) to build a computer for the mathematical Centre. They were about five years older than I, had been close friends since the age of 12, and had already worked on their machinery for quite a few years when I joined them. When I came their machine did not work, but they had learned a lot; among the things they had learned was to recognise that their first effort contained too many technological mistakes to be salvageable, and that they should start afresh. So, early in the game I became involved in the design of "the next machine", something that would be repeated a number of times. In retrospect, the most noteworthy aspect of that cooperation is that we accepted without any discussion a discipline that seems rare today: the very first thing we did was to write a programmer's manual for the next machine, including its *complete* functional description. (For some reason, I was always the one that wrote that document.) This document then served as a contract between them and me: they knew what they had to build and I knew that I could build upon. And about a year later, we would be both ready: they had built their machine and I had all the basic software finished. (Not tested, of course, but that turned always out to be superfluous. At the time I did not realise how valuable this training was of programming for a machine that did not exist yet; now I know.)

I designed my first non trivial algorithms. The algorithm for The Shortest Path was designed for the purpose of demonstrating the power of the ARMAC at its official inauguration in 1956, the one for the Shortest Spanning Tree was designed to minimise the amount of copper in the backplane wiring of the X1. In retrospect, it is revealing that I did not rush to publish these two algorithms: at that time, discrete algorithms had not yet acquired mathematical respectability, and there were no suitable journal. Eventually they were offered in 1959 to "Numerische Mathematik" in an effort of helping that new journal to establish itself. For many years, and in wide circles, The Shortest Path has been the main pillar for my name and fame, and then it is a strange thought that it was designed without pencil and paper, while I had a cup of coffee with my wife on a sunny cafe terrace in Amsterdam, only designed for a demo...

Life became serious when Bram and Carel embarked on the design of the X1, which after a while they conceived with a real-time interrupt. I was frightened out of my wits, only being too much aware that this would confront me with a to all intents and purposes nondeterministic machine. For about three months I have been able to delay the decision to include the interrupt, until eventually, they flattered me out of my resistance. (Their plea was along the lines "Yes, we understand that this makes your task much harder, but surely someone like you must be up to it, etc.".) The real-time interrupt handler became my thesis topic; I earned my Doctorate in late 1959.

In the mean time, the ALGOL project was underway. I had attended a few conferences outside the Netherlands, but it was ALGOL that introduced me to frequent foreign travel. Those trips were very tiring, for my English was untrained. (I am very grateful to Mike Woodger from the National Physical Laboratory, who quickly started correcting my English, but in the beginning only the errors he had heard me make more than once.) In the summer of 1959 I had discovered how to implement subroutines that could call themselves, by the end of the year I saw how to use a stack for the evaluation of expressions and how to translate expressions from the usual infix notation into "reverse Polish" (only we did not call it that). For several months, Harry D. Huskey had been working at the Mathematical Centre, having been invited by van Wijngaarden in the hope that he would teach us how to write what was then called "an algebraic translator". Van Wijngaarden wanted us to base our ALGOL 60 compiler on Huskey's work which was a mess (Huskey being an experimental programmer); I had seen that work and knew that we had to ignore it, but in order to convince van Wijngaarden I had to bang my fist on his table. (I remember this, because this has been the only time that I did so.) Late December I started with J.A. Zonneveld on an implementation of ALGOL 60 for the X1. We coded the translator in duplo, fully aware that we were making history: each evening each of us took his own copy of the manuscript home so that in case of a fire at the Mathematical Centre our translator would not be lost. (There was no fire.) In August 1960 our implementation was working, more than a year before the one of our nearest competitor, the group of Peter Naur in Copenhagen. In May 1960 I submitted the short article "Recursive Programming" to "Numerische Mathematik", which published it quickly; here I introduced the term "stack", while describing the quintessence of our implementation. For many years I thought that that paper had been ignored in the USA, but later I learned that it had established my name among America's scholarly computing scientists.

The ALGOL compiler had been a major effort on which I had worked day and night for 8 months, and I needed to recover; more than anything I needed to be free from the responsibility for a large project. For a few years I did not build something sophisticated and ambitious, but thought, mainly about the logical intricacies of multiprogramming. This was when I formulated the Mutual Exclusion Problem, thought about synchronizing primitives and designed "semaphores" for that purpose. I did not publish that last invention because the company "Electrologica", for which I was a consultant, was going to incorporate them in their next machine, but I freely lectured about the topic. In 1961, at Brighton, Peter Naur and I lectured on ALGOL 60 and its implementation; at that occasion, Tony Hoare and I met for the first time.

At the end of summer of 1962 I attended the IFIP Congress in Munich, where thanks to van Wijngaarden I was given the opportunity to deliver an invited speech. My speech was well received, and the congress was the occasion at which Niklaus Wirth and I met each other for the first time. When I came back from Munich, it was September, and I was Professor of Mathematics at the Eindhoven University of Technology.

Later I learned that I had been the Department's third choice, after two numerical analysts had turned the invitation down; the decision to invite me had not been an easy one, on the one hand because I had not really studied mathematics, and on the other hand because of my sandals, my beard and my "arrogance" (whatever that may be). I have also been told that I should be very grateful to the Department because it had taken a great risk by appointed me. For two years I gave the courses in Numerical Analysis because they had no one else to do so; at the same time I began building a little group of computing scientists. In the years that followed, I built with Cor Ligtmans, Piet Voorhoeve, Nico Habermann and Frits Hendriks what became known as the "THE Multiprogramming System" for the X8 (the next machine of Electrologica, of which the university had ordered a copy); machine and operating system served the computation centre of the university very well for about half a decade, until the X8 had to be replaced by a more powerful engine.

With the THE Multiprogramming System we also knew we were making history, because it was the world's first operating system conceived (or partitioned) as a number of loosely coupled, explicitly synchronized, co-operating sequential processes, a structure that made proofs of absence of the danger of deadlock and proofs of other correctness properties feasible. I have presented the quintessence of this design in Teddinton, in Edinburgh, and in Gatlinburg, Tennessee (at an ACM Conference on Operating Systems Principles), and it got all the attention and recognition it deserved. The Department saw it differently and disbanded the little group, in reaction to which most of them left the University. It was obviously too difficult for the Department to honour what had been achieved: as classical mathematicians they were not used to group efforts at all, and further more it had no connection to the kind of mathematics they were used to. In addition, some began to fear that, instead of a strong component of the Department, Computing Science could become too powerful a competitor for the true mathematics that were the Department's real concern. For me, life became a little bit difficult.

For more than half a year I suffered from a rather severe depression, but things became better when I realised that I was frightened by the next challenge, viz. the study of difficulty as such, and of how to overcome it. How do we do difficult things? My previous more ambitious projects acquired the status of finger exercises, and for therapeutic reasons I wrote "Notes on Structured Programming"; as far as my depression was concerned, the therapy worked.

Of the "Notes on Structured Programming" I sent only about 20 copies to friends abroad, but by that time the copier had become ubiquitous, and (without my knowledge) the text spread like wildfire: later I met people in faraway places who treasured a fifth or sixth generation copy! It was not a great text and its spontaneous wide distribution surprised me when I heard of it; the only explanation I have been able to think of is that it was the first text that openly took the programming challenge very seriously. It has certainly played a role in the founding the IFIP Working Group 2.3 on "Programming Methodology", a topic that would only in the 70s become a subject of academic concern.

Late one afternoon in 1972, the telephone in my office at the University rang; I picked up the receiver, and this was Franz L. Alt of the ACM, calling from the USA to tell me that I had won the ACM Turing Award. It took me by complete surprise for I had never considered myself as a potential recipient of that most prestigious award, and I could hardly believe it, but Alt convinced me that it was really true, and then we disconnected. When I realised that my six predecessors had all been native English speakers working at famous institutes in the USA or the UK, I got even a bit overwhelmed. When I had calmed down I went to the office of the Department's Chairman to inform him if he was still in. He was, and I told him. His first reaction was "In the world of computing, they are rather lavish with prizes, aren't they?". When in the fall of that year, another colleague, N.G. de Bruijn, launched a plan in which a massive undergraduate course in programming would earn the money for the graduate program in "genuine mathematics", I knew it was time to go. After having given Philips the first right of refusal, I became Burroughs Research Fellow on the 1st of August 1973.

That same summer I designed the first so-called "self-stabilizing systems". It was a new problem of which no one knew whether it could be solved at all, and the solutions were at the time a tour de force. I submitted a short article to the Communications of the ACM, which published it in 1974; it was ignored until Leslie Lamport drew attention to it in 1984, and since then it has added a new dimension to ways of dealing with error recovery.

In the fall of 1973 I designed "predicate transformers" as a tool for defining program semantics in a way that would provide a suitable basis for a calculus for the derivation of programs. In the late 60s, Tony Hoare had provided a basis for this work: he taught us the relevance of the predicate calculus for reasoning about programs and showed us how inference rules could be tied to the formal syntax of the programming language, but it was at the time not clear to which extent his "axiomatic basis" defined the programming language semantics. I tightened that up and introduced in passing nondeterminacy. This formed the starting point for the book "A Discipline of Programming", which kept me busy for the next few years. Prentice-Hall published it in 1976, it is still for sale, it has been quite influential; the organisation that publishes the Science Citation Index has identified it as a "Citation Classic".

I did the "On-the-fly garbage collection", which at the time was quite a feat of fine-grained interleaving, and the distributed termination detection of diffusing computation, which emerged as a fundamental problem. In 1981 I designed a surprising sorting algorithm.

On the whole, my years as Burroughs Research Fellow have been a wonderful time. I would visit the company in the USA a few times a year and my further assignment was to do my own thing, which I did in the smallest Burroughs research outfit in the world, viz. my study on the second floor of our house in Nuenen. I maintained my contracts by mail, by going to the University once a week, and by travelling all over the world. I lectured all over the world, but foreign travel lost what little attraction it had (I never was a very good tourist:

I managed to visit Moscow without being dragged to the Kremlin). Whenever I was in the country, I would work on Friday with Carel S. Scholten; others I saw as regularly were W.H.J. Feijen, A.J.M. van Gasteren and Alain J. Martin.

It was too good to last, and it didn't. When Michael W. Blumenthal took over, Burroughs Corporation changed from an enlightened computer manufacturer into a financial organisation, and precisely the kind of people I had established relations with were the first to leave the company: I lost my audience there. Whereas Burroughs's intellectual horizon was shrinking, my own one was widening, as my interests shifted from computing and programming methodology to mathematical methodology in general. A return to the University Campus seemed appropriate, and when the University of Texas at Austin offered me the Schlumberger Centennial Chair in Computer Sciences, the offer was actually welcome. The offer was not so surprising, nor my acceptance, for I knew UT and they knew me. All through my years at Burroughs I had made it a rule to visit, where ever I went, the local university and since the late 70s Burroughs had an Austin Research Centre, a visit to which was included in most of my trips to the USA.

In my interest in mathematical methodology I had been encouraged by a grant that I received thanks to the efforts of Dr. Donald W. Braben of the Venture Research Unit of BP, a grant that enabled me to offer a position to Netty van Gasteren (who thus became the first BP Venture Research Fellow). Another strong stimulus came from my efforts in which I would soon be joined by Carel S. Scholten to give my "predicate transformer semantics" of the 70s the necessary mathematical underpinning; these efforts became an exciting experience when we discovered (after few notational innovations) the surprising utility of calculational proofs. In 1988, Netty van Gasterned successfully defended her Ph.D. thesis "On the shape of mathematical arguments".

In the summer of 1984 we moved to Austin, Texas, where we have been happy ever since, partly by the simple device of having warned ourselves beforehand that The University of Texas at Austin would not be Heaven-on-Earth. (It is not.) With the prolonged recession, the pressures to do the wrong things, always already strong, are mounting, and part of the university gives in. And in computing, the excessive preoccupation with speed is, of course, a little bit vulgar.

Austin, 29 November 1993

Joseph Goguen

Joseph Goguen is Professor of Computer Science & Engineering at the University of California, San Diego, and Director of the Meaning and Computation Lab there, from 1996. He was previously the Professor of Computing Science, Fellow of St. Anne's College, and Director of the Centre for Requirements and Foundations at Oxford University. Prior to that he was Senior Staff Scientist at SRI International and a Senior Member of the Center for the Study of Language and Information at Stanford, and before that a full professor at UCLA. He has also taught at Berkeley and Chicago, and held fellowships at IBM Research and Edinburgh University. Professor Goguen's Bachelor degree is from Harvard and PhD from Berkeley, and he has been a distinguished lecturer at the Universities of Syracuse and Texas. He is best known for his research on abstract data types and specification; his current intersts include requirements engineering, user interface design, and a large Japanese sponsored project to build an industrial strength environment for a new version of his OBJ specification language.

Tossing Algebraic Flowers down the Great Divide

Introduction

Computer science today is extraordinarily successful; chips are reproducing and evolving far faster than humans, and millions of humans are exchanging email, visiting websites, and discussing HTML, Java, and high speed modems in cybercafes across the world. Computers are the only significant commodity to ever get progressively cheaper as they get better, throughout their entire history.

But computer science is in deep crisis, expanding, fragmenting and specializing faster than any other discipline, faster than anyone can understand, let alone predict. Moreover, computer science is increasingly seen as marginal to its applications, and this is particularly true of theoretical computer science.

Information is the life blood of modern society, and much of it is managed and distributed by computer systems; they control cash machines, factories, nuclear reactors, telecommunication networks, and ballistic missiles, as well as arcade games, the family car brakes, and an almost unimaginable variety of databases, e.g. for health records, country music hits, vehicle registrations, stock transactions, and DNA sequences.

Nobody knows where all this is going or what it will mean for people's lives; those in government who are responsible for overseeing technology and its impact are often remarkably ignorant. This is due not only to the unprecedented growth of information technology, but also to the unprecedented nature of its relationship to society: Information only has value insofar as it has *meaning*, i.e., is about something, whether money, braking distances, transaction costs, or genetics. Otherwise, it is just *data*, patterns of bits, strings of characters, etc. Data can only become *information* when people care about it for some reason and are able to interpret it. This means that information technology, and thus computer science, is bound up with the social at a very basic level having to do with the nature of information itself.

Most work in theoretical computer science ignores all this. I would never suggest that theory has no value, but I do suggest that this "great divide" between theory and practice helps explain the ever declining interest shown in theory. The rest of this paper explores this theme in various ways, beginning with a classification of the fragments – we might say cultures – of computing, and then moving to more detailed consideration of the interplay between theory and practice in certain areas.

This paper can also be considered a survey of some attempts to bridge the great divide between theory and practice in computing, mainly using various kinds of algebra, as well as to bridge the "even greater divide" between technical and social aspects of computing, which in turn is but a small part of the huge rift between science and technology on one side, and society on the other (see [6] for more on this).

From another perspective, this paper can be considered a diary from a very personal journey[1], moving from a mathematical view of computing, through a process of questioning why it wasn't working as hoped, to a wider view that tries to integrate the technical and social dimensions of computing. This journey has required a struggle to acquire and apply a range of skills that I could never have imagined would be relevant to computer science. Always I have sought to

[1] I hope the large number of citations that signpost stages along this journey won't be thought too egomaniacal; the truth is that this essay evolved out of one of those dreadful documents that (many) professors have to write to get a salary increase, so that it was easier to leave all these citations in than to take them out. And who knows, someone might find some of them useful.

discover things of beauty – "flowers" – and present them in a way that could benefit all beings, though of course I don't expect that very many people will share my aesthetics or my ethics. I also hope that this piece may help younger researchers to see more of the process and the human context of research; for that reason I have tried to bring out attitudes as well as facts. Although this paper is by no means intended as a general survey, I have still tried to be as fair as possible to everyone; but through the years, I have discovered that sins of omission and misattribution are inevitable, especially in such a very personal path through such a broad landscape, and so I apologize and invite readers to send me corrections and additions that they consider important.

Five Cultures of Computing

For present purposes, we may identify five cultural fragments of computing, each involving a different group of people, characterized by different goals and different activities:

1. **Computer hardware** Hardware engineers have been phenomenally successful. As this is written, an ordinary high-end PC runs at 333 MHz, has 128 Megabytes of RAM, and 8 gigabits of hard disc (which is probably more computing power than all the computers in the world of twenty-five years ago combined), along with peripherals that would once have been astonishing, including FAX, CD quality digital stereo, and real time audio and video over the internet; by the time you read this, there will most likely have been further advances. Twenty-five years ago, the arpanet had only a few hundred users, and for the most part we all knew each other; now the internet has millions of users, including many we would probably not care to know.
2. **Computer software** Software engineers have been less fortunate than hardware engineers in their choice of profession. The fantastic improvements in hardware have fueled escalating expectations for software that are not being met. Huge "legacy" systems, often written in obsolete and/or obscure languages, with little documentation and generations of superimposed "patches," are exceedingly difficult to maintain, but are surprisingly common. And numerous expensive software failures have been reported, at the Federal Aviation Agency, the Internal Revenue Service, the European Space Agency, and more, as discussed in Section "State of the Software Arts".
3. **Shrinkwrap computing** Consumers are beneficiaries of the great success of hardware as well as victims of the doubtful state of software. There is an immense popular culture of computing, with dozens of magazines on the racks of supermarkets, record stores, and drug stores; many newspapers have weekly features on computing, often a special pullout section; there are also very many popular books, and even TV and radio shows. But shrinkwrap software is typically badly documented, full of bugs, hard to understand, and hard to use. It is also notoriously quickly replaced by a "new and improved" version, or even an entirely new system, often requiring more powerful hardware, and of course providing new bugs.

4. **Sociology of computing** The study of computing by psychologists, sociologists and anthropologists is a fairly new and still relatively small phenomenon. The relevance of such work to user interface design and team management is clear, but I will argue in Section "Some Social Aspects of Computing". that it also has a much broader relevance to computing.
5. **Theoretical computer science** Unfortunately, the explosive growth of computing practice has had little effect on theoretical computer science, which continues to decline in relative importance within industrial and even academic circles, despite many impressive advances in its own terms.

It seems that these five cultures[2] are rather isolated from each other, having quite distinct conferences, journals, methods, and goals. Of course the boundaries are somewhat vague and overlapping, but it does seem to me that they are moving away from each other at an increasing rate, and that each one itself is becoming increasingly fragmented. As a result, there is a great need for communication and even unification between these cultures as well as within them. My major theses are (1) that computer science has not paid sufficient attention to social issues, and (2) that theoretical computer science could play a key role in a reunification that would yield significant benefits to both society and education.

Software Engineering

Beginning with a closer look at the state of software engineering, the subsections below sketch some ways to apply algebra to problems related to software engineering. These include: general system and sheaf theories; abstract data types; specification languages; logical programming and institutions; parameterized programming; order sorted algebra; hidden algebra; specification libraries and reuse; and distributed cooperative engineering. In each case I try to indicate the original motivation, and how I now see this in terms of larger concerns about relations between technology and society.

State of the Software Arts

Large complex software systems fail much more often than seems to be generally recognized. Perhaps the most common case is that a project is simply cancelled before completion. This may be due to time and/or cost overruns or other management difficulties that seem insurmountable; it may be due to politics; it might even be due to purely technical difficulties. One highly visible example is the cancellation by the US Federal Aviation Agency of an $8 billion contract from IBM to build the next generation air traffic control system for the entire country [158]. This is perhaps the largest default in history, but there are many more examples, including cancellation by the US Department of Defense of a $2 billion contract with IBM to provide modern information systems to

[2] A wag might want to call them hardware, software, shrinkware, wetware and airware, respectively.

replace myriads of obsolete, incompatible systems. Other highly publicized failures include IBM software to deliver real time sports data to the media at the 1996 Olympic Games in Atlanta, the $2 billion loss of the European Ariane 5 satellite, and the failure of theUnited Airlines baggage delivery system at Denver International Airport, delaying its opening by one and a half years [30].

What these examples have in common is that they were hard to hide. Anyone who has worked in the software industry has seen numerous examples of projects that were over time, over cost, or failed to meet crucial requirements, and hence were cancelled, curtailed, diverted, replaced, or released anyway, sometimes with dire consequences, and sometimes with a loud declaration of success, even though the system was never used, and may well have been unusable. For obvious reasons, the organizations involved usually try to hide their failures, but experience suggests that half or more of large complex systems fail in some significant way, and that the frightening list in the previous paragraph is just the tip of an enormous iceberg. Much more information about computer system failures can be found in the Risks Forum run by Peter Neumann (see `http://www.csl.sri.com/neumann.html` and [154]).

Experience shows that many failures are due to a mismatch between the social and technical aspects of a supposed solution. It is understandable that software engineering has been biased towards a formal view of information, because computer programs consist of precise instructions that manipulate formally defined structured representations of data, and this is what software engineers are trained to deal with, as opposed to relatively more messy social situations. But we now know that ignoring the situated, social aspect of information can be fatal in designing and building software systems.

Category and General System Theories

In the late 1960s, I greatly admired the smooth way that category theory captured many important general concepts in mathematics (e.g., see [128, 129]), and I greatly regretted the lack of a similar apparatus for engineering. My first attempt to use categories was in my thesis, which gave axioms for fuzzy set theory (see Section "Fuzzy Logic and Information Theory"). Because of this enthusiasm, I wrote several introductions to category theory for computer scientists, beginning in the early 1970s with the "ADJ"[3] report series [107–109], illustrating basic categorical concepts mainly with examples from automata and formal languages, which were the focus of theoretical computer science at that time. I've been told that many East Europeans of that generation learned both basic category theory and theoretical computer science from these reports. Our original goal was to write one or more comprehensive books, something like what

[3] The ADJ group, {Goguen, Thatcher, Wagner, Wright}, was formed during my tenure as Research Fellow in the Mathematical Sciences at IBM Research, Yorktown Heights, initially to study the relationship between category theory and computer science; see [51] for many historical details; for some reason, I wrote all of the initial reports, but in compensation, had very little role in some of the final reports.

Bourbaki did for mathematics, but ADJ fell apart before we got any further than these reports. Later I wrote "A Categorical Manifesto" [56] to provide for each basic concept of category theory a "doctrine" of how to use it in practice, and still later, [53] developed category theory from scratch while proposing a general theory of unification.

Following clues from the systems engineering, general systems theory (hereafter **GST**) and cybernetics of the late 1960s, I decided that the most general concepts of engineering might be system, behavior, and interconnection, formalized in such a way as to include hierarchical whole/part relationships. Systems were taken to be diagrams in a category, behaviors were given by their limits, and interconnections were given by colimits of diagrams; some very general laws about interconnection and behavior hold in this setting [35, 37, 78]. The most complete exposition is in [62], which has full proofs of all results.

One especially nice feature of this approach is that it does not build in any notion of causality, and therefore can capture the sort of mutual interdependence that occurs, for example, in electrical circuits. This contrasts with models like automata in which a causal dependence of the next state on the current state and current input is built into the model. It is also consistent with philosophical ideas like mutual causation and interdependent origination[4], which go beyond naive reductionist causality. But it was (and still is) disappointing to me that so few people felt any need for concepts and theories of such generality; they seem happy to have (more or less) precise ideas about specific systems or small classes of systems, with little concern for what concepts like system, behavior and interconnection might actually mean. Still, this categorical GST has had a significant indirect impact on computer science: its application to the Clear and OBJ module systems influenced some important programming languages, including Ada, ML and C++ (see Section "Parameterized Programming and Generic Modules").

A general theory of objects based on sheaf theory [40] arose from this work, and has been applied (for example) to the semantics of concurrent systems, including concurrent object based languages [62], hardware description languages [166], and semantics for object based concurrent information systems [23]. Sheaves can express the kind of local causality combined with global nondeterminism that characterizes many different kinds of model, from partial differential equations to automata. They can capture not only variation over time, but also over space and over space-time [62]; and they can embrace the incompleteness of observation that is characteristic of all real empirical work. This can be helpful at the philosophical level in dispelling the illusion that models fully capture reality (see the discussion in Section "Realism and Idealism"). It can

[4] The concept of interdependent origination goes back over 2,500 years to the Buddha; in the Pali language, it is called *paṭicca-samuppāda* [12]. This kind of thinking can also be found in much contemporary AI, e.g., the robotics of Rodney Brooks, which is intended to be fast and cheap, because it doesn't require any central control (cf. the wonderful documentary movie *Fast, Cheap and out of Control*, directed by Errol Morris).

also be useful technically, for example in capturing the way that the behavior of distributed concurrent systems depends only on local interactions. There is now a slow but steady stream of research on sheaf theory in computer science, though there is not as yet a coherent community.

In the early 1970s, I formulated the minimal realization of automata as an adjoint functor [38]; this soon evolved into much more general results about the minimal realization of machines in categories, which gave a neat unification of system theory (in the sense of electrical engineering) with automaton theory [36]. I consider this a major vindication of the categorical approach to systems.

Abstract Data Types and Algebraic Semantics

The history of programming languages, and to a large extent of software engineering as a whole, can be seen as a succession of ever more powerful abstraction mechanisms. The first stored program computers were programmed in binary, which soon gave way to assembly languages that allowed symbolic codes for operations and addresses. FORTRAN began the spread of "high level" programming languages, though at the time it was strongly opposed by many assembly programmers; important features that developed later include blocks, recursive procedures, flexible types, classes, inheritance, modules, and genericity. Without going into the philosophical problems raised by abstraction (which in view of the discussion of realism in Section "Philosophy of Computing" may be considerable), it seems clear that the mathematics used to describe programming concepts should in general get more abstract as the programming concepts get more abstract. Nevertheless, there has been great resistance to using even abstract algebra, let alone category theory.

One of the most important features of modern programming is abstract data types (hereafter, **ADT**s), which encapsulate some data within a module, providing access to it only through operations that are associated with the module. This idea seems to have been first suggested by David Parnas [155, 156] as a way to make large programs more manageable, because changes will be confined to the inside of the module, instead of being scattered throughout the code. For example, if dates had been encapsulated as an ADT, the so-called "year two thousand problem" would not exist. Not all increases in abstraction make programming easier; an abstraction must match the way programmers think, or it won't help. This may explain why ADTs have been more successful than higher order functions.

In the early 1970s, there was no precise semantics for ADTs, so it was impossible to verify the correctness of an implementation for a module, or even to formulate what correctness means. Initial algebra semantics provided the first rigorous formulation of these problems, with solutions that were useful, although they have been improved (see Section "Hidden Algebra"). Initial algebra semantics was born in [41], which (among other things) formulated (Knuthian) attribute semantics as a homomorphism from an initial many sorted syntactic algebra generated by a context free grammar, to a semantic algebra[5]. The step to ADTs was facilitated by my realization that Lawvere's characterization of the natural numbers as an initial algebra [132] could be extended to other data structures [105, 106]. As a young researcher at this time, I was really shocked by the attempts of certain senior colleagues to reconfigure the history of this period to their own advantage; this is why I wrote the paper [51].

What really pleased me was the neat parallel established between Emmy Noether's insight that algebra is the study of sets with structure given by operations, and David Parnas's insight that modules should encapsulate data with operations; more than that, the algebraic approach established an equivalence between abstractness in ADTs and abstractness in algebra[6]; furthermore, equations among operations came into specifications of ADTs the same way as in abstract algebra.

It also seemed splendid that computability properties worked out so well: an algebraic version of the Turing-Church thesis says that an algebra is computable iff it is a reduct of a finitely presented initial algebra with an equationally definable equality; these are also the algebras for which so-called "inductionless induction" proofs are valid [47]. Moreover, an algebra is: semicomputable iff it is a reduct of a finitely presented initial algebra; cosemicomputable iff it is a reduct of a finitely presented final algebra; and computable iff a reduct of a finitely presented algebra that is both initial and final. The reason that the computability notion associated with Scott-style denotational semantics doesn't work for algebras is explained in [46]. This field was pioneered in a series of papers by Jan Bergstra and John Tucker, surveyed in [150], which also explains basic many sorted algebra and abstract machines. Some conjectures from [150] were solved with Meseguer and Moss in [153]. The computational side of ADTs includes term rewriting, which featured in early drafts of [106]. Initial algebra semantics has also been used in linguistics, to explicate to the notion of compositionality [122].

Meseguer and I studied the rules of deduction for many sorted algebra in [94]. This paper surprised the community by showing that the naive generalization of the usual unsorted rules (as previously used in the ADT literature) is unsound. We gave a sound and complete set of rules, and showed that the unsorted rules did work for certain signatures; the difficulties involve implicit universal quantification over empty sorts. Complete rules of deduction for many sorted conditional equational logic are given in [92]. The first rigorous proof of correctness for the inductionless induction proof technique that was originally suggested by David Musser, is given in [47]. The formulation of inductionless

[5] This built on an approach to many sorted algebra developed for my course Information Science 329, Algebraic Foundations of Computer Science, first taught in 1969 at the University of Chicago, including the now familiar use of indexed sets, the word 'signature' with symbol Σ, and its formal definition.

[6] This is because any two initial objects in a category are isomorphic; hence we speak of "the" abstract data type of a specification in exactly the same way that we speak of "the" initial algebra of a variety. Each is determined up to isomorphism, and the fact that each an "abstract algebra" and an "abstract data type" are isomorphism classes of algebras expresses their independence of representation.

induction in [47] is more general than in some later work, which was restricted to just constructors; [47] also pointed out that the essence of inductionless induction is "proof by consistency," and gave the simple but fundamental result that for an equational specification that is canonical (i.e., terminating and Church-Rosser) as a set of rewrite rules, the normal forms of ground terms form an initial algebra. This result justifies term rewriting as an operational semantics for initial algebra semantics.

ADJ later extended initial algebra semantics to continuous algebras (at about the same time as Maurice Nivat) and continuous algebraic theories [110], and then to rational algebraic theories [173]. This inspired Meseguer and me to generalize to an arbitrary category, getting *initial model semantics* (see Section "Logical and Multi-Paradigm Programming").

Specification

In the early seventies, most theoretical research concerned the semantics of programs and programming languages, and the verification of programs. There was little or no work on specification, modules, or verification at these levels. But we now know, and even then many suspected, that this is not where the leverage lies for real applications; in fact, most debugging effort goes into fixing errors in requirements, specifications and designs, and very little into fixing errors in coding (around 5%) [5]. Moreover, the problems that arise for large programs are qualitatively very different from those that arise for small programs. It is *not* just as easy to find specifications and invariants for the flight control software of a real airplane as it is for a sorting algorithm; in fact, finding specifications and invariants is not an important activity in real industrial work. On the contrary, it turns out that finding requirements (i.e., determining what kind of system to build), structuring the system (modular design), understanding what has already been done (reading documentation and talking to others), and organizing the efforts of a large team, are all much more important for a large system development effort. As Tony Hoare said about his research (largely on program correctness), "It has turned out that the world just does not suffer significantly from the kind of problem that our research was originally intended to solve" [120].

I thought that since we knew how to do ADTs as theories, the next step (according to categorical GST) should be to interconnect these theories using colimits; then a description of such an interconnection would be a design for a system. Of course, things weren't entirely straightforward – it seems they never are! – but Rod Burstall and I succeeded in designing the Clear specification language [8–10], which integrated initial and loose semantics with generic modules using "data constraints"[7] by extending an idea of Horst Reichel [161]. Clear seems to have been the first specification language with a rigorous semantic definition, and its modules seemed promising as a way to handle large systems.

[7] It is interesting to notice that these must be *morphisms*, rather than just theory inclusions.

I badly wanted to execute specifications, because I had noticed that it is all too easy to write them incorrectly, and I also thought it could be very helpful in teaching. Around 1974, I conceived the OBJ language for this purpose, using order sorted algebra[8] with mixfix syntax, and with term rewriting as operational semantics [42]; the goal was to make specifications as readable and testable as possible. The final OBJ3[9] [114] version of OBJ was implemented by a team led by José Meseguer, including contributions by Kokichi Futatsugi, Jean-Pierre Jouannaud, Claude and Hélène Kirchner, David Plaisted, Joseph Tardo, and others [104, 100, 27, 114]; this system provided both loose and initial semantics, rewriting modulo equations, generic modules, order sorted algebra with retracts, and user definable builtins[10]; OBJ2 was heavily used in designing OBJ3 [125], and I think greatly speeded up this effort, by facilitating team communication and documenting interfaces. Many other languages have followed OBJ's lead, including ACT ONE [24], which was used in the well known LOTOS hardware description language.

Although I never thought program verification had much practical value, I do think it has educational value, and in 1996, Grant Malcolm and I published a book on verifying imperative programs using OBJ3, based on a course we taught at Oxford [88]. Unlike other books on this topic, every program proof in the book is executable, and students can do all their homework on a computer; this produced a large improvement in both motivation and understanding, presumably in part due to the addictive quality of programming [141]. The use of algebra avoids the awkwardness and/or lack of rigor of techniques like predicate transformers and three valued logic that are found in some other books on this subject.

It is unhealthy to confuse a formal notation with a formal method. A *method* should say how to do something, whereas a *notation* allows one to say something [141]; thus OBJ is a notation, but using it as described in [88] gives a method for proving properties of imperative programs. Methods are rarely completely mechanical, because some of the problems that must be faced are usually uncomputable (e.g., finding loop invariants). Nevertheless, the method of [88] is surprisingly effective, in part because OBJ does all the routine work mechanically, and even provides hints to help with doing much of the non-routine work. For me, the frontier of research in this area is the use of systems like OBJ3 for theorem proving, e.g., in first order logic with equalities as atoms [71], or for verifying distributed concurrent systems (see [89] and see Section "Hidden Algebra").

OBJ later developed a whole family of extensions, some of which are discussed in Section "Parameterized Programming and Generic Modules"), and then it

---

[8] Actually, a precursor called error algebra, motivated by the importance of error handling in real systems.

[9] "OBJ" refers to the general design, while "OBJ3" refers to a specific implementation.

[10] These were originally intended for providing builtin data structures like numbers, but were later used in implementing complex systems on top of OBJ, since they allow access to the underlying Lisp system [114].

spawned a next generation, which is even now under construction. Some of the most important members of this next generation are the following:

1. SpecWare is a product from Kestrel Institute that has been very successful in synthesizing a wide range of scheduling algorithms, some of which are in daily practical use. SpecWare has a top level command `colimit` which computes the colimit of a collection of theories and theory morphisms [165].
2. Maude [148] is an extension of OBJ to rewriting logic [147], which is particularly suited to specifying concurrency. This project is led by Dr. José Meseguer at SRI International, where most of the original OBJ3 development was done, with support from the Office of Naval Research. Maude also has an efficient implementation and a number of interesting new features, including a logical and operational foundation in membership equational logic [149].
3. The CafeOBJ project [29] aims to make algebraic formal methods accessible to practicing software engineers. The CafeOBJ language is similar to OBJ, but enriched with features for both rewriting logic (as in Maude) and hidden algebra (see Section "Hidden Algebra"), to help specify and verify distributed concurrent systems. The CafeOBJ consortium includes several large Japanese companies, and is supported by MITI (the Japanese Ministry of Industry and Technology); more information on this project can be obtained from `http://ldl-www.jaist.ac.jp:8080/cafeobj` and from [20].
4. CoFI is another large effort to design and build an algebraic specification language. It is a highly collaborative multinational project with a distinctively European flavor, much influenced by the success of OBJ; see `http://www.brics.dk/Projects/CoFI`.

The most recent information on the OBJ family of systems and its relatives can be obtained from the website `http://www.cs.ucsd.edu/goguen/sys/obj.html`.

The motivation for all this work is of course to provide tools to formalize and verify the meaning of software and hence improve programming practice. From a purely theoretical point of view, there has been, and still is, a great deal of progress; but this has only made it more painfully clear that social issues play a dominant role in the transition to practical applications. There is an old saying that if you invent a better mouse trap, the world will beat a path to your door. But it's not true. You need to do field testing, file an environmental impact statement, get a designer label, mount a large advertising campaign – and then you need to train the mice!

Institutions

Because Clear is based on colimits of theories, Burstall and I were able to give it a very general semantics independent of the underlying logic in which theories are expressed, provided that logic has certain simple and very usual properties, which constitute the notion of **institution** [73, 75]. The basic feature of institutions is a duality between models and the (logical) sentences used in specifications arising through a relation of satisfaction that is parameterized by the signature involved. In the traditional cases of equational and first order logic, models are algebras and first order structures, respectively, but for LILEANNA (see Section "Parameterized Programming and Generic Modules"), models are given by Ada programs. So institutions give a way to deal with issues in programming and specification languages (as well as databases and other kinds of system) independently of their underlying logic; I really love this kind of generality.

The theory of institutions was developed further in [74], showing how to generate institutions from the simpler structures of charters and parchments, how to put institutions together, and how to greatly generalize them; in particular, morphisms of sentences are introduced to support proof theory. The notion of inclusion system was introduced in [21] to axiomatize the notion of inclusion morphism in categorical terms, and then used to study some mathematical properties of specification modules, including the relation between pushouts preserving (various kinds of) conservative extension, Craig style interpolation properties, and some distributive laws for information hiding. There is now a rather large literature on institutions, with applications to many different areas, e.g., [17] concerns multi-institutional specification. However, it did take nine years (!) for the basic paper on institutions to be published in journal form [75]; this is the longest refereeing and editorial delay of which I ever heard.

Parameterized Programming and Generic Modules

Parameterized programming [48, 52, 27, 28] makes the advantages of the Clear module system available for real programming languages, as well as for more practical specification languages. In addition to semantic interfaces for generic modules (where axioms describe when the module will behave as advertised), parameterized programming provides module expressions to describe systems as interconnections, and views both to describe module bindings for instanting generics, and to serve as global assertions about semantic properties of subsystems; default views greatly simplify module expressions, and multiple inheritance for modules arises in a natural way. These ideas were first implemented in OBJ3. Although I'm happy that some of this influenced the languages Ada, ML, and C++, it can still be distressing to see the compromises involved.

LIL [50] extends parameterized programming to handle programs and specs together, by giving each module a specification "header" as well as implementations. LIL provides "two dimensional" module composition following the "CAT" ideas [72], where vertical structure refers to the layering of software to use capabilities from lower layers, while horizontal structure refers to a single layer. LIL has been implemented as LILEANNA [168, 169], which uses Ada for code and ANNA [135] for specs; it has been used to build helicopter navigation software. New features of LILEANNA include operations to add, delete and modify module functionality, at both the code and spec levels. A formal semantics is given for all this in [111], using a concrete set theoretic exposition of institutions; some general "laws of software engineering" are given showing how various module operations relate. Hyperprogramming [55] extends the idea of organizing information around a specification header to support requirements as well as specs and code, with traceability, controlled evolution, and management of configurations, versions, families, documentation, etc., as well as system generation and software reuse. All this is of course intended to ease the development of large systems, and in particular, to make reuse more effective in practice.

An approach to software architecture based on these ideas is given in [67], where module expressions provide a module connection (also called an architecture description) language. Any mixture of architectural styles can be supported, and modules can involve information hiding. Detailed design and coding are unnecessary if a suitable database of specifications and relationships among them is available, because executing a module expression yields an executable system, constructed by manipulating and linking implementation modules. The underlying ADT of the database is called a **module graph**; it includes specs, source code, compiled code, and many kinds of relationship, mostly among specs. So far there has been little interest in these ideas and even some antagonism; perhaps the community is too fragmented to accept a combination of formal semantics, mixed architectural styles, and generating systems from designs.

Order Sorted Algebra

Real software has many features that are difficult or impossible to treat with ordinary many sorted algebra. These include the raising and handling of exceptions, overloaded operators, subtypes, inheritance, coercions, and multiple representations. Error algebra [42] was a first crack at some of these problems, although it didn't work out. The second try was order sorted algebra (hereafter, **OSA**) [44], which reached fruition in joint work with Meseguer [99, 151]. This approach provides a partial ordering relation on sorts, interpreted semantically as subset inclusion among model carriers. Meseguer and I proved [151] that many simple ADTs have no adequate many sorted equational specification, because (what we call) the constructor-selector problem can't be solved in this setting. OSA is only slightly more difficult than many sorted algebra, and essentially all results generalize without much fuss; in particular, there are initial models and an efficient operational semantics. Because OSA is strongly typed, many terms that intuitively should be well formed because they evaluate to well formed terms, are actually ill formed; [99] introduced *retracts* to handle this problem. Sort constraints [99] extend OSA to support equational definitions of bounded data structures and partial operations. There are now many different variants and extensions of OSA, too numerous to mention here, although Meseguer's membership equational logic [149] should not be omitted. Although successful in this sense within the algebraic specification community, OSA seems to have had little influence elsewhere, and retracts have not been taken up anywhere.

Semantics of Logic Programming

Eqlog [95, 15] combines (equality based) functional programming with (predicate based) logic programming, by combining their logics, which are equational and Horn clause logic respectively, to get Horn clause logic with equality (actually the order sorted version). Eqlog's operational semantics also combines those of functional and logic programming, using both term rewriting and unification with backtracking; moreover, the design permits efficient special purpose algorithms for builtin types like numbers and lists, as well as narrowing to solve equations over user defined ADTs. Eqlog introduced several features that were new for logic programming, including user definable ADTs, strong typing with subsorts and overloading, generic modules, and a "wide spectrum" integration of coding with specification, design and prototyping. A good deal of theory was done to support Eqlog, including: complete rules of deduction, and Herbrand and initiality theorems, both for order sorted Horn clause logic with equality [79, 96]; order sorted unification [152]; order sorted narrowing and resolution [79, 152]; correctness criteria for builtin algorithms; and an initial model semantics generalizing and subsuming the traditional Herbrand universe construction. I still can't understand why the logic programming community prefers fixpoint semantics to this elegant algebraic approach.

Logical and Multi-Paradigm Programming

Major programming styles or "paradigms" that have emerged in addition to traditional imperative programming include functional programming, logic programming, and object oriented programming. Each can be considered a kind of logical programming [49], where a *logical programming language* is characterized as follows:

- its statements are sentences in some logic;
- its computation is deduction in that logic; and
- its denotational semantics is given by models in the logic.

For example, higher order functional programming is (or can be) based on higher order equational logic, and OBJ is based on order sorted first-order conditional equational logic. Similarly, logic programming is based on Horn clause logic. This approach can be made precise using institutions [96, 74, 49], and it has been enriched and extended by Meseguer with his theory of "general logics" [146].

In this setting, it is natural to generalize initial algebra semantics to *initial model semantics* [95, 96], using initiality in an arbitrary category; this often yields simpler proofs of general results by avoiding details of construction and representation. The initial model idea appears in a new guise in [96], to handle the semantics of logic programming over builtin types and algorithms as *free extensions* of the given builtin model. This idea also features in the elegant semantics for so called *constraint based* programming developed by Diaconescu [15, 16]. However initiality is not the right semantics for every logical language.

Programming paradigms can be combined by viewing them as logical programming languages and then combining their logics [49, 74, 96]. Thus Eqlog combines functional and logic programming [95, 96], and FOOPS combines object oriented and functional programming, by using reflective order sorted conditional equational logic [97, 115], since features of the object paradigm can be obtained by reflectivizing other logics; FOOPlog combines all three paradigms [97] by reflectivizing Horn clause logic with equality. All implementations of these systems were built on top of OBJ3. This research direction went so far as designing and prototyping special purpose hardware, the Rewrite Rule Machine, for executing declarative languages efficiently, based on term rewriting chips [98, 133, 57]. Despite all the interest once shown in declarative programming, the Fifth Generation, etc., there seems to be little interest in precise explications for declarative and logical programming and reflection, or in general purpose declarative architectures. See [11] for the latest on reflective logic and its applications.

Hidden Algebra

While (order sorted) equational logic works well for unchanging (immutable or "Platonic") data types like the numbers, it can be awkward for software modules having an internal state that changes over time. In 1982, Meseguer and I developed a theory of abstract machines [92] for this purpose, and proved minimal (final) and initial realization theorems for them; this theory naturally generalizes algebraic ADTs as well as classical automata. The minimal realization adjunction for automata [38] helped inspire this work, and it was also pleasant to realize that many intuitions from John Guttag's early work could be vindicated [116, 117].

In collaboration with Răzvan Diaconescu, Rod Burstall, and most recently especially Grant Malcolm, this work has developed into a new *hidden algebra* approach [54, 58, 77, 7, 87, 140, 89], intended to facilitate proving properties of *designs*, as opposed to code, and in particular, to facilitate refinement proofs, that one level of design is correctly realized by another. The main contribution is *hidden coinduction* techniques for (relatively) easy proofs of *behavioral properties*; this is important for software engineering because many practical implementation techniques provide the desired behavior, without realizing it as in the specification; hidden theories capture what have elsewhere been called behavioral types. This constitutes a *method* for using notations like OBJ3 and CafeOBJ to verify software designs; it is especially suitable for the object paradigm. The distinction between hidden and visible sorts allows the latter to be used for immutable data types (typically given by initial semantics). Hidden algebra has also opened intriguing new perspectives on nondeterminism and concurrency: nondeterminism arises naturally simply by *not specifying* some behaviors [89]; and concurrency is described by an elegant universal[11] construction on hidden theories [77]. A hidden Herbrand theorem which unifies the object and logic

[11] In the sense of category theory; i.e., it is a construction defined purely in terms of morphisms.

paradigms at the logical level is proved in [90]. There is now an excellent hidden group at Oxford, including Grant Malcolm, Corina Cîrstea, James Worrell, and Simone Veglioni [139, 13, 14, 172]; the recent proof that categories of hidden algebras are topoi [142] is especially exciting. There is also an exciting flurry of hidden activity around the CafeOBJ project at the Japan Institute of Science and Technology, which includes Răzvan Diaconescu, Kokichi Futatsugi, Shusaku Iida, Dorel Lucanu and Michihiro Matsumoto [18–20, 121].

Distributed Cooperative Engineering with TATAMI

Typical software engineering projects have multiple workers with multiple tasks that interleave in complex ways, often working at multiple sites with different schedules, so that it is difficult to share information and coordinate tasks; documentation is often hard to find, out of date, incomplete, or non-existent; requirements change; specifications change; personnel change; and it is hard to determine which parts of the system most need attention (e.g., see [137], especially its hypergraph model of evolution). Despite all this, most formal methods and tools to support them take a single user with a single unchanging task as their (usually implicit) model of interaction.

Recent work at the UCSD Meaning and Computation Lab seeks to address the distributed cooperative aspect of software engineering with an environment called TATAMI [80, 102] for CafeOBJ, having the capability (but *not* the necessity) for complete formal verification. Formality provides a discipline for both designing TATAMI and using it; however, the most practical use exploits the task structure of formality without requiring logical completeness, to ensure that all relevant dependencies are known, and that documentation, test cases, etc. are in predictable locations. Confidence values in the unit interval are associated with project tasks instead of Boolean truth values; this fuzzy logic (following [33]) allows critical path analysis to aid task allocation, taking account of different levels of formality and criticality.

TATAMI is supported by a truth maintenance protocol that resolves inconsistencies while updating local databases, allowing multiple versions at multiple sites, including incomplete and even incorrect proofs. The Kumo[12] tool generates websites for project documentation and assists with verification; it is now being used to populate a library with tutorial examples; see `http://www.cs.ucsd.edu/groups/tatami`. The output of Kumo can be read by any web browser, and in addition to the proof itself, provides for proof execution on remote servers, as well as animation and informal explanations. User interface design for this system uses algebraic semiotics (see Section "Algebraic Semiotics"), and its requirements were driven by my own participant observation as a software engineer. We are also using ideas from narratology (the study of stories) and even cinema to organize project websites to facilitate navigation and comprehension; since TATAMI supports multiple development lines, it also supports multiple narrative lines [101]. This group includes Kai Lin, Akira Mori,

[12] This is a Japanese word for spider.

Grigore Roşu, and Akiyoshi Sato. Lin is the heroic implementer of the Kumo system, and Roşu has given a hidden algebraic correctness proof for the TATAMI protocol.

Preliminary experiments using Kumo and TATAMI have been encouraging. This project is struggling to make earlier theoretical work more relevant to practice; however it is well beyond the capability of one small research group to build a truly industrial strength environment. So far this research has been greeted largely with incomprehension.

Discussion

Although my early work may have an austere kind of beauty from its abstraction and generality, it seems difficult for engineers to integrate such results into practice. Much of my later research has tried to understand what is required to bridge this divide, and in particular, my excursions into the social, described in the next section, have this motivation.

Looking back over Section "Software Engineering" above, the extent to which I have failed to rise above the rather partisan divisions that are so much part of the present research scene is embarrassing, and has compromised my goal of evaluating ideas from a broad social perspective. I can only say that I have done my best at this time, and hope to have more perspective in the future.

Some Social Aspects of Computing

As already noted several times, social issues can dominate computer technology, e.g., in the construction of real systems that must be used in some social context, and in the propagation of new ideas into practice. This section reviews some approaches to reconciling computer technology with its social context. Sometimes algebraic methods are applied to the social, and sometimes social aspects of computation theory are considered.

Discourse Analysis

My earliest adventures into the social sciences were in discourse analysis, an area of sociolinguistics concerned with the large grain structure of language. Several types of discourse have a definite regular structure, including planning [134], explanation [113] and command and control discourse [82, 85]. The latter was developed in studies supported by NASA, on statistical properties of aviation discourse in emergency situations, for application to aviation safety. Later work on multi-media instruction supported by the Office of Naval Research involved naturalistic experiments and ideas from semiotics, for application to human-computer interface design [84]; this later led to the work discussed in Section "Algebraic Semiotics. Two other discourse types are stories [126] and jokes [162], which are interesting because they embed values of the speaker and audience, and can therefore be used to study those values [69, 86]. This was applied in

[81] as part of a study to determine requirements for a system to computerize a small headhunting firm, by collecting stories and jokes told during breaks and at lunch, and then collating them into a "value system tree" for the firm, as described in [60, 86]; this work also showed how to extract dataflow diagrams from task oriented discourse. For some reason, linguistics seems stuck on the syntax of sentences, despite the fact that there are important applications at higher levels. On the other hand, it must be admitted that there are a great variety of approaches to discourse analysis, the procedures tend to be time consuming, and at least the techniques used in my own work to describe discourse structure involve an unfamiliar formalism. So this work remains largely unknown. A critical overview of techniques for gathering information about social situations is given in [83]; in general, the greater the accuracy, the greater the difficulty.

Requirements Engineering

Case studies and experience suggest that flawed requirements may be the most significant source of errors in system development; moreover, it has been shown that requirements errors are the most expensive to correct at later stages [5]. Case studies and experience also suggest that social, political and cultural factors are very often responsible for the flaws in requirements; however, this area has been little studied. It follows that studying social aspects of requirements engineering has great leverage. But (using our ongoing metaphor) requirements engineering involves problems deep down inside the great divide.

The Centre for Requirements and Foundations was founded in 1991 at Oxford University to work towards a scientific basis for requirements engineering taking adequate account of social issues, as well as advanced technology; another goal was to develop appropriate new methods and tools for capturing and analyzing requirements, to help build systems that are better for users, as well as less stressful for those involved in the building process. When I left Oxford, the Centre had completed two main projects, supported by a large grant from BT. The first was a case study in requirements elicitation using techniques from sociology and linguistics, particularly interaction (video) analysis, but also discourse and conversation analyses [118, 136]. The second was a classification of methods and tools used for requirements [4]. The first project built on early work with Linde on requirements elicitation [81], which was followed by a critical survey of elicitation methods [83]. The second project was in part inspired by work of Lyotard [138] on post-modernism. Other work from the Centre included a book [124] consisting of (revised) papers from a workshop organized by the Centre, some more papers [64, 123], and the TOOR object oriented tool for tracing requirements [160]. The view that the essence of requirements engineering is to reconcile social and technical aspects of system design was proposed in [65] and elaborated in later work [66, 69]; it amounts to saying that requirements engineering consists of building bridges across the divide. Hence there is no disciplinary home for this area, and thus despite its great economic importance, there is little academic effort devoted to it; in particular, I don't know of any degree programs in this area.

Algebraic Semiotics

Semiotics is defined as the theory of signs and their meanings, a very difficult area deep inside the social side of the great divide. Unfortunately there seem to be at least as many approaches to semiotics as there are authors who have considered it, and many other fields with different names also cover the same or closely related ground, e.g., cognitive linguistics [127, 26, 170]. Algebraic semiotics [70, 68] is one more: it tries to combine insights from both sides of the divide, to obtain a precise formulation of certain problems about meaning and to allow the construction of supporting technology. Among the main insights are: (1) signs mediate meaning (emphasized by Charles Saunders Peirce [157]); (2) signs come in structured systems (made clear and studied in detail for language by Ferdinand de Saussure [164]); (3) structure preserving maps ("morphisms") are often at least as important as structures[13]; and (4) discrete abstract structures can be described by algebraic theories (see Section "Abstract Data Types and Algebraic Semantics). In algebraic semiotics, *sign systems* are algebraic theories with extra structure, and *semiotic morphisms* are used to study representations, metaphors, translations, meanings, etc.; because these map signs in one system to signs in another, rather than mapping individual signs, Saussure's insight is raised from sign systems to their morphisms. In [70, 68], techniques are also given for comparing the quality of semiotic morphisms, and a new version of categorical colimits, developed in collaboration with Grigore Roşu, is used for combining meanings and for studying the effect of context on meaning; this includes "blends" in the sense of [26]. The potential to connect diverse areas on both sides of the great divide and to enter new application areas, such as user interface design, seems very exciting.

When applied to user interface design, algebraic semiotics can model the content and structure of information through its representation [68, 80, 101, 102]; it is now being used to design interfaces for the TATAMI system (see Section "Distributed Cooperative Engineering with TATAMI"). Although traditional ergonomics, HCI (human computer interface), and cognitive science are very good for issues like keyboard layout, color choice, font size, and window layout, they are less useful for more semantic problems. Examples using algebraic semiotics are accessible over the web at `http://www.cs.ucsd.edu/groups/tatami`, and a "semiotic zoo" of ordinary design choices that are bad for interesting reasons is available at `http://www.cs.ucsd.edu/users/goguen/zoo`, with algebraic explanations. The zoo illustrates how algebraic semiotics can be applied to syntax, understanding stories, and much more.

---

13 This is an insight from mathematics. The journey through this paper has already encountered several cases where morphisms are important: initial extensions for constraint programming (in Section); data constraints (in Section); and inclusion systems (Section "Logical and Multi-Paradigm Programming"). Eilenberg and Mac Lane [25] gave this insight a more definite and systematic form with the invention of category theory.

Fuzzy Logic and Information Theory

Many people have had the intuition that mathematical logic fails to capture the imprecision and robustness of practical reasoning. Fuzzy sets and logic [174] are a successful move in this direction, though it is far from capturing the richness and complexity of ordinary human reasoning. In the late 1960s, it was fashionable to give axioms on the category of things having some structure (with structure preserving morphisms) to characterize that category up to natural equivalence. This is what I did for fuzzy sets in my thesis [32, 34, 39]; earlier papers concerned other aspects of fuzzy sets, e.g., extending them to more general values than the unit interval [31, 33]. This was the first foundational work on fuzzy sets and fuzzy logic. Later I set up the Fuzzy Robot Users Group at UCLA, and did some early empirical work on fuzzy algorithms, though it was never properly published (but see [103] for an abstract). Some applications to philosophy and the social sciences, and some limitations of fuzzy set theory were discussed in [45]. Today fuzzy sets and fuzzy logic are very popular areas. But in the late 1960s, work in this area was bitterly opposed by more traditional parts of engineering, such as classical control theory, to the extent that I found it impossible to get funding to continue my research in this area, and had to abandon it.

It is (or should be) a scandal that in the middle of a period called the "information age" and characterized by an astonishing expansion of information technology, there is no adequate theory of information, nor even any adequate definition of information. In the late 1960s, I taught a course at the University of Chicago on traditional Shannon-style information theory [76], and encountered great difficulty trying to extend it to human situations; in fact this theory does not apply to meaning, but rather to data compression and transmission – after all, it was developed for the (then) Bell Telephone Company – because it ignores the crucial human aspects that underlie meaning. This motivated using categorical GST for a complexity based information theory, which was then applied to music [43]; here the information content of a behavior is the minimum value of the sum of weights of (hierarchical) interconnected components whose behavior is the given one. It was startling that the classical equalities and inequalities of information theory still hold in this exceedingly general setting; moreover it did capture some insights about the structure of music. Several Oxford students did experiments in this area for their MSc theses, but again it became clear that no purely formal approach, however abstract and general, could deal with human meaning in any deep sense [63] (see also [45]).

Recently I returned to this area, but from the opposite side of the great divide, defining information in social terms [69]: *An item of information is an interpretation of a configuration of signs for which members of some social group are accountable.* The goal is to get a theory of information adequate for understanding and designing systems that process information. This research draws on ideas from ethnomethodology [163], semiotics, logic, and the sociology of science. The paper [69] also presents some case studies and makes the perhaps surprising argument that because of its social situatedness, information has an intrinsic ethical dimension.

Philosophy of Computing

It is usually thought that philosophy has little to do with the practical engineering concerns that dominate computer science today. But philosophies are just coherent expositions of particular approaches to defining, approaching and (maybe) solving problems[14]. Nothing is more practical than a good philosophy: our approach to a situation has an effect, often decisive, on what happens. As Wesley Phoa [159] puts it,

> All practical work is based on philosophical presuppositions: they may be conscious or unconscious, innocuous or fatal. ... In any case, we might as well be aware of them, and aware of the alternatives. There is a more positive reason to be interested in philosophy though: philosophical reflection can be a potential source of practical ideas.

This section discusses some philosophies that seem especially relevant to today's computer science.

Realism and Idealism

Realism is the view that certain things (e.g., numbers) really exist; one can be "realistic" in this sense about many different things. For pure mathematics, realism is called *Platonism*. Here it seems harmless, perhaps even charming; but it can become dangerous when extended further into applied mathematics. Within computer science, systems analysis, systems engineering, requirements engineering, etc., realism can be the view that there really is such a "thing" as "the system," and (going beyond realism into confusion) that some particular model built for some particular purpose completely "captures" the system. This easily can (and often does) lead to incorrect decisions based on some weak area of the model, that should have been reevaluated and strengthened, but was not because of the erroneous belief that the model *really is* the system. I encountered some examples of this in the area of computer security, in connection with the so-called Bell-LaPadula security model [3, 91, 93].

It is easier to understand the attraction of realism when it is considered in the context of its opposition to *idealism*, the view that things (either some particular class of things, or in a more extreme form, all things) only exist in our minds; the most extreme form of this view is *solipsism*, that nothing exists outside oneself. The problem with idealism is that it seems to negate science (as well as religion and all metaphysics); so realists can be seen as trying to save "reality" for science (and/or religion).

My own view is that this is a false duality, based on an unexamined presupposition, namely dualism. Simplifying quite a bit, *dualism* asserts a rigid separation between a material realm and a spiritual realm; such a view was advocated by Descartes in order to gain a ground for science that was free from interference

[14] Actually, even the notion that "problems" exist and should be "solved" is a philosophical presupposition that is open to question!

by religion. Today dualism is largely considered untenable, for example because there is no empirical evidence for a separate realm of the spiritual. So within this historical context, idealism can be seen as the attempt to reduce everything to the spiritual (though not usually in the religious sense) and realism as the attempt to reduce everything to the material – which is what most science is about anyway. In my view, both are crude attempts to eliminate Cartesian duality, which after all does reflect some aspects of everyone's experience. Instead of trying to collapse the duality to one of its poles, it seems preferable to *transcend* duality in a way that denies the actual existence of two realms, and still adequately explains our experience. My preference is for *phenomenology*, which starts with experience, disallowing any external distinctions; it then proceeds with a careful analysis of experience and its ground; however, this paper isn't the place to consider such issues; some further discussion and references may be found in [63].

Modernism

Modernism can be defined as belief in the adequacy of hierarchy, formalization and control to achieve desired ends without error. The success of computer science, along with technology as a whole, has long been seen as establishing the correctness of such a philosophy, so much so that it is usually not even explicitly formulated. But increasing experience with ever more ambitious projects has led to more awareness of the limitations of modernism. A classic example in software engineering is the so-called waterfall model of software development, which requires a rigid top-down hierarchical control of the development process, by dividing it into strictly sequential stages, typically something like requirements, design, specification, build and test. But such an organization of the work process makes it very difficult to correct errors in earlier stages, to learn from errors, and to respond to the plethora of changes in the surrounding context of technology, organization, law, etc. that are inevitable and unending in today's fast paced environment. Of course, now there are many process models that improve upon the waterfall model, but I think there is still insufficient appreciation of the importance of alternative techniques like rapid prototyping and user participation in design.

Theoretical computer scientists need not look so far afield for examples of excessive modernism. What might be called the "error free" school of formal methods aims for programs that have no bugs at all. For example, [22] claims

> we have ... "a calculus" for a formal discipline – a set of rules – such that, if applied successfully: (1) it will have derived a correct program; and (2) it will tell us that we have reached such a goal.

From a narrow point of view, this effort succeeded, modulo certain technical difficulties which can however be corrected[15]. However, there is a fundamental

[15] These include the following: (1) there is a gap in the logical foundations, in that the first order logic used for expressing conditions is not actually sufficiently expressive

difficulty, namely that it attempts to control the programming process by imposing a rigid top-down derivation sequence, working backwards from the initial top level specification (the "postcondition") to the final code, in which each step is derived by applying a "weakest precondition" (hereafter, "wp") formula.

It is not coincidental that the "wp calculus" requires significant human invention at exactly the most difficult points, namely the loops. And for most programs that go much beyond the trivial, the insights needed to write the loop invariants are tantamount to already knowing how to write the program; moreover, these insights are more difficult to achieve when using wp than they would be in a more conventional setting. When I was at Oxford, I saw several very good students who had been taught that the wp calculus was the right way to program, become so discouraged over the difficulties they experienced, that they came to believe they could never learn how to program and should therefore seek a different profession. This is a great pity! In general, rigid top-down ideologies inhibit experimentation and make it hard to explore tradeoffs. Moreover, they can be harmful to students, wasteful of time, reinforcing of an inflexible view of life, and inhibiting to intuition, learning and creativity. Finally, as noted in the first paragraph of Section "Specification", correctness of code is the wrong problem to solve. (An overview of some recent debates on philosophical foundations of formal methods may be found in [1].)

A further difficulty with formal methods is that they tend to be very brittle, in the sense that slight changes in a specification can lead to drastic changes in what must be done to achieve that spec. We can see the importance of this difficulty by noting that the very rapid rate of change of requirements, which is so typical of large projects, implies an even more rapid rate of change for specifications. This makes many formal methods very difficult, perhaps even impossible, to apply in practice [137]. Let me be clear that I am not criticizing formal methods as such – in fact, I believe they can be very useful in practice, *especially* for large programs, by focusing on the large grain structure of software (see Section "Parameterized Programming and Generic Modules" and [137]).

Jean-François Lyotard [138] defines *modernism* more broadly than at the beginning of this section, as any approach that justifies its claims to universality through a "grand unifying" story, which he calls a *meta-narrative*. For example, the realist approach to systems theory discussed in Section "Realism and Idealism" tells the story of unique pre-existing systems really out there to be captured. By contrast, Lyotard says *postmodernism* is characterized by multiple "local language games" that cannot necessarily be unified, or even neatly classified (this notion of language game comes from late work of Wittgenstein). A great deal has been written about postmodernism, and though much of it is

– something like the infinitary logic proposed by Erwin Engeler in the 1960s for this purpose is necessary; (2) many important programming features are not treated, including procedures, blocks, modules, and objects – in general, all large grain features are omitted; and (3) data structures and types are treated loosely, and variables that range over different kinds of item are not carefully distinguished.

trash, there is a general agreement that we may now be entering a postmodern period.

We have already seen the issue of whether there are many or just one language game in the discussion of requirements engineering in Section "Requirements Engineering"; my view is that requirements engineers must deal with multiple "stakeholders" who may have not just different viewpoints on the "same system," but may actually have completely different ways to conceptualize their experience. A related philosophical approach is the *deconstructionism*[16] of Derrida, which can be seen as a strategy for undermining meta-narratives. Phoa [159] applies deconstruction to aspects of theorem proving, software engineering and artificial intelligence; in particular, Phoa's view of requirements is remarkably similar to that presented here. It can also be argued that the approach to distributed cooperative theorem proving and software engineering taken in the TATAMI system (Section "Distributed Cooperative Engineering with TATAMI") is postmodern, or even Derridean, because it supports multiple points of view, including views that are mutually inconsistent, or even self inconsistent, and because it uses multiple narratives as an organizational principle.

Autopoiesis

Maturana and Varela [144] define an *autopoietic system* to be

> ... a network of processes of production of components that produces the components that: (i) through their interactions and transformations continuously regenerate the network of processes that produced them; and (ii) constitute it as a concrete unity in the space in which they exist by specifying the topological domain of its realization as such a network.

(See [2, 167, 143, 145] for more on this area, and see [171, 112] for some possibly ill-advised attempts to formalize this notion; also cf. footnote 4. Sorry for the confusing prose in this quote.)

The relevance of this to software engineering is discussed in [42]: Anyone familiar with large software projects knows that there is a sense in which they "have a life of their own," in that some projects seem healthy and vibrant from the start, and overcome even unexpected obstacles with enthusiasm and intelligence, while others always seem disorganized and depressed, suffering from such symptoms as unrealistic goals, inadequate equipment, poor planning, (seemingly) insufficient funding, faulty communication, indecisive leadership, frequent reorganizations, and/or deep rifts between internal factions.

A software development project is not a formal mathematical entity. Perhaps it is usefully seen as an autopoietic process, an evolving organization of informational structures, continually recreating itself by building, modifying, and reusing its structures; in the language of Maturana, this is "development through mutually recursive interactions among structurally plastic systems" [143]. For

16 See also the hilarious Woody Allen film *Deconstructing Harry*.

example, an unhealthy project may struggle for survival by reassigning responsibilities, redefining subprojects, and even trying to reconstrue the conditions that define its success. On the other hand, a healthy project may develop new tools to enhance its productivity.

In this view, computers, printouts, compilers, editors, design tools, and even programmers, can be seen as supporting substrates, just as body parts are supporting substrates for a person[17]. Autopoietic systems are about as far as we know how to get from rigid top-down hierarchical goal-driven control systems; autopoietic systems thrive on error, and reconstruct themselves on the basis of what they learn from their mistakes. Autopoiesis can be considered an implementation technique for postmodernism. See [59] for more on attitudes towards errors in computer science.

Ethics

One aspect of the great divide that seems especially troubling in the late twentieth century is the separation of technology and science from ethics. A positive sign is the increasing availability of courses on professional ethics in engineering schools; on the other hand, ethical behavior seems to be on the decrease. Of course, there is a huge philosophical literature on ethics, but little of it directly addresses technology. My own concern is to bridge the gap between technology and ethics on intellectual grounds. In [69], I argue for an inherent ethical dimension to information (but not mere data), through its being embedded in a context of concern by some social group (see the definition of information in Section "Fuzzy Logic and Information Theory"). This is not an appropriate place for details, but we should recall that understanding values can be crucial for getting requirements that match user needs. Other perspectives on the "great divide" can be found in the book [6] in which the paper [69] appears. A more radical view appears in [119], where Heidegger considers Western civilization to be fundamentally entangled with a separation of technology from ethics, based on an untenable instrumental conception of technology; see also the discussions in [61] and [63]. Heidegger [119] further claims that by questioning deeply enough into the essence of technology, perhaps in desperation, we may find what we need to go beyond the current impasse.

But no one should think that postmodernism, deconstructionism or phenomenology is going to answer every question concerning technology, the modern world, or the philosophy of computation. The issues that must be faced are extremely complex and diverse, and it appears to me that we are only at the beginning of some new ways of thinking. If our civilization survives, there will doubtless be many profound changes ahead in how we think about technology and its relation to society.

[17] Of course, this does not mean that groups have moral or spiritual priority over individuals, or that people should be viewed as components of systems in anything like the same way that Ada packages can be.

Some Concluding Remarks

I'm afraid that the reader may have found this paper rather a long strange trip[18], starting from the practice of software engineering, then going to category theory, and eventually ethics, passing through topics like equational deduction, various programming and specification paradigms, semiotics, theorem proving, requirements engineering and philosophy. Nevertheless, theoretical computer science has been a constant theme, for example appearing in the application of ADTs and theory morphisms to semiotics and user interface design, and in the philosophical exploration of difficulties that arise in computing practice. I believe that the kind of concept developed in theoretical computer science, especially its algebraic and categorical branches, is very well suited to combating the fragmentation of computer science as a whole, and hope to have given some salient examples; this seems especially important for education.

It seems valuable to use algebraic theories to model the kind of situated abstract data types that appear in semiotics, and it also seems valuable to study the trajectories of projects that involve theory. Both kinds of research straddle the great divide, and both are risky. It is much safer to stay within the confines of a single discipline, and if possible work on a single well defined problem of recognized practical importance. Few people want to read a paper like [68] in a serious way, because it requires familiarity with diverse areas of philosophy, mathematics, linguistics, literature and computer science; so the author isn't likely to get much recognition, and it will be hard to publish. But the safe way can also be a painful way. Working on a narrow well defined problem of recognized importance will almost guarantee intense competition. Of course this is what academic departments like, because it makes comparison and measurement easier. In this way, departmental boundaries are inimical to interdisciplinary work, as well as to innovative and/or very abstract work. On the other hand, a community is necessary to develop tools and methods and get them actually used; transfering either theory or technology is by definition a social process [130], so that no lone researcher or small group can hope to get very far. But this is not to say that fame and fortune are the only measures of success; on the contrary, quite different measures like aesthetics and coherence seem to be at least as important for theoretical computer science.

In this paper, I've tried to share some of the hopes and disappointments I've felt for various ideas and projects; these remarks are not intended as bragging or complaining; at this stage in my life, I don't take any of them seriously (though I must admit that some did seem very serious at the time). The goal of these remarks is to help younger researchers. Most work of most researchers never has very much influence, every career has difficulties, and innovative and/or interdisciplinary work is especially likely to be difficult. Moreover, although the conventions of scientific writing hide it, every researcher has some emotional involvement with his/her work, and this is something that should be dealt with, not just ignored. Those interested in the sociology of technology may also find

---

[18] With a nod of the head to the Grateful Dead song *Truckin'*.

it interesting to see how ideas have sometimes flowed from one area to another. Our civilization needs to heal the wound between the social and the technical-scientific world views, and I hope to have given some hints about how this might occur; see also the charming documentary novel *Aramis, or the Love of Technology*, by Bruno Latour [131].

After discussing this paper with a student, he asked if I felt "melancholy" looking back over a whole career of tossing algebraic flowers down the great divide. Not really, I replied; to be sure, this is a sad metaphor, but is not all life similar? Of course there are difficulties, and perhaps there are successes. But commitment to our art and craft, to our profession, to our own integrity, and to other people, are far more important; life occurs as it is lived, and to live it fully is to appreciate joys and pains as they are, rather than in the light of ambitions that can only intensify pain and confusion.

Acknowledgements

I enjoy working with others, and much of my favorite research has been joint work with Rod Burstall, Răzvan Diaconescu, Kokichi Futatsugi, Charlotte Linde, Grant Malcolm, José Meseguer, David Plaisted, James Thatcher, Will Tracz, Eric Wagner, Timothy Winkler, Jesse Wright, and personnel at the UCSD Meaning and Computation Lab, including Kai Lin, Eric Livingston, Akira Mori, Grigore Roşu, and Akiyoshi Sato. Especially I thank Rod Burstall, Peter Landin, Saunders Mac Lane, John McCarthy, and Christopher Strachey, whose pioneering research inspired my interests in semantics, algebra and logic. Burstall and Mac Lane have done even more, they have encouraged and guided me, and my gratitude and debt are very great; I have also learned much from Charlotte Linde about sociolinguistics and discourse analysis. I wish to thank the organizations that have supported my research, including the National Science Foundation, the Office of Naval Research, the National Aeronautics and Space Administration, British Telecom, Fujitsu, several European Union projects, and two MITI (Japan) projects, including the CafeOBJ project.

References

1. Jon Barwise. Mathematical proofs of computer system correctness. Technical Report CSLI-89-136, Center for the Study of Language and Information, Stanford University, August 1989.
2. Gregory Bateson. *Mind and Nature: A Necessary Unity.* Bantam, 1980.
3. D.E. Bell and L.J. LaPadula. Secure computer systems: Mathematical foundations and model. Technical report, MITRE, 1974.
4. Matthew Bickerton and Jawed Siddiqi. The classification of requirements engineering methods. In Stephen Fickas and Anthony Finkelstein, editors, *Requirements Engineering '93*, pages 182–186. IEEE, 1993.
5. Barry Boehm. *Software Engineering Economics.* Prentice-Hall, 1981.
6. Geoffrey Bowker, Leigh Star, William Turner, and Les Gasser. *Social Science, Technical Systems and Cooperative Work: Beyond the Great Divide.* Erlbaum, 1997.

7. Rod Burstall and Răzvan Diaconescu. Hiding and behaviour: an institutional approach. In A. William Roscoe, editor, *A Classical Mind: Essays in Honour of C.A.R. Hoare*, pages 75–92. Prentice Hall, 1994.
8. Rod Burstall and Joseph Goguen. Putting theories together to make specifications. In Raj Reddy, editor, *Proceedings, Fifth International Joint Conference on Artificial Intelligence*, pages 1045–1058. Department of Computer Science, Carnegie-Mellon University, 1977.
9. Rod Burstall and Joseph Goguen. The semantics of Clear, a specification language. In Dines Bjorner, editor, *Proceedings, 1979 Copenhagen Winter School on Abstract Software Specification*, pages 292–332. Springer, 1980. Lecture Notes in Computer Science, Volume 86; based on unpublished notes handed out at the Symposium on Algebra and Applications, Stefan Banach Center, Warsaw, Poland, 1978.
10. Rod Burstall and Joseph Goguen. An informal introduction to specifications using Clear. In Robert Boyer and J Moore, editors, *The Correctness Problem in Computer Science*, pages 185–213. Academic, 1981. Reprinted in *Software Specification Techniques*, Narain Gehani and Andrew McGettrick, editors, Addison-Wesley, 1985, pages 363–390.
11. Manuel Clavel and José Meseguer. Reflection and strategies in rewriting logic. In José Meseguer, editor, *Proceedings, First International Workshop on Rewriting Logic and its Applications*. Elsevier Science, 1996. Volume 4, *Electronic Notes in Theoretical Computer Science.*
12. Steven Collins. *Selfless Persons.* Cambridge, 1983.
13. Corina Cîrstea. A semantic study of the object paradigm, 1996. Transfer thesis, Programming Research Group, Oxford University.
14. Corina Carstea. Coalgebra semantics for hidden algebra: parameterized objects and inheritance, 1997. Paper presented at 12th Workshop on Algebraic Development Techniques.
15. Răzvan Diaconescu. *Category-based Semantics for Equational and Constraint Logic Programming.* PhD thesis, Programming Research Group, Oxford University, 1994.
16. Răzvan Diaconescu A category-based equational logic semantics to constraint programming. In Magne Haveraaen, Olaf Owe, and Ole-Johan Dahl, editors, *Recent Trends in Data Type Specification*, pages 200–222. Springer, 1996. Lecture Notes in Computer Science, Volume 389.
17. Răzvan Diaconescu Extra theory morphisms for institutions: logical semantics for multi-paradigm languages. Technical Report IS-RR-96-0024S, Japan Institute of Science and Technology, 1996. To appear, *J. Applied Categorical Structures.*
18. Răzvan Diaconescu Foundations of behavioural specification in rewriting logic. In José Meseguer, editor, *Proceedings, First International Workshop on Rewriting Logic and its Applications.* Elsevier Science, 1996. Volume 4, *Electronic Notes in Theoretical Computer Science.*
19. Răzvan Diaconescu and Kokichi Futatsugi. Logical semantics for CafeOBJ. Technical Report IS-RR-96-0024S, Japan Institute of Science and Technology, 1996.
20. Răzvan Diaconescu and Kokichi Futatsugi. *CafeOBJ Report: The Language, Proof Techniques, and Methodologies for Object-Oriented Algebraic Specification*, volume 6. World Scientific, 1998. To appear, AMAST Series in Computing.
21. Răzvan Diaconescu, Joseph Goguen, and Petros Stefaneas. Logical support for modularisation. In Gerard Huet and Gordon Plotkin, editors, *Logical Environments*, pages 83–130. Cambridge, 1993.

22. Edsger Dijkstra. Guarded commands, nondeterminacy and formal derivation of programs. *Communications of the Association for Computing Machinery*, 18:453–457, 1975.
23. Hans-Dieter Ehrich, Joseph Goguen, and Amilcar Sernadas. A categorial theory of objects as observed processes. In Jaco de Bakker, Willem de Roever, and Gregorz Rozenberg, editors, *Foundations of Object Oriented Languages*, pages 203–228. Springer, 1991. Lecture Notes in Computer Science, Volume 489.
24. Hartmut Ehrig, Werner Fey, and Horst Hansen. ACT ONE: An algebraic specification language with two levels of semantics. Technical Report 83–03, Technical University of Berlin, Fachbereich Informatik, 1983.
25. Samuel Eilenberg and Saunders Mac Lane. General theory of natural equivalences. *Transactions of the American Mathematical Society*, 58:231–294, 1945.
26. Gilles Fauconnier and Mark Turner. Conceptual projection and middle spaces. Technical Report 9401, University of California at San Diego, 1994. Dept. of Cognitive Science.
27. Kokichi Futatsugi, Joseph Goguen, Jean-Pierre Jouannaud, and José Meseguer. Principles of OBJ2. In Brian Reid, editor, *Proceedings, Twelfth ACM Symposium on Principles of Programming Languages*, pages 52–66. Association for Computing Machinery, 1985.
28. Kokichi Futatsugi, Joseph Goguen, José Meseguer, and Koji Okada. Parameterized programming in OBJ2. In Robert Balzer, editor, *Proceedings, Ninth International Conference on Software Engineering*, pages 51–60. IEEE Computer Society, March 1987.
29. Kokichi Futatsugi and Ataru Nakagawa. An overview of Cafe specification environment. In *Proceedings, ICFEM'97*. University of Hiroshima, 1997.
30. W. Wyat Gibbs. Software's chronic crisis. *Scientific American*, pages 72–81, September 1994.
31. Joseph Goguen. L-fuzzy sets. *Journal of Mathematical Analysis and Applications*, 18(1):145–174, 1967.
32. Joseph Goguen. *Categories of Fuzzy Sets*. PhD thesis, Department of Mathematics, University of California at Berkeley, 1968.
33. Joseph Goguen. The logic of inexact concepts. *Synthese*, 19:325–373, 1968–69.
34. Joseph Goguen. Categories of V-sets. *Bulletin of the American Mathematical Society*, 75(3):622–624, 1969.
35. Joseph Goguen. Mathematical representation of hierarchically organized systems. In E. Attinger, editor, *Global Systems Dynamics*, pages 112–128. S. Karger, 1971.
36. Joseph Goguen. Minimal realization of machines in closed categories. *Bulletin of the American Mathematical Society*, 78(5):777–783, 1972.
37. Joseph Goguen. Categorical foundations for general systems theory. In F. Pichler and R. Trappl, editors, *Advances in Cybernetics and Systems Research*, pages 121–130. Transcripta Books, 1973.
38. Joseph Goguen. Realization is universal. *Mathematical System Theory*, 6:359–374, 1973.
39. Joseph Goguen. Concept representation in natural and artificial languages: Axioms, extensions and applications for fuzzy sets. *International Journal of Man-Machine Studies*, 6:513–561, 1974.
40. Joseph Goguen. Objects. *International Journal of General Systems*, 1(4):237–243, 1975.
41. Joseph Goguen. Semantics of computation. In Ernest Manes, editor, *Proceedings, First International Symposium on Category Theory Applied to Computation and*

Control, pages 151–163. Springer, 1975. (San Fransisco, February 1974.) Lecture Notes in Computer Science, Volume 25.

42. Joseph Goguen. Abstract errors for abstract data types. In Eric Neuhold, editor, *Proceedings, First IFIP Working Conference on Formal Description of Programming Concepts*, pages 21.1–21.32. MIT, 1977. Also in *Formal Description of Programming Concepts*, Peter Neuhold, Ed., North-Holland, pages 491–522, 1979.
43. Joseph Goguen. Complexity of hierarchically organized systems and the structure of musical experiences. *International Journal of General Systems*, 3(4):237–251, 1977.
44. Joseph Goguen. Order sorted algebra. Technical Report 14, UCLA Computer Science Department, 1978. Semantics and Theory of Computation Series.
45. Joseph Goguen. Fuzzy sets and the social nature of truth. In M.M. Gupta and Ronald Yager, editors, *Advances in Fuzzy Set Theory and Applications*, pages 49–68. North-Holland, 1979.
46. Joseph Goguen. Some ideas in algebraic semantics. In Ken Hirose, editor, *Mathematical Logic and Computer Science.* IBM Japan, 1979. Proceedings, Third IBM Symposium on Mathematical Foundations of Computer Science.
47. Joseph Goguen. How to prove algebraic inductive hypotheses without induction, with applications to the correctness of data type representations. In Wolfgang Bibel and Robert Kowalski, editors, *Proceedings, Fifth Conference on Automated Deduction*, pages 356–373. Springer, 1980. Lecture Notes in Computer Science, Volume 87.
48. Joseph Goguen. Parameterized programming. *Transactions on Software Engineering*, SE–10(5):528–543, September 1984.
49. Joseph Goguen. One, none, a hundred thousand specification languages. In H.-J. Kugler, editor, *Information Processing '86*, pages 995–1003. Elsevier, 1986. Proceedings of 1986 IFIP Congress.
50. Joseph Goguen. Reusing and interconnecting software components. *Computer*, 19(2):16–28, February 1986. Reprinted in *Tutorial: Software Reusability*, Peter Freeman, editor, IEEE Computer Society, 1987, pages 251–263, and in *Domain Analysis and Software Systems Modelling*, Rubén Prieto-Díaz and Guillermo Arango, editors, IEEE Computer Society, 1991, pages 125–137.
51. Joseph Goguen. Memories of ADJ. *Bulletin of the European Association for Theoretical Computer Science*, 36:96–102, October 1989. Guest column in the 'Algebraic Specification Column.' Also in *Current Trends in Theoretical Computer Science: Essays and Tutorials*, World Scientific, 1993, pages 76–81.
52. Joseph Goguen. Principles of parameterized programming. In Ted Biggerstaff and Alan Perlis, editors, *Software Reusability, Volume I: Concepts and Models*, pages 159–225. Addison Wesley, 1989.
53. Joseph Goguen. What is unification? A categorical view of substitution, equation and solution. In Maurice Nivat and Hassan Aït-Kaci, editors, *Resolution of Equations in Algebraic Structures, Volume 1: Algebraic Techniques*, pages 217–261. Academic, 1989.
54. Joseph Goguen. An algebraic approach to refinement. In Dines Bjorner, C.A.R. Hoare, and Hans Langmaack, editors, *Proceedings, VDM'90: VDM and Z – Formal Methods in Software Development*, pages 12–28. Springer, 1990. Lecture Notes in Computer Science, Volume 428.
55. Joseph Goguen. Hyperprogramming: A formal approach to software environments. In *Proceedings, Symposium on Formal Approaches to Software Environment Technology.* Joint System Development Corporation, Tokyo, Japan, January 1990.

56. Joseph Goguen. A categorical manifesto. *Mathematical Structures in Computer Science*, 1(1):49–67, March 1991.
57. Joseph Goguen. Semantic specifications for the Rewrite Rule Machine. In Aki Yonezawa and Takayasu Ito, editors, *Concurrency: Theory, Language and Architecture*, pages 216–234, 1991. Proceedings of a U.K.–Japan Workshop; Springer, Lecture Notes in Computer Science, Volume 491.
58. Joseph Goguen. Types as theories. In George Michael Reed, Andrew William Roscoe, and Ralph F. Wachter, editors, *Topology and Category Theory in Computer Science*, pages 357–390. Oxford, 1991. Proceedings of a Conference held at Oxford, June 1989.
59. Joseph Goguen. The denial of error. In Christiane Floyd, Heinz Züllighoven, Reinhard Budde, and Reinhard Keil-Slawik, editors, *Software Development and Reality Construction*, pages 193–202. Springer Verlag, 1992.
60. Joseph Goguen. The dry and the wet. In Eckhard Falkenberg, Colette Rolland, and El-Sayed Nasr-El-Dein El-Sayed, editors, *Information Systems Concepts*, pages 1–17. Elsevier North-Holland, 1992. Proceedings, IFIP Working Group 8.1 Conference (Alexandria, Egypt).
61. Joseph Goguen. Hermeneutics and path. In Christiane Floyd, Heinz Züllighoven, Reinhard Budde, and Reinhard Keil-Slawik, editors, *Software Development and Reality Construction*, pages 39–44. Springer Verlag, 1992.
62. Joseph Goguen. Sheaf semantics for concurrent interacting objects. *Mathematical Structures in Computer Science*, 11:159–191, 1992.
63. Joseph Goguen. Truth and meaning beyond formalism. In Christiane Floyd, Heinz Züllighoven, Reinhard Budde, and Reinhard Keil-Slawik, editors, *Software Development and Reality Construction*, pages 353–362. Springer Verlag, 1992.
64. Joseph Goguen. Social issues in requirements engineering. In Stephen Fickas and Anthony Finkelstein, editors, *Requirements Engineering '93*, pages 194–195. IEEE, 1993.
65. Joseph Goguen. Requirements engineering as the reconciliation of social and technical issues. In Marina Jirotka and Joseph Goguen, editors, *Requirements Engineering: Social and Technical Issues*, pages 165–200. Academic, 1994.
66. Joseph Goguen. Formality and informality in requirements engineering. In *Proceedings, International Conference on Requirements Engineering*, pages 102–108. IEEE Computer Society, April 1996.
67. Joseph Goguen. Parameterized programming and software architecture. In *Proceedings, Reuse'96*, pages 2–11. IEEE Computer Society, April 1996.
68. Joseph Goguen. Semiotic morphisms. Technical Report CS97–553, UCSD, Dept. Computer Science & Eng., 1997. Early version in *Proc., Conf. Intelligent Systems: A Semiotic Perspective, Vol. II*, ed. J. Albus, A. Meystel and R. Quintero, Nat. Inst. Science & Technology (Gaithersberg MD, 20–23 October 1996), pages 26–31.
69. Joseph Goguen. Towards a social, ethical theory of information. In Geoffrey Bowker, Leigh Star, William Turner, and Les Gasser, editors, *Social Science, Technical Systems and Cooperative Work: Beyond the Great Divide*, pages 27–56. Erlbaum, 1997.
70. Joseph Goguen. An introduction to algebraic semiotics, with applications to user interface design. In Chrystopher Nehaniv, editor, *Proceedings, Computation for Metaphors, Analogy and Agents*. University of Aizu, 1998. Aizu–Wakamatsu, Japan, 6–10 April 1998.
71. Joseph Goguen. *Theorem Proving and Algebra*. MIT, to appear.

72. Joseph Goguen and Rod Burstall. CAT, a system for the structured elaboration of correct programs from structured specifications. Technical Report Report CSL-118, SRI Computer Science Lab, October 1980.
73. Joseph Goguen and Rod Burstall. Introducing institutions. In Edmund Clarke and Dexter Kozen, editors, *Proceedings, Logics of Programming Workshop*, pages 221–256. Springer, 1984. Lecture Notes in Computer Science, Volume 164.
74. Joseph Goguen and Rod Burstall. A study in the foundations of programming methodology: Specifications, institutions, charters and parchments. In David Pitt, Samson Abramsky, Axel Poigné, and David Rydeheard, editors, *Proceedings, Conference on Category Theory and Computer Programming*, pages 313–333. Springer, 1986. Lecture Notes in Computer Science, Volume 240.
75. Joseph Goguen and Rod Burstall. Institutions: Abstract model theory for specification and programming. *Journal of the Association for Computing Machinery*, 39(1):95–146, January 1992.
76. Joseph Goguen and Lee Carlson. Axioms for discrimination information. *IEEE Transactions on Information Theory*, pages 572–574, September 1975.
77. Joseph Goguen and Răzvan Diaconescu. Towards an algebraic semantics for the object paradigm. In Hartmut Ehrig and Fernando Orejas, editors, *Proceedings, Tenth Workshop on Abstract Data Types*, pages 1–29. Springer, 1994. Lecture Notes in Computer Science, Volume 785.
78. Joseph Goguen and Susanna Ginali. A categorical approach to general systems theory. In George Klir, editor, *Applied General Systems Research*, pages 257–270. Plenum, 1978.
79. Joseph Goguen, Jean-Pierre Jouannaud, and José Meseguer. Operational semantics of order-sorted algebra. In Wilfried Brauer, editor, *Proceedings, 1985 International Conference on Automata, Languages and Programming*. Springer, 1985. Lecture Notes in Computer Science, Volume 194.
80. Joseph Goguen, Kai Lin, Akira Mori, Grigore Roşu, and Akiyoshi Sato. Distributed cooperative formal methods tools. In Michael Lowry, editor, *Proceedings, Automated Software Engineering*, pages 55–62. IEEE, 1997. Lake Tahoe CA, 3–5 November 1997.
81. Joseph Goguen and Charlotte Linde. Cost-benefit analysis of a proposed computer system. Technical report, Structural Semantics, 1978.
82. Joseph Goguen and Charlotte Linde. Linguistic methodology for the analysis of aviation accidents. Technical report, Structural Semantics, December 1983. NASA Contractor Report 3741, Ames Research Center.
83. Joseph Goguen and Charlotte Linde. Techniques for requirements elicitation. In Stephen Fickas and Anthony Finkelstein, editors, *Requirements Engineering '93*, pages 152–164. IEEE, 1993. Reprinted in *Software Requirements Engineering (Second Edition)*, ed. Richard Thayer and Merlin Dorfman, IEEE Computer Society, 1996.
84. Joseph Goguen, Charlotte Linde, and Tora Bikson. Optimal structures for multimedia instruction. Technical report, Computer Science Lab, SRI International, July 1985.
85. Joseph Goguen, Charlotte Linde, and Miles Murphy. Crew communication as a factor in aviation accidents. In E. James Hartzell and Sandra Hart, editors, *Papers from the 20th Annual Conference on Manual Control.* NASA Ames Research Center, 1984.
86. Joseph Goguen and Luqi. Formal methods and social context in software development. In Peter Mosses, Mogens Nielsen, and Michael Schwartzbach, editors,

Proceedings, Sixth International Joint Conference on Theory and Practice of Software Development (TAPSOFT 95), pages 62–81. Springer, 1995. Lecture Notes in Computer Science, Volume 915.

87. Joseph Goguen and Grant Malcolm. Proof of correctness of object representation. In A. William Roscoe, editor, *A Classical Mind: Essays in Honour of C.A.R. Hoare*, pages 119–142. Prentice Hall, 1994.
88. Joseph Goguen and Grant Malcolm. *Algebraic Semantics of Imperative Programs.* MIT, 1996.
89. Joseph Goguen and Grant Malcolm. A hidden agenda. Technical Report CS97–538, UCSD, Dept. Computer Science & Eng., May 1997. To appear in *Theoretical Computer Science*. Early abstract in *Proc., Conf. Intelligent Systems: A Semiotic Perspective, Vol. I*, ed. J. Albus, A. Meystel and R. Quintero, Nat. Inst. Science & Technology (Gaithersberg MD, 20–23 October 1996), pages 159–167.
90. Joseph Goguen, Grant Malcolm, and Tom Kemp. A hidden Herbrand theorem, to appear.
91. Joseph Goguen and José Meseguer. Security policies and security models. In Marvin Schafer and Dorothy D. Denning, editors, *Proceedings, 1982 Symposium on Security and Privacy*, pages 11–22. IEEE Computer Society, 1982.
92. Joseph Goguen and José Meseguer. Universal realization, persistent interconnection and implementation of abstract modules. In M. Nielsen and E.M. Schmidt, editors, *Proceedings, 9th International Conference on Automata, Languages and Programming*, pages 265–281. Springer, 1982. Lecture Notes in Computer Science, Volume 140.
93. Joseph Goguen and José Meseguer. Unwinding and inference control. In Dorothy D. Denning and Jonathan K. Millen, editors, *Proceedings, 1984 Symposium on Security and Privacy*, pages 75–86. IEEE Computer Society, 1984.
94. Joseph Goguen and José Meseguer. Completeness of many-sorted equational logic. *Houston Journal of Mathematics*, 11(3):307–334, 1985.
95. Joseph Goguen and José Meseguer. Eqlog: Equality, types, and generic modules for logic programming. In Douglas DeGroot and Gary Lindstrom, editors, *Logic Programming: Functions, Relations and Equations*, pages 295–363. Prentice Hall, 1986. An earlier version appears in *Journal of Logic Programming*, Volume 1, Number 2, pages 179–210, September 1984.
96. Joseph Goguen and José Meseguer. Models and equality for logical programming. In Hartmut Ehrig, Giorgio Levi, Robert Kowalski, and Ugo Montanari, editors, *Proceedings, 1987 TAPSOFT*, pages 1–22. Springer, 1987. Lecture Notes in Computer Science, Volume 250.
97. Joseph Goguen and José Meseguer. Unifying functional, object-oriented and relational programming, with logical semantics. In Bruce Shriver and Peter Wegner, editors, *Research Directions in Object-Oriented Programming*, pages 417–477. MIT, 1987. Preliminary version in *SIGPLAN Notices*, Volume 21, Number 10, pages 153–162, October 1986.
98. Joseph Goguen and José Meseguer. Software for the Rewrite Rule Machine. In Hideo Aiso and Kazuhiro Fuchi, editors, *Proceedings, International Conference on Fifth Generation Computer Systems 1988*, pages 628–637. Institute for New Generation Computer Technology (ICOT), 1988.
99. Joseph Goguen and José Meseguer. Order-sorted algebra I: Equational deduction for multiple inheritance, overloading, exceptions and partial operations. *Theoretical Computer Science*, 105(2):217–273, 1992. Drafts exist from as early as 1985.

100. Joseph Goguen, José Meseguer, and David Plaisted. Programming with parameterized abstract objects in OBJ. In Domenico Ferrari, Mario Bolognani, and Joseph Goguen, editors, *Theory and Practice of Software Technology*, pages 163–193. North-Holland, 1983.

101. Joseph Goguen, Akira Mori, and Kai Lin. Algebraic semiotics, ProofWebs and distributed cooperative proving. In Yves Bartot, editor, *Proceedings, User Interfaces for Theorem Provers*, pages 25–34. INRIA, 1997. Sophia Antipolis, 1–2 September 1997.

102. Joseph Goguen, Akira Mori, Kai Lin, and Akiyoshi Sato. Formal tools for distributed cooperative engineering, 1998. In preparation.

103. Joseph Goguen and Efraim Shaket. Fuzzy sets at UCLA. *Kybernetes*, 8:65–66, 1979.

104. Joseph Goguen and Joseph Tardo. An introduction to OBJ: A language for writing and testing software specifications. In Marvin Zelkowitz, editor, *Specification of Reliable Software*, pages 170–189. IEEE, 1979. Reprinted in *Software Specification Techniques*, Nehan Gehani and Andrew McGettrick, editors, Addison Wesley, 1985, pages 391–420.

105. Joseph Goguen and James Thatcher. Initial algebra semantics. In *Proceedings, Fifteenth Symposium on Switching and Automata Theory*, pages 63–77. IEEE, 1974.

106. Joseph Goguen, James Thatcher, and Eric Wagner. An initial algebra approach to the specification, correctness and implementation of abstract data types. In Raymond Yeh, editor, *Current Trends in Programming Methodology, IV*, pages 80–149. Prentice Hall, 1978.

107. Joseph Goguen, James Thatcher, Eric Wagner, and Jesse Wright. A junction between computer science and category theory, I: Basic concepts and examples (part 1). Technical report, IBM Watson Research Center, Yorktown Heights NY, 1973. Report RC 4526.

108. Joseph Goguen, James Thatcher, Eric Wagner, and Jesse Wright. An introduction to categories, algebraic theories and algebras. Technical report, IBM Watson Research Center, Yorktown Heights NY, 1975. Report RC 5369.

109. Joseph Goguen, James Thatcher, Eric Wagner, and Jesse Wright. A junction between computer science and category theory, I: Basic concepts and examples (part 2). Technical report, IBM Watson Research Center, Yorktown Heights NY, 1976. Report RC 5908.

110. Joseph Goguen, James Thatcher, Eric Wagner, and Jesse Wright. Initial algebra semantics and continuous algebras. *Journal of the Association for Computing Machinery*, 24(1):68–95, January 1977.

111. Joseph Goguen and William Tracz. An implementation-oriented semantics for module composition, 1997. Draft manuscript.

112. Joseph Goguen and Francisco Varela. Systems and distinctions; duality and complementarity. *International Journal of General Systems*, 5:31–43, 1979.

113. Joseph Goguen, James Weiner, and Charlotte Linde. Reasoning and natural explanation. *International Journal of Man-Machine Studies*, 19:521–559, 1983.

114. Joseph Goguen, Timothy Winkler, José Meseguer, Kokichi Futatsugi, and Jean-Pierre Jouannaud. Introducing OBJ. In Joseph Goguen and Grant Malcolm, editors, *Algebraic Specification with OBJ: An Introduction with Case Studies*. Academic, to appear. Also Technical Report SRI-CSL-88-9, August 1988, SRI International.

115. Joseph Goguen and David Wolfram. On types and FOOPS. In Robert Meersman, William Kent, and Samit Khosla, editors, *Object Oriented Databases: Analysis, Design and Construction*, pages 1–22. North Holland, 1991. Proceedings, IFIP TC2 Conference, Windermere, UK, 2–6 July 1990.
116. John Guttag. Abstract data types and the development of data structures. *Communications of the Association for Computing Machinery*, 20:297–404, June 1977.
117. John Guttag, Ellis Horowitz, and David Musser. Abstract data types and software validation. *Communications of the Association for Computing Machinery*, 21(12):1048–1064, 1978.
118. Christian Heath, Marina Jirotka, Paul Luff, and Jon Hindmarsh. Unpacking collaboration: the interactional organisation of trading in a city dealing room. In *European Conference on Computer Supported Cooperative Work '93*. IEEE, 1993.
119. Martin Heidegger. The question concerning technology. In *Basic Writings*, pages 283–217. Harper and Row, 1977. Translated by David Krell; original from 1953.
120. C.A.R. Hoare. Unification of theories: A challenge for computing science. In Magne Haveraaen, Olaf Owe, and Ole-Johan Dahl, editors, *Recent Trends in Data Type Specification*, pages 49–57. Springer, 1996. Lecture Notes in Computer Science, Volume 389.
121. Shusaku Iida, Michihiro Matsumoto, Răzvan Diaconescu, Kokichi Futatsugi, and Dorel Lucanu. Concurrent object composition in CafeOBJ. Technical report, Japan Institute of Science and Technology, 1997.
122. Theo Janssen. Compositionality. In Johan van Benthem and Alice ter Meulen, editors, *Handbook of Logic and Language*, pages 417–473. Elsevier/MIT, 1997.
123. Marina Jirotka. Ethnomethodology and requirements engineering. Technical Report PRG-TR-92-27, Centre for Requirements and Foundations, Oxford University Computing Lab, 1991.
124. Marina Jirotka and Joseph Goguen. *Requirements Engineering: Social and Technical Issues*. Academic, 1994.
125. Claude Kirchner, Hélène Kirchner, and Aristide Mégrelis. OBJ for OBJ. In Joseph Goguen and Grant Malcolm, editors, *Algebraic Specification with OBJ: An Introduction with Case Studies*. Academic, to appear.
126. William Labov. The transformation of experience in narrative syntax. In *Language in the Inner City*, pages 354–396. University of Pennsylvania, 1972.
127. George Lakoff and Mark Johnson. *Metaphors we Live by*. Chicago, 1980.
128. Saunders Mac Lane. *Categories for the Working Mathematician*. Springer, 1971.
129. Saunders Mac Lane. *Mathematics: Form and Function*. Springer, 1986.
130. Bruno Latour. *Science in Action*. Open, 1987.
131. Bruno Latour. *Aramis, or the Love of Technology*. Harvard, 1996.
132. F. William Lawvere. An elementary theory of the category of sets. *Proceedings, National Academy of Sciences, U.S.A.*, 52:1506–1511, 1964.
133. Sany Leinwand, Joseph Goguen, and Timothy Winkler. Cell and ensemble architecture of the Rewrite Rule Machine. In Hideo Aiso and Kazuhiro Fuchi, editors, *Proceedings, International Conference on Fifth Generation Computer Systems 1988*, pages 869–878. Institute for New Generation Computer Technology (ICOT), 1988.
134. Charlotte Linde and Joseph Goguen. Structure of planning discourse. *Journal of Social and Biological Structures*, 1:219–251, 1978.
135. David Luckham, Friedrich von Henke, Bernd Krieg-Brückner, and Olaf Owe. ANNA*: A Language for Annotating Ada Programs*. Springer, 1987. Lecture Notes in Computer Science, Volume 260.

136. Paul Luff, Marina Jirotka, Christian Heath, and David Greatbatch. Tasks and social interaction: the relevance of naturalistic analyses of conduct for requirements engineering. In Stephen Fickas and Anthony Finkelstein, editors, *Requirements Engineering '93*, pages 187–190. IEEE, 1993.
137. Luqi and Joseph Goguen. Formal methods: Problems and promises. *IEEE Software*, 14(1):73–85, 1997.
138. Jean-François Lyotard. *The Postmodern Condition: a Report on Knowledge.* Manchester, 1984. Theory and History of Literature, Volume 10.
139. Grant Malcolm. Behavioural equivalence, bisimilarity, and minimal realisation. In Magne Haveraaen, Olaf Owe, and Ole-Johan Dahl, editors, *Recent Trends in Data Type Specifications.* Springer, 1989. Lecture Notes in Computer Science, Volume 389.
140. Grant Malcolm and Joseph Goguen. Proving correctness of refinement and implementation. Technical Report Technical Monograph PRG-114, Programming Research Group, University of Oxford, 1994. Submitted for publication.
141. Grant Malcolm and Joseph Goguen. An executable course on the algebraic semantics of imperative programs. In Michael Hinchey and C. Neville Dean, editors, *Teaching and Learning Formal Methods*, pages 161–179. Academic, 1996.
142. Grant Malcolm and James Worrell. Toposes of abstract machines with observational semantics, 1997. Draft, Oxford University Computing Laboratory.
143. Humberto Maturana. Biology of language: The epistemology of reality. In George Miller and Eric Lenneberg, editors, *Psychology and Biology of Thought and Language: Essays in Honor of Eric Lenneberg*, pages 27–64. Academic, 1978.
144. Humberto Maturana and Francisco Varela. *Autopoiesis and Cognition: The Realization of the Living.* Reidel, 1980.
145. Humberto Maturana and Francisco Varela. *The Tree of Knowledge.* Shambhala, New Science Library, 1987.
146. José Meseguer. General logics. In H.-D. Ebbinghaus et al., editors, *Proceedings, Logic Colloquium 1987*, pages 275–329. North-Holland, 1989.
147. José Meseguer. Conditional rewriting logic: Deduction, models and concurrency. In Stéphane Kaplan and Misuhiro Okada, editors, *Conditional and Typed Rewriting Systems*, pages 64–91. Springer, 1991. Lecture Notes in Computer Science, Volume 516.
148. José Meseguer. A logical theory of concurrent objects and its realization in the Maude language. In Gul Agha, Peter Wegner, and Aki Yonezawa, editors, *Research Directions in Object-Based Concurrency.* MIT, 1993.
149. José Meseguer. Membership algebra as a logical framework for equational specification, 1997. Draft manuscript. Computer Science Lab, SRI International.
150. José Meseguer and Joseph Goguen. Initiality, induction and computability. In Maurice Nivat and John Reynolds, editors, *Algebraic Methods in Semantics*, pages 459–541. Cambridge, 1985.
151. José Meseguer and Joseph Goguen. Order-sorted algebra solves the constructor selector, multiple representation and coercion problems. *Information and Computation*, 103(1):114–158, March 1993. Revision of a paper presented at LICS 1987.
152. José Meseguer, Joseph Goguen, and Gert Smolka. Order-sorted unification. *Journal of Symbolic Computation*, 8:383–413, 1989.
153. Lawrence Moss, José Meseguer, and Joseph Goguen. Final algebras, cosemicomputable algebras, and degrees of unsolvability. *Theoretical Computer Science*, 100:267–302, 1992. Original version from March 1987.

154. Peter G. Neumann. *Computer-Related Risks.* ACM (Addison-Wesley), 1995.
155. David Parnas. Information distribution aspects of design methodology. *Information Processing '72*, 71:339–344, 1972. Proceedings of 1972 IFIP Congress.
156. David Parnas. A technique for software module specification. *Communications of the Association for Computing Machinery*, 15:330–336, 1972.
157. Charles Saunders Peirce. *Collected Papers.* Harvard, 1965. In 6 volumes; see especially Volume 2: Elements of Logic.
158. Tekla Perry. In search of the future of air traffic control. *IEEE Spectrum*, 34(8):18–35, August 1997.
159. Wesley Phoa. Should computer scientists read Derrida? Technical Report 24/5/93, University of New South Wales, 1993. School of Computer Science and Engineering.
160. Francisco Pinheiro and Joseph Goguen. An object-oriented tool for tracing requirements. *IEEE Software*, pages 52–64, March 1996. Special issue of papers from ICRE'96.
161. Horst Reichel. Initially restricting algebraic theories. In Piotr Dembinski, editor, *Mathematical Foundations of Computer Science*, pages 504–514. Springer, 1980. Lecture Notes in Computer Science, Volume 88.
162. Harvey Sacks. An analysis of the course of a joke's telling in conversation. In Richard Baumann and Joel Scherzer, editors, *Explorations in the Ethnography of Speaking*, pages 337–353. Cambridge, 1974.
163. Harvey Sacks. *Lectures on Conversation.* Blackwell, 1992. Edited by Gail Jefferson.
164. Ferdinand de Saussure. *Course in General Linguistics.* Duckworth, 1976. Translated by Roy Harris.
165. Y.V. Srinivas and Richard Jüllig. SpecWare language manual, version 2.0. Technical report, Kestrel, 1996.
166. Victoria Stavridou, Joseph Goguen, Steven Eker, and Serge Aloneftis. FUNNEL: A CHDL with formal semantics. In *Proceedings, Advanced Research Workshop on Correct Hardware Design Methodologies*, pages 117–144. IEEE, 1991. Turin.
167. William Irwin Thompson, editor. *Gaia: a Way of Knowing.* Lindisfarne, 1987.
168. William Tracz. LILEANNA: a parameterized programming language. In *Proceedings, Second International Workshop on Software Reuse*, pages 66–78, March 1993. Lucca, Italy.
169. William Tracz. *Formal Specification of Parameterized Programs in* LILLEANNA. PhD thesis, Stanford University, 1997.
170. Mark Turner. *The Literary Mind.* Oxford, 1997.
171. Francisco Varela and Joseph Goguen. The arithmetic of closure. *Journal of Cybernetics*, 8:125ff, 1978. Also in *Progress in Cybernetics and Systems Research*, Volume 3, edited by R. Trappl, George Klir and L. Ricciardi, Hemisphere Publishing Co., 1978.
172. Simone Veglioni. *Integrating Static and Dynamic Aspects in the Specification of Open, Object-based and Distributed Systems.* PhD thesis, Oxford University Computing Laboratory, 1998.
173. Jesse Wright, James Thatcher, Eric Wagner, and Joseph Goguen. Rational algebraic theories and fixed-point solutions. In Michael J. Fischer, editor, *Proceedings, Seventeenth Symposium on Foundations of Computing*, pages 147–158. IEEE, 1976.
174. Lotfi Zadeh. Fuzzy sets. *Information and Control*, 8:338–353, 1965.

Helmut Jürgensen

Born in Rieseby, a village some 50 km north of Kiel, Germany, in 1942, Helmut Jürgensen attended school in Rieseby and in the town of Eckernförde. He studied ancient languages, ancient literature and mathematics at the universities in Kiel and Tübingen, Germany. After a doctorate with a thesis on classical Greek rhetorics in 1968, he joined the Department of Mathematics at Kiel University. In 1976 he received the degree of Dr. rer. nat. habil. at Kiel university with a thesis on semigroup extensions. From 1968 until 1976 he taught Computer Science and Mathematics at Kiel University. From 1977 until 1983, he was a Professor of Theoretical Computer Science at the Darmstadt University of Technology. In 1983 he joined the Department of Computer Science of the University of Western Ontario in London, Canada, where he also holds an honorary professorship with the Department of Mathematics. Since 1997 he is also with the Department of Computer Science of Potsdam University, Germany.

Jürgensen's research interests include: automata and formal languages; combinatorics of words; descriptional complexity; codes; communication; semigroups; circuit testing; multi-purpose document specification; communication for the handicapped.

Towards Computer Science

Σοφώτατόν τοι κἀμαθέστατον χρόνος.
Euenos

In my contribution to this volume I shall try to share stories about the early days of computer science in West Germany. I shall follow the traces of my own footsteps into computer science; not that I find them particularly important, but they provide me with a means for structuring my thoughts. Moreover, the path that led me into computer science was sufficiently unorthodox to provide the occasional humorous turn.

I shall recount events as I remember them. I have tried to verify all dates and facts; I have consulted my letter files dating back to about 1965 and also asked some friends and colleagues from the early days for confirmation. Despite this, I cannot exclude that I shall err in some details. It was occasionally very difficult to dig up the facts. This experience, for me, re-inforces the need for a volume like this one. Much of the folklore about the early days of computer science will soon be forgotten; museums of computer history cannot and will not capture the atmosphere of a night spent on a cot beside a computer nor the excitement of finding an undocumented computer instruction nor the frustration of hunting for a program error in the system software that was leading to irreproducible deadlocks. Indeed, it might be already too late to begin building an archive of early computer programs, folklore and memories.

While the term *Computer Science* appears in the title of this contribution, I shall rather use the term *Informatics* in the sequel. No physicist would make the hammer the main topic of his science and, consequently, call it *Hammer Science*. Or, what about calling dentistry *Drill Science?* The choice of the term *Computer Science* is, in my view, very unfortunate as it tends to convey to the outsider and the student that computer science is about using computers. I have to admit, however, that using the term *Informatics* does not *entail* a better understanding of the field: My announcement of a public lecture on randomness and descriptional complexity at Potsdam University a while ago was met with quite some surprise by some colleagues in the Faculty of Science: that this would be topic of informatics? I shall address this issue again later in this paper.

While I emphasize this matter of names, I admit that, to a very large extent, this contribution talks about computers. The direction of early research in informatics was, of course, highly influenced by our *playing* with those mysterious new machines, the potential of which was completely unexplored.

First Encounters

When I began my studies at Christian-Albrechts-Universität in Kiel, Germany, in the summer of 1961, while I might have read about computers in novels of science fiction, I certainly did not really believe in their existence, let alone dream of ever touching one: I wrote my first programs in the summer of 1962. The computer was a *Zuse Z22,* a room-filling machine consisting of two large steel cabinets and a desk with a special keyboard, control lights, a teletype machine serving as a printer and also as a five-bit tape puncher for output, and a five-bit paper tape reader.[1] The machine had a magnetic core memory of 25 words and an 8K-word drum, each word with 38 bits. About 500 vacuum tubes and about 2400 diodes made it work – when it was not too warm. Many times did I sit

---

[1] A photograph of the Z22 is shown in [1], p. 152. The book [1] is an absolutely outstanding source concerning the early days of computing. I am grateful to its author, W. de Beauclair, and to F. Genser, who has joined him in the preparation of a successor volume, for the permission to reproduce the picture of the Z22 and, below, the picture of the X1. In the book [1], one also finds close-up photographs of the electronic components and the periphery of early computers.

beside the technician watching the oscilloscope and trying to find which broken part had ruined the computation of a whole night. When your only output media are teletype print on paper and punched paper tape, punched and read at about 15 to 25 characters per second, you don't want to write a back-up very often, if ever.

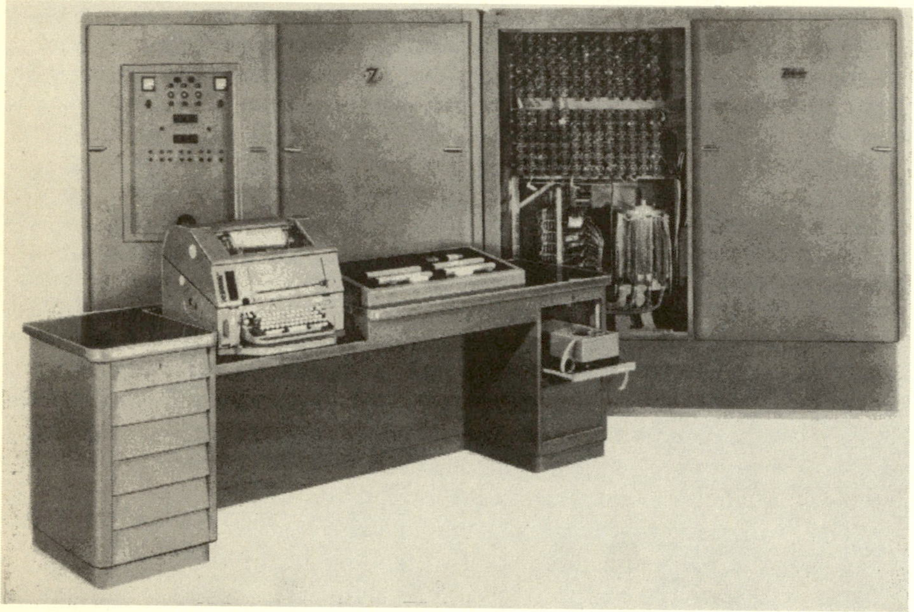

The Z22: On the desk in the foreground is a teletype machine used for output on paper or punched paper tape; beside it is the console and, a bit lower on a drawer, the mechanical paper tape reader. Later, we also had an additional faster optical tape reader. In the open cabinet one sees the drum in the lower part; the upper half is filled with plug-in circuits, each typically consisting of a vacuum tube and some resistors and capacitors.

Most users of the Z22 would write programs in machine code – this would be called assembly code today: the instruction set was fully *analytical,* that is, each bit in the operation code part of an instruction had its own meaning and could be combined with every other bit. This allowed us to write very compact, sophisticated and hard-to-read programs.[2] With the *assembler* and the run-time support for input, output, and arithmetic taking up $1\frac{1}{2}$K words on the drum, space was the major concern. Many of the programming techniques which we invented – and probably re-invented – would be called hacking today. Of course,

[2] I was very much reminded of this when, some 15 years ago, I wrote programs for the TI-59, a pocket computer with programs limited to at most 1000 instructions and a magnetic card reader for program or data input.

also speed was important – an integer addition took 0.6 ms, an integer multiplication took about 10 ms, and an integer division could take even 60 ms; for a very short cycle with a known number of rounds we would load the instructions of the cycle into the core memory and then exploit the structure of the address register to iterate these instructions and get back to the program on the drum correctly; when the number of iterations was unknown at the time of programming, we would actually compute some of the instructions at run-time, thus creating small self-modifying and data-dependent programs. Another important way to save time was to organize the program and the data on the drum in such a way that the drum rotation was taken into account: for example, storing an array of integers in consecutive words on the drum could be very bad, because the program might have to wait for the time of an entire drum rotation, about 10 ms, to access the next array element whereas clever spacing of the array elements could reduce the access time to significantly less than 1 ms. The idea that such kind of optimization could ever be carried out automatically would have been considered utopian.

There was also ALGOL 60 on that machine taking up 4K of drum space for the compiler and run-time system. Nevertheless, there were some scientists, who dared to use ALGOL. I particularly recall a researcher in nuclear physics, an astronomer, and also a meteorologist spending many nights *mit der*[3] *Zuse*. I learned ALGOL by helping them to debug their programs. The degree to which programming tools have improved since then can probably be best appreciated when one reads a typical error message of those days:

```
Syntax error after line 176
```

Interestingly, while compiler tools today (1997) give incomparably better error diagnoses, the *theory* of such error diagnosis for parsers and compilers seems not to have developed much since then, as shown in [7].

To use the Z22 one reserved time, usually for several hours and at night for large projects. The computer room was furnished with a small bed. Throughout the long nights, the computer's loudspeaker would be on. With some experience one could hear from the pitch patterns what the computer was doing and whether it had just gone *astray*. A change in the pitch pattern would usually wake one up – too late of course if there had really been a mistake, but sometimes soon enough to salvage at least part of the work.

Among the many small peculiarities I remember about using the Z22 – I think I could still program it today – was the debugging problem. In about 15 minutes you had been able to enter your program by paper tape. Now, single-stepping through the program – without a monitor to show what was happening – you suspected a mistake: to make data visible, you would key in instructions, bit-by-bit, to save the accumulator, to fetch the relevant data item and to display its bits. To patch a program you would key in instructions bit-by-bit – modifying and re-entering the whole program would necessitate changing the paper tape

---

[3] Note the affectionate feminine in German.

(sometimes cutting and pasting[4] would work, though) and reading that paper tape again. Sure, we were hacking; however, at the same time constraints of this kind made us program very carefully; correcting a mistake would require a major effort.

At that time I was enrolled in Greek and Latin and – as a compromise with my parents, who believed I would never make a living with these topics and wanted me to study something more *useful* – mathematics. As there was no clearly defined curriculum, after fulfilling the very basic requirements I could go on as many tangents as I wanted during the next few years: philosophy (Apel), Sanskrit and Hindi (Lienhard, Hofmann), Hittite, linguistics (Hofmann), music (Gudewill), logic (Schütte). In early 1963 I completed a small thesis on the expression of fear in the tragedies of Aischylos [9] that opened the way into the doctoral student seminar for Greek, Latin and ancient history; for that thesis, I used the Z22 to extract and compute some language statistics, a revolutionary computer application to me; only years later would I learn about the computer linguistics research carried out at the same time in many places throughout the world.

In late 1963 I proposed a topic for my doctoral thesis to Professor H. Diller, a leading German scholar of Greek specializing in ancient medicine.[5] At about the same time I joined the research group of J. Neubüser, then at Kiel University, later at Aachen University in Germany; Neubüser was a pioneer of computer algebra: he investigated the structure of finite groups by computer. Among the many people in that group I remember, in particular, V. Felsch, with whom I spent many days and nights at various computers debugging our programs and the system software on which it was meant to run – but this was later.

My first project in Neubüser's group was to write a program for the multiplication of group elements represented as words, the group being given by an abstract presentation. My first attempt was fundamentally wrong – at that time everything surrounding the *word problem* was still arcane and certainly not taught; the proof of the undecidability of the word problem for groups was only about 10 years old and its consequences were not understood by even the leaders in group theory.

In early 1964 I took both, my thesis topic and the computer algebra problem, with me to Eberhard-Karls-Universität Tübingen. The short time spent there, one term, had several implications on the direction I would take, but also allowed me to re-evaluate my options. I took courses in numismatics from the enthusiastic archeologist H. Hommel, attended a seminar by the famous E. Zinn, sat at the feet of W. Jens (on medieval rhetorics) and E. Bloch (on philosophy), enjoyed the lectures and seminar of *Arrius Nurus, poeta laureatus,* (H. Schnur) conducted in Latin on modern Latin poetry, and learnt about the influence of classical literature on modern American novels in lectures by a visiting professor

[4] Using *real* scissors and *real* glue!

[5] 25 years later, on another continent, in London, Ontario, I met a professor of ancient medicine and his wife, both former doctoral students of Professor Diller.

from the distant America. I took a single course in mathematics, on coding and information theory, from W. Vogel, and this experience is one of the origins of my ongoing interest in these areas.

Returning to Kiel in the fall of 1964, I brought with me a complete program for the multiplication of words in solvable groups and a program tracer, a program that would allow me to trace any program and print out its step-by-step actions. This was written for the legendary ELECTROLOGICA X1 (Kiel University had just acquired an X1 as an addition to the Z22, probably in 1962). Both programs had never been tested before I returned, but worked right away – debugging without a computer and proving a program correct are certainly safer than test runs; but few people understood this issue.

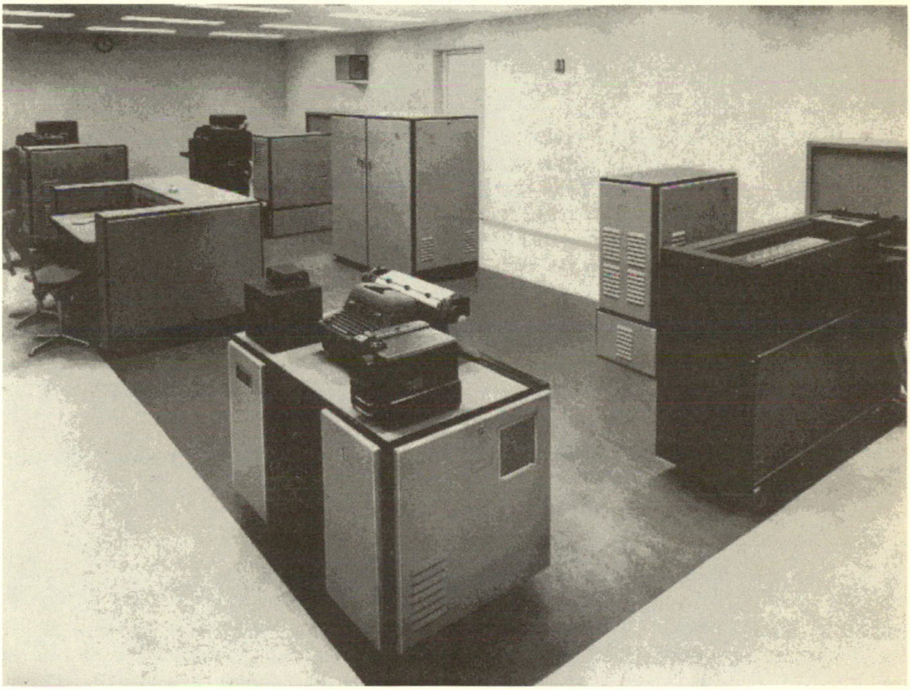

The X1: On the front cabinet there is a paper tape puncher and a type writer for output and a paper tape reader for input. Behind it to the left is the console. The dark cabinet at the front right seems to be a line printer. Our installation also included magnetic tapes and a card reader and puncher.

The program for word multiplication was of the self-modifying kind: from the given group presentation it would generate a multiplication program; using it, further relations would be computed and, on their basis, a new multiplication program would be computed. That one was then used by the actual group theory programs [11]. The main algorithm has still been used in the *Cayley* and *SOGOS* systems until the early eighties [6], [21].

The X1 was revolutionary in many ways.[6] One of its most striking points was

[6] The photograph of the X1 is reproduced from [1], p. 187; see also footnote 1.

that it had true interrupts with hardware support for semaphores. The core of its run-time system was wired in the upper parts of its magnetic core memory, including the sophisticated interrupt programs for paper tape input and output due to E. Dijkstra.[7] Many years later, when it was decided to use the X1 as a sattelite to our new X8, we had to design the hardware for a new interrupt on the X1 and to find the space for *one* additional instruction in Dijkstra's program. Proudly, V. Felsch and I detected one redundant instruction after many days of analysis, and we could re-wire the upper memory of the machine.

ELECTROLOGICA announced a whole series of new machines, starting with the X8, the architecture of which was based on some modern insights in computer programming and algorithms at that time. The architecture of these machines deviated significantly from that of IBM type computers which had already started to dominate the market. For example, the X8 incorporated an ingenious floating point arithmetic and some then highly innovative – by modern standards rather rudimentary – instructions for environment and task switching. In retrospect I still find it very sad that this innovative approach to computer architecture was unsuccessful economically and was finally aborted when ELECTROLOGICA was sold.

The X8 changed much of our programming behaviour. In the group theory project we continued to use assembly code, because none of the "high"-level languages available – ALGOL or FORTRAN[8] – would support the data structures we needed without an enormous overhead in time and space. Our machine was large for that time, 32K of 27-bit words main memory and 512K words of drum memory, but not large enough for our purposes to accept that overhead. However, we started to experiment with new implementation and debugging methods. My X1-tracer was re-written as an X8-tracer that could also handle interrupts. The group theory programs were gradually transferred to the X8. However, I continued to work on the Z22 as well for another related project.

Part of the group theory project was that Hasse diagrams of the lattices of subgroups would be drawn – by hand. Some of these diagrams were huge, involving several hundreds of subgroups so that the drawing of a single diagram could take weeks; moreover, a small mistake in lay-out decisions could require us to re-do the work of the past week; or a single incorrect line could force us to start all over. At this time the university had just acquired a plotter, the *Zuse Graphomat Z64*, a plotter with an accuracy of up to $\frac{1}{32}$mm. Not knowing about the *travelling salesman problem,* I started writing a program for drawing Hasse diagrams of partially ordered sets – on the Z22 – that actually ended up being used for all the drawings in the project [10], [27]; the program employed what would now be called a greedy algorithm. This program was later improved in [13]. Since then, significantly more sophisticated plotting programs for Hasse

[7] For more information about the X1 see Dijkstra's contribution to this volume.

[8] The FORTRAN compiler was written by H. Felsch as part of his doctoral thesis; compiler writing and code optimization were still in the early stages. Tools like LEX and YACC did not exist, and most of a compiler was written in assembly code.

diagrams have been developed, especially [34], which exploit the structure of the partially ordered set under consideration.

While all this was going on, while I learnt about Frege, Wittgenstein and Russell, read the *Song of Ullikummi,* learnt about Beethoven's string quartets, read Premchand and Kalidasa, while I played second violin in a lay string quartet, while I wrote programs to plot geodesics on a four-dimensional ellipsoid – as a research assistant to a mathematician, Reinhild Rumberger then, Reinhild Jürgensen since 1968 –, while I catalogued books in the library of the Department of Indology, the thesis matured and was submitted in early 1968 [12]. After the *rigorosum,* an examination in Greek, Latin and Mathematics, I received two job offers: join a team working on the thesaurus of classical Greek to computerize that work; join the Department of Mathematics with the aim of helping to build a, still to be founded, Department of Informatics.

Growing up

In 1968, informatics had just started to grow into a discipline of its own, finally separating itself from its parents, mathematics or electrical engineering. In West Germany, several far-sighted scientists – including F. L. Bauer, W. Händler, G. Hotz, K. H. Weise, to name only the few I remember – developed a plan to found departments of informatics at selected universities – under Weise's direction, Kiel University was one of these – to serve as foci for the emerging activities in the new field; they were also instrumental in founding the GI *(Gesellschaft für Informatik).*

The field of informatics was still so compact that a researcher could easily be aware of and contributing to, so seemingly distant areas as compiler building, coding theory and semigroups of automata. Moreover, the number of scientists interested in the field was small enough that meaningful discussions across the boundaries of the subareas could be held at conferences still representing informatics in its entirety. Of course, the IFIP congresses were already huge (I attended the one in Ljubljana in 1971 just after having been at an informatics conference in Haifa[9]); yet, there were also many smaller meetings among which the Oberwolfach meetings between 1969 and 1975 and the early ICALP conferences and their predecessors were particularly fruitful. The 1969 Oberwolfach meeting on *Formal Languages and Automata,* could be said to establish, finally, informatics as a field of its own in West Germany; it also forged contacts among researchers in Europe, most of whom were literally pioneers; the list of people planning to attend the meeting[10] includes a few people who were famous then, but many more names which were completely unknown then, but are recognized today:

---

[9] In Ljubljana I heard D. E. Knuth on the complexity of concrete algorithms; in Haifa I met J. A. Brzozowski, Sh. Greibach, A. Paz, D. Simovici and M. Yoeli for the first time.

[10] I could not find the final list of attendants in my notes; but for Oberwolfach meetings, the pre-conference list is usually accurate to more than 90%.

R. Albrecht, U. Augustin, S. Backes, J. Bečvář, J. Berstel, G. Bertram, K. H. Böhling, J. F. Böhme, W. Brauer, J. R. Büchi, V. Claus, K. Čulík I, H. Curry, P. Deussen, K. Döpp, J. Dörr, J. Duske, B. Eggers, H.-D. Ehrich, J. Eickel, H. Feldmann, M. Fliess, F. Gross, W. Händler, I. Havel, G. Hotz, H. Jürgensen, P. Kandzia, L. Kalmár, W. Knödel, J. Král, V. Kudielka, H. Langmack, C. Lenormand, J. J. C. Loeckx, W. Menzel, J. Merkwitz, G. C. Moisil, G. H. Müller, M. Nivat, W. Oberschelp, R. Ochranová, W. Oettli, J.-F. Perrot, M. Rosendahl, B. Schlender, A. Schmitt, H. J. Schneider, C.-P. Schnorr, A. Schönhage, D. Siefkes, M. Sintzoff, J. Specht, P. Strnad, W. Stucky, H. Thiele, L. A. M. Verbeek, R. Vollmar, A. P. J. van der Walt, H. Walter, D. Wotschke, K. Zeller, K. Zuse.

The proceedings [5] of this meeting reveal the diversity and energy of this emerging field of research, calling it explicitly *Informatics.* The lists of participants of the next seven Oberwolfach meetings until 1975, when the series was replaced by more specialized conferences, include, for instance and beyond the names listed above:

K. Alber, H. Alt, D. A. Alton, G. Ausiello, D. B. Benson, W. Bibel, A. Blikle, M. Blum, L. Boasson, E. Börger, J. Brzozowski, B. Buchberger, J. Černý, R. Cori, W. Coy, A. B. Cremers, O. J. Dahl, R. P. Daley, W. Dörfler, K. Ecker, H. Ehrig, S. Eilenberg, J. Engelfriet, P. van Emde Boas, M. J. Fischer, P. Flajolet, F. W. von Henke, K. Indermark, L. S. Jutting, L. Kalmár, T. Kameda, M. Karpinski, R. Kemp, W. Knödel, W. Kuich, I. Kupka, K. Lagally, J. Małuszynski, H. Maurer, O. Mayer, K. Mehlhorn, A. R. Meyer, E. Mincozzi, B. Monien, R. P. Nederpelt, M. Nielsen, A. Nijholt, H. Noltemeier, M. S. Patterson, T. Postelnicu, L. Priese, Th. Ottmann, W. G. Paul, B. Reusch, J. Riguet, G. Rozenberg, A. Salomaa, J. E. Savage, B. Schinzel, E. M. Schmidt, W. Schwabhäuser, H. Schwichtenberg, O. Spaniol, S. Skyum, W. Stucky, G. Thierrin, K. Weihrauch, M. Yoeli.

One notices the gradual widening of the horizon – across the Atlantic – and, when one goes through the lists by year, – sadly – also the decrease in or lack of contact with the East.

For audacious pioneers only the sky is the limit. In 1968, I started a research project on grammatical inference for natural languages, that is, designing an algorithm for the construction of a grammar approximating a given natural language L, when examples of correct sentences are provided; as the language paradigm to use I chose the one I was deeply involved with at the time, the Hittite language! Of course, the project produced no direct results; it was far too ambitious then and still would be so today. However, it had a few important and lasting side-effects:

- I needed an implementation language and decided in favour of LISP of which I knew absolutely nothing. The rather entertaining story of LISP at Kiel University is mentioned on p. 252 of [32]; here is a more complete account:

  > With no idea how LISP would work and how one would program it, I decided to start an implementation. Fortunately, there was a published LISP system available (in print) for the PDP-1 [4]. I contacted DEC and received,

from them, a manual for the PDP-1 and a punched paper tape containing the assembler for the PDP-1. I then prepared a punched paper tape of the published LISP system and created a simulator of the PDP-1 on the X8 – this was rather tricky due to differences in the handling of interrupts. To get a running – albeit very slow – LISP on the X8 *only* the following steps were needed: load the simulator; read the PDP-1 assembler; simulate the PDP-1 to run the assembler to read the LISP system; simulate the PDP-1 to run the LISP system. Once DEC had sent me their information, this was completed in about two weeks.

This implementation allowed me and a few students to experiment at least and to learn hands-on how LISP worked. A few months later, I received a genuine LISP implementation for the X8 from W. L. van der Poel, which we then augmented – adding secondary storage management with *concurrent* garbage collection [31], adding a compiler [20] and various other items. Since then, LISP has been my preferred prototyping language, not because it is particularly well-suited, but because it is so very simple to use.

- As another side-effect of the inference project I began to study syntactic congruences of languages. The need to find an algorithmic – or at least a mathematical – expression of the idea of *adequate grammar* for a given language – the goal of the language inference project explicitly included the adequacy of the grammar, a notion quite fashionable at the time when the Latin-based grammar model everybody knew from school had just been proclaimed inadequate for many languages – lead me to the work of S. Marcus and M. Novotný on modelling grammatical categories.[11] Their work was largely ignored in the West and still has not found its due recognition.

The work of two scientists, in particular, whom I first met at Oberwolfach, had a thorough influence on my own research directions. There was L. A. M. Verbeek, who had just discovered a new type of semigroup extensions – *union extensions* – and derived the rudiments of a theory for them in his dissertation [33]; and there was J.-F. Perrot, who arrived at amazingly elegant algebraic descriptions of combinatorial properties of prefix codes via their syntactic monoids [30].

The grammar inference problem was soon laid aside by me because it was too hard then – and judging by the state of learning theory today, it is still too hard to attack in any meaningful fashion, as we simply do not understand learning mechanisms for languages well enough – and, aside from extending the

---

[11] I still have a large collection of papers on mathematical linguistics and contextual grammars written by S. Marcus or his students, mostly in Romanian; when looking around my office I see his *Poetica matematică* [24], *Introduction mathématique à la linguistique structurale* [25], and the collection *Metode distribuţionale algebrice în lingvistică* [26]. I also treasure a set of course notes by Novotný on syntactic categories of 1967, [28], and a book [29] on how to find a grammar from a language and how to recover the language from the grammar.

computer algebra work from groups to semigroups (see [15] and the survey [14]), I turned to the extension theory of semigroups and groupoids, taking Verbeek's ideas to the limits (see [16] and [17] and the references provided there), and to the study of syntactic congruences of languages – the syntactic class of a word being the idealized representative of its syntactic category. The latter topic and that of codes and information – omitting a few excursions – seem to have held my interest since then [19].

One of the main events – for me – in the early seventies was a visit to J. Szép in Budapest in 1974, at the Karl Marx University of Economics, as it was called then. Szép, a well-known group-theorist, had started to investigate the structure of *finite* semigroups, using some ideas also found in E. S. Ljapin's then already classical book [23],[12] and had also initiated a publication series of his institute with some of his work in the early issues. He invited me to join a project for a monograph on finite semigroups. Together with F. Migliorini of Siena, Italy, we finally completed this project – omitting the word *finite* – with the publication of a small book *Semigroups* in 1991, comprising ideas developed over those more than 15 years [18]. This first visit to Hungary – by plane to West Berlin, by bus to East Berlin, by plane to Budapest – was also my first encounter with Eastern Europe. It was followed by many more, fondly remembered visits to Hungary establishing friendships, in particular with J. Szép, I. Peák and F. Gécseg.

In some sense, the iron curtain had holes for informatics at that time – at least from our point of view. While extremely severe travel and publication restrictions were in place in many of the Eastern countries or their universities – I recall one East German colleague telling me in confidence many years ago that the letter he had received from me months earlier had been stamped *not to be replied to* when he found it on his desk, opened of course; and I know of many incidents of a similar or more severe kind – we could still meet in the East at conferences and exchange ideas. Romania became an exception as contacts rapidly collapsed in the early eighties. For theory researchers, the many workshops organized by V. Topentscharov and his group, specifically also by K. Peeva, in Bulgaria offered valuable opportunities for encounter and discussion.

In 1974, I visited A. Salomaa in Århus, Denmark, and then again in 1976 in Turku, Finland.[13] This started a lasting friendship, re-inforced probably by the fact that I moved to one of his favourite places, London in Canada, where he had spent several years in the late sixties.

Informatics in those growing-up years was a field of fast-moving focus. Informatics then and even today reminds me of the game of *Clue* – or the corresponding movie –, in which a Mr. Body has been murdered and the remaining group

---

[12] Ljapin's book contains a huge amount of knowledge about semigroups, much of which has become unknown as semigroup theory took a different overall direction. I remember, vividly, the days Ljapin spent at our home in 1978 in the context of an Oberwolfach conference on semigroups. He entertained our daughters, at the age of grand- or great-grand-children to him, by drawing cartoons of crocodiles eating chairs.

[13] Due to this visit in Turku, our older daughter Astrid, who had then just turned 5, still knows how to count to 100 in Finnish.

of people investigates by running around in a panic from one room to another and back again – occasionally leaving someone behind. Once the easy problems of a field have been solved, most people move on to the next fashionable field, possibly even proclaiming the old field dead. For instance, the areas of *formal languages* and of *automata* have been proclaimed dead by several former leaders in these fields; despite this, they are very much alive, as many of their core questions are still unsolved. In this context it is interesting to note the rather recent revival and applications of the theory of descriptional complexity – in the sense of Kolmogorov or Chaitin – once considered buried by most people, unjustified as this may have been.

After 7 years at Technische Hochschule Darmstadt, from 1977 to 1983, we moved to London in Canada. Our connection to Canada was initiated by G. Thierrin in London, who invited me in 1976[14] and many more times after that, by D. Wood, who invited me to McMaster in Hamilton, and by J. A. Brzozowski and K. Culik, with whom I spent a sabbatical at Waterloo in 1982; Salomaa was at Waterloo at the same time and Wood had just moved there. Our move to Canada was rather accidental: At a party in Thierrin's house – the kitchen, to be precise, where Canadian parties tend to end up – someone commented on their having a hard time finding new faculty, to which I jokingly replied: Why don't you take me? The next day I received a phone call from the chairman, who had not been present at the party, and an offer less than a year later.

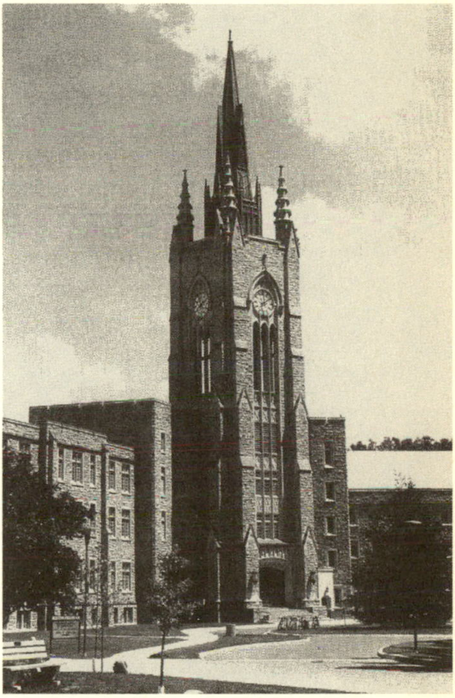

The University of Ontario, Middlesex College (1959), housing the Department of Mathematics, the School of Journalism, the Department of Computer Science and the Graduate Students' Club.

In London since 1983, I have seen the department grow – not just in numbers, but also in vision and achievement. A. Lindenmayer became a frequent visitor and a friend. In one of our papers, originally intended for the Scientific American, we tried to explain the main ideas about L systems to a general readership [22]; we often fought over single words or commas, but loved

[14] This was my first visit to North America; I also spent a few days in Waterloo meeting J. A. Brzozowski; in Bowling Green, Ohio, meeting Satyanarayana; at Purdue University meeting M. Drazin and J. R. Büchi; and at Pennsylvania State University meeting G. Lallement.

every moment of it; how much do I regret that Aristid could not see the finished work!

Our move to London also brought about a very close co-operation with J. A. Brzozowski (Waterloo); Waterloo and London are only about 100 km apart, and Brzozowski and I have been visiting each other on a regular basis over the past 12 years, trying to solve some of the most intriguing mathematical problems in circuit testing: modelling faults realistically, computing tests and determining test complexity [2], [3].

Maturing?

After initial enthusiasm about and initial support for, informatics in West Germany, its development as a discipline had virtually halted at many West German universities in the early eighties. There were not enough faculty positions compared to the number of students – a class size exceeding 300 was quite common; there was no money to replace outdated computer technology and little understanding of the fact that our equipment was outdated. In Darmstadt, students and faculty had to share computer terminals and to line up for them. Still in 1981, the university administration recounted to me – then Dean of Informatics – the story of *that professor* hired around 1975 who, in his negotiations, *had asked for a terminal in his bedroom,* – that is, who had simply wanted to have a modem connection from his home – as something totally absurd and unheard-of.[15] Electronic mail was used for everyday-business in most computer science departments in North America in the early eighties; access to electronic mail was also one of *our ridiculous* wishes in Darmstadt at that time. Informatics, where it was established, was fighting an up-hill battle against the traditional disciplines. Moreover, informatics itself was not sure of where it wanted to go with sometimes rather fierce battles between the various scientific camps. The momentum had been lost.

Potsdam, Neues Palais (about 1760). Part of the university resides in the *Communs* building and the stables of *Neues Palais* in *Park Sanssouci*. The Informatics Department is located just across the street in two rather modest buildings.

15 *That professor* was D. Parnas, who had already left Darmstadt before 1977 and whom I only met much later in London, Canada. To dampen enthusiasm, modem connections from home to the university are still a rather exotic item to expect in the German setting of 1997.

Some years ago, J. Gruska and I made an attempt to formulate our understanding and vision of informatics [8]. As I now look back on more than 30 years, I find the field being far less coherent than when we started. The characteristic ambivalence of informatics as a field bridging engineering, science, linguistics, psychology – and probably a few others – that made such a spicy contrast to established fields seems to have become devisive, as resources get scarcer. Informatics as a whole has not defined its discipline – or rather, many of those educated or educating in informatics have no vision of or do not agree on a vision for the field.

In view of this, it is not surprising that informatics is still looked down upon with some suspicion by the established fields. In the first few months of my present experiment of returning to Germany after 15 years in London, I learnt very soon that informatics still will have to do a lot of convincing. There was even a serious proposal to close down the informatics department. Still, the basic misunderstanding abounds that informatics is about *using computers* and this undermines the credibility of our field.

To convince others, *we* ought to agree on what informatics is about; this seems to be more difficult now than it was some thirty years ago. In this context the rôle of theoretical informatics is by no means clear. The gap between the experimental research in informatics and theory is huge – on the one hand, because theory is still trying to solve old hard and fundamental problems while technology has changed; on the other hand, because the experimental side very often lacks the mathematical and modelling skills to deal with the relevant issues in a scientific fashion. In fact, we seem to be moving further and further away from a consensus about the essence of the field and the rôles of its components.

I conclude this paper of memories with thanks to and special mention of two scientists who were of particular influence on my way into informatics: H. Diller, my *Doktorvater;* and K.-H. Weise, who provided me with the opportunity to enter the field.

References

1. W. de Beauclair: *Rechnen mit Maschinen.* Friedr. Vieweg & Sohn, Braunschweig, 1968.
2. J. A. Brzozowski, H. Jürgensen: An Algebra of Multiple Faults in RAMs. *Journal of Electronic Testing, Theory and Applications,* **8** (1996), 129–142.
3. R. David, J. A. Brzozowski, H. Jürgensen: Testing for Bounded Faults in RAMS. *Journal of Electronic Testing, Theory and Applications,* **10** (1997), 197–214.
4. L. P. Deutsch, E. C. Berkeley: The LISP Implementation for the PDP-1 Computer. In: E. C. Berkeley, D. G. Bobrow (eds.): *The Programming Language LISP; Its Operation and Applications.* MIT Press, Cambridge, 1974, 199–252.
5. J. Dörr, G. Hotz (eds.): *Tagungsbericht: Automatentheorie und formale Sprachen. Berichte aus dem Mathematischen Forschungsinstitut Oberwolfach* **3**, Bibliographisches Institut, Mannheim, 1970.

6. V. Felsch: A Machine Independent Implementation of a Collection Algorithm for the Multiplication of Group Elements. In: R. D. Jenks (ed.): *SYMSAC'76, Proceedings of the 1976 ACM Symposium on Symbolic and Algebraic Computation.* ACM, New York, 1976, 159–166.
7. G. George: *Natural-Language Processing: Context-Dependent Error Correction in English Text.* M. Sc. Thesis, The University of Western Ontario, London, Ontario, Canada, 1997.
8. J. Gruska, H. Jürgensen: Maturing of Informatics. *Images of Programming, Dedicated to the Memory of A. P. Ershov.* edited by D. Bjørner, V. Kotov, North-Holland, Amsterdam, 1991, I-55–I-69.[16]
9. H. Jürgensen: *Die Darstellung der Furcht in den Chorliedern des Aischylos.* Kiel, 1963, 53 pp.
10. H. Jürgensen, K. Ferber: A Programme for the Drawing of Lattices. *Computational Problems in Abstract Algebra, Proceedings of a Conference, Oxford, 1967,* edited by J. Leech. Pergamon Press, Oxford, 1969, 83–87.
11. H. Jürgensen: Calculation with the Elements of a Finite Group Given by Generators and Defining Relations. *Computational Problems in Abstract Algebra, Proceedings of a Conference, Oxford, 1967,* edited by J. Leech. Pergamon Press, Oxford, 1969, 47–57. (Russian translation in: *Computations in Algebra and Number Theory,* edited by B. Venkov, D. Fadeev. Moskva, 1976, 32–46.)
12. H. Jürgensen: *Der antike Metaphernbegriff.* Dissertation, Kiel, 1968.
13. H. Jürgensen, J. Loewer: Drawing Hasse Diagrams of Partially Ordered Sets. In: G. Kalmbach, *Orthomodular Lattices.* Academic Press, London, 1983, 331–345.
14. H. Jürgensen: Computers in Semigroups. *Semigroup Forum* **15** (1977/78), 1–20.
15. H. Jürgensen: Transformationendarstellungen endlicher abstrakt präsentierter Halbgruppen. *Computing* **21** (1979), 333–342.
16. H. Jürgensen: Zur Charakterisierung von Vereinigungserweiterungen von Halbgruppen durch partielle Morphismen. *Acta Scientiarum Mathematicarum (Szeged)* **38** (1976), 73–78.
17. H. Jürgensen: Fastideale in vollständig 0-einfachen Halbgruppen *Mathematica Slovaca* **26** (1976), 73–76.
18. H. Jürgensen, F. Migliorini, J. Szép: *Semigroups.* Akadémiai Kiadó, Budapest, 1991, 121 pp.
19. H. Jürgensen, S. Konstantinidis: Codes. *Handbook of Formal Language Theory I,* edited by G. Rozenberg and A. Salomaa, Springer-Verlag, Berlin, 1997, 511–607.
20. B. Kalhoff: *Ein Kompiler für das System LISP1.5 X8 Kiel.* Diplomarbeit, Universität Kiel, 1972.
21. R. Laue, J. Neubüser, U. Schönwälder: Algorithms for Finite Soluble Groups and the SOGOS System. In: M. D. Atkinson (ed).: *Computational group the-*

[16] A much more extensive and ambitious manuscript of about 80 pages on this topic is still waiting in our files for completion.

ory, Proceedings of the London Mathematical Society Symposium on Computational Group Theory. Academic Press, London, 1984, 105–135.
22. A. Lindenmayer, H. Jürgensen: Grammars of Development: Discrete State Models for Growth, Differentiation, and Gene Expression in Modular Organisms. *Lindenmayer Systems, Impacts on Theoretical Computer Science, Computer Graphics, and Developmental Biology,* edited by G. Rozenberg and A. Salomaa. Springer-Verlag, Berlin, 1992, 3–21.
23. E. C. Ляпин: Полугруппы. Moskow, 1960. English translation: E. S. Ljapin: *Semigroups.* Revised edition, American Mathematical Society, Providence, 1968.
24. S. Marcus: *Poetica matematică.* Editura Academiei Republicii Socialiste România, Bucharest, 1970.
25. S. Marcus: *Introduction mathématique à la linguistique structurale.* Monographies de linguistique mathématique **1**, Dunod, Paris, 1967.
26. S. Marcus (ed.): *Metode distribuţionale algebrice în linguistică.* Editura Academiei Republicii Socialiste România, Bucharest, 1977.
27. J. Neubüser: *Die Untergruppenverbände der Gruppen der Ordnungen* ≤ 100 *mit Ausnahme der Ordnungen 64 und 96.* Habilitationsschrift, Universität Kiel, 1967, 370 pp.
28. M. Novotný: *Einführung in die algebraische Linguistik.* IIM-Skriptum 40/67, Rheinisch-Westfälisches Institut für Instrumentelle Mathematik, Bonn, 1967.
29. M. Novotný: *S algebra od jazyka ke grammatice a zpět.* Československá akademie věd, Prague, 1988.
30. J.-F. Perrot: *Contribution à l'étude des monoïdes syntactiques et de certains groupes associés aux automates finis.* Thèse, Paris VI, 1972.
31. F. Simon: *Sekundärspeicher in LISP.* Diplomarbeit, Universität Kiel, 1970.
32. H. Stoyan: *LISP – Anwendungsgebiete, Grundbegriffe, Geschichte.* Akademie-Verlag, Berlin, 1980.
33. L. A. M. Verbeek: *Semigroup Extensions.* Thesis, Delft, 1968.
34. R. Wille: Lattices in Data Analysis; How to Draw Them with a Computer. In: I. Rival (ed.): *Algorithms and Order. Proceedings of the NATA Advanced Study Institute held in Ottawa, Canada, May 31–June 13, 1987.* NATO ASI Series C, Mathematical and Physical Sciences, **255**, Kluwer Academic Publishers, Dordrecht, 1989, 33–58.

Richard M. Karp

Professor Richard M. Karp was born in Boston, Massachusetts in 1935 and was educated at the Boston Latin School and Harvard University, where he received the Ph.D. in Applied Mathematics in 1959. From 1959 to 1968 he was a member of the Mathematical Sciences Department at the IBM Thomas J. Watson Research Center. From 1968 to 1994 he was a Professor at the University of California, Berkeley, where he had appointments in Computer Science, Mathematics and Operations Research. From 1988 to 1995 he was also associated with the International Computer Science Institute in Berkeley. In 1994 he retired from Berkeley and was named University Professor (Emeritus). In 1995 he moved to the University of Washington, where he is a Professor of Computer Science and Engineering and an Adjunct Professor of Molecular Biotechnology.

The unifying theme in Karp's work has been the study of combinatorial algorithms. His most significant work is the 1972 paper "Reducibility Among Combinatorial Problems," which shows that many of the most commonly studied combinatorial problems are disguised versions of a single underlying problem, and thus are all of essentially the same computational complexity. Much of his subsequent work has concerned the development of parallel algorithms, the probabilistic analysis of combinatorial optimization problems, and the construction of deterministic and randomized algorithms for combinatorial problems. His current research is concerned with strategies for sequencing the human genome, the physical mapping of large DNA molecules, the analysis of regulatory pathways in cells, and other combinatorial problems arising in molecular biology.

Professor Karp has received the U.S. National Medal of Science, the Harvey Prize (Technion), the Turing Award (ACM), the Centennial Medal (Harvard), the Fulkerson Prize (AMS and Math. Programming Society), the von Neumann Theory Prize(ORSA-TIMS), the Lanchester Prize (ORSA) the von Neumann Lectureship (SIAM) and the Distinguished Teaching Award (Berkeley). He is a member of the National Academy of Sciences, the National Academy of Engineering and the American Philosophical Society, as well as a Fellow of the American Academy of Arts and Sciences. He holds four honorary degrees.

The Mysteries of Algorithms

This article is about the key events and decisions in my career as a computer scientist. I belong to the first generation that came to maturity after the invention of the digital computer. My school years coincided with enormous growth in the support of science in the United States as a consequence of the great scientific breakthroughs during World War II, followed by the Sputnik era of the late 50's. I have been able to work in two great research universities and a great industrial research laboratory. No previous generation had access to such opportunities. If I had been born a few years earlier I would have had a very different, and undoubtedly less satisfying, career.

In the course of my career I have had to make a number of critical choices. As a student, I had to decide where to go to school and which subjects to study. Later, there were decisions about where to work and which professional responsibilities to undertake. Most importantly, there were decisions about entering or initiating new research areas. I have been guided through this maze of decisions by three principles:

Understand what you are good at and what you like to do, and choose accordingly. In the words of Socrates, "Know thyself."

Disregard the fashions of the day and search for new areas of research that are about to become important. In the words of the great hockey player and philosopher Wayne Gretzky, "Skate to where the puck is gonna be."

To find exciting problems, look at the interfaces between disciplines.

My work has touched on many fields, including computer science, mathematics, operations research, statistics, engineering and molecular biology, but the unifying theme has been the study of combinatorial algorithms. After more than four decades and exposure to countless algorithms, I continue to find the subject full of surprises and mysteries.

Getting Educated

Mathematics was my favorite subject in school. Between the ages of ten and fourteen I developed some skill at mental arithmetic, culminating in the ability to entertain my friends by multiplying four-digit numbers in my head. At the age of 13 I was exposed to plane geometry and was wonderstruck by the power and elegance of formal proofs. I recall feigning sickness in order to stay home from school and solve geometry problems. I was fortunate to receive a solid classical education at Boston Latin School, the second-oldest school (after Harvard) in the country. Next to mathematics my favorite subject was Latin, which I studied for six years. I took special pleasure in diagramming Latin sentences, a pursuit not very different from the solution of mathematical puzzles.

Having sailed through Boston Latin with little difficulty I had developed an inflated sense of my own ability. This was quickly dispelled at Harvard, where I found that I actually had to work in order to earn good grades, and that there were many students who equalled or surpassed my ability. I discovered that my writing ability was no better than workmanlike, and that laboratory science was definitely not the field for me.

The early mathematics courses were easy enough, but in the second half of my junior year I faced a tougher challenge in a course based on Halmos' monograph "Finite-Dimensional Vector Spaces," which developed linear algebra from the point of view of operator theory. The course coincided with the ending of an ill-fated sophomore romance. My unhappy love life shattered my morale, and for a time I retreated from the world, spending my afternoons at the Boylston Chess Club in Boston. As a result I spent virtually no time on mathematics, failed to master the Halmos text, and did poorly in the course.

In my senior year I greatly enjoyed a course in probability theory from Hartley Rogers, who encouraged me very strongly to pursue a mathematical career. On the other hand, I felt overmatched in Ahlfors' graduate course in Complex Analysis, where the students included a future Nobel Prize winner, a future Fields Medalist and a mathematical prodigy who was taking the course as a freshman.

By the middle of my senior year I had concluded that a career in pure mathematics was not for me. Being reluctant to leave Cambridge, and even more reluctant to work for a living, I decided to become a Ph.D. student at the Harvard Computation Lab, where Howard Aiken had built the Mark I and Mark IV computers. A faculty member at the Lab advised me to drop Prof. Ahlfors' Complex Analysis course immediately in favor of a solid introductory course in Accounting. Being an obedient young man I accepted this advice and approached Prof. Ahlfors for permission to drop the course. At first he was reluctant, but when he heard that I was giving up his course for one in Accounting he must have decided that there was no hope for me, as he gave his consent at once. At that point the die was cast - I was not to become a pure mathematician.

The Computation Lab

When I entered the Comp Lab in 1955 there were no models for a curriculum in the subject that today is called computer science. The young faculty offered courses in numerical analysis, switching theory, data processing, computational linguistics and operations research, and outside the Lab I took a variety of courses in applied mathematics, electrical engineering, probability and statistics. My performance was spotty, but I seemed to have a special feel for those topics that involved probability and discrete mathematics, and my successes in those areas produced a feeling of confidence. Sputnik in 1957 led to boom times in technical fields, and summer jobs became plentiful. Productive summers with M.I.T. Lincoln Lab and General Electric further fortified my sense that I might amount to something after all. A major turning point was Tony Oettinger's numerical analysis seminar in the fall of 1957, where I had the opportunity to give some talks, and discovered the pleasure of teaching.

Don Knuth has called attention to a breed of people who derive great aesthetic pleasure from contemplating the structure of computational processes. I still recall the exact moment when I realized that I was such a person. It was when a fellow student, Bill Eastman, showed me the Hungarian Algorithm for solving the Assignment Problem. I was fascinated by the elegant simplicity with which the algorithm converged inexorably upon the optimal solution, performing no arithmetic operations except addition and subtraction.

My Ph.D. dissertation was based on the idea that the flow of control in a computer program can be represented by a directed graph, and that graph theory algorithms can be used to analyze programs. In a loose sense this work was a precursor of the later development of the field of code optimization. Tony Oettinger was my supervisor, and my other readers were Ken Iverson and Gerry Salton.

The Comp Lab's old boy network was already operative by the time I finished my dissertation, and Fred Brooks, who had preceded my by two years at the Lab and had become a major player at IBM, set me up with a wide range of interviews. I accepted a position in the Mathematical Sciences Department within IBM's Research Division.

Nirvana on the Hudson

In January, 1959 I reported for work at the Lamb Estate, a former sanitarium for wealthy alcoholics that was the temporary home of the fledgling IBM Research Division. There was a diverse group of applied mathematicians under the direction of Herman Goldstine, John von Neumann's long-time collaborator, and an exciting group under Nat Rochester, doing what would today be called cognitive science. The atmosphere was informal; a high point of each day was the lunchtime frisbee game on the vast lawns that surrounded the Lamb Estate.

I was assigned to work on algorithms for logic circuit design under the direction of the topologist Paul Roth, who had made fundamental contributions to the subject. It was this work that first brought me up against the harsh realities of combinatorial explosions. While some of the algorithms our group devised scaled well with increasing problem size, others were essentially enumerative, and their running time escalated exponentially as the number of variables increased.

To my great good fortune IBM Research became a mecca for combinatorial mathematicians during the early sixties. Although computers were primitive by today's standards, they could already be used to solve logistical problems of significant size. Splendid algorithms for linear programming and network flow problems had been discovered, and the field of combinatorial algorithms was in a stage of rapid development. In the summer of 1960 the leaders of the field came together at the Lamb Estate for an extended period. Among the visitors were Richard Bellman, George Dantzig, Merrill Flood, Ray Fulkerson, Ralph Gomory, Alan Hoffman, Ed Moore, Herb Ryser, Al Tucker and Marco Schutzenberger. Soon thereafter IBM brought in Ralph Gomory and Alan Hoffman to build a combinatorics research group.

Alan Hoffman became my mentor. He is a virtuoso at linear algebra, linear programming and algebraic graph theory, and has a talent for explaining mathematical ideas with clarity, precision and enthusiasm. Although my interests were more algorithmic than his, his style of exposition became a model for my own. I gained an understanding of the theory of linear programming and network flows, and came to appreciate how special their structure was compared to nastier problems such as integer programming and the traveling-salesman problem. During this period Mike Held and I developed the 1-tree heuristic, which remains the best method of computing a tight lower bound on the cost of an optimal traveling-salesman tour. With a bit of help from Alan Hoffman and Phil Wolfe we realized that our heuristic was a special case of an old method called Lagrangian relaxation; this connection motivated many other researchers to apply Lagrangian relaxation to difficult combinatorial optimization problems.

I also visited the National Bureau of Standards to work with Jack Edmonds on network flow problems. We pointed out, perhaps for the first time, the distinction between a *strongly polynomial algorithm*, whose running time (assuming unit-time arithmetic operations) is bounded by a polynomial in the dimension of the input data, and a *polynomial-time algorithm*, whose running time is bounded by a polynomial in the number of bits of input data. We gave the first strongly polynomial algorithm for the max-flow problem. For the min-cost flow problem we introduced a scaling technique that yielded a polynomial-time algorithm, but we were unable to find a strongly polynomial algorithm. The first such algorithm was obtained by Eva Tardos in the early 80's.

A few years before our collaboration began Edmonds had published a magnificent paper entitled "Paths, Trees and Flowers"which gave an algorithm for constructing a matching of maximum cardinality in any given graph. The paper began by introducing the concept of a "good algorithm," Edmonds' term for what is now called a polynomial-time algorithm. He showed that his algorithm for matching was a good one, and, even more importantly, raised the possibility that, for some combinatorial optimization problems, a good algorithm might not exist. This discussion by Edmonds was probably my first exposure to the idea that some standard combinatorial optimization problem might be intractable in principle, although I later learned that Alan Cobham and Michael Rabin had thought along similar lines, and that the possibility had been discussed extensively in Soviet circles.

IBM had a strong group in formal models of computation under the leadership of Cal Elgot. Through my contacts with that group I became aware of developments in automata theory, formal languages and mathematical logic, and followed the work of pioneers of complexity theory such as Rabin, McNaughton and Yamada, Hartmanis and Stearns, and Blum. Michael Rabin paid an extended visit to the group and became my guide to these subjects. From Hartley Rogers' splendid book "Theory of Recursive Functions and Effective Computability" I became aware of the importance of reducibilities in recursive function theory, but the idea of using subrecursive reducibilities to classify combinatorial problems did not yet occur to me.

My own work on formal models centered around parallel computation. Ray Miller, Shmuel Winograd and I did work that foreshadowed the theory of systolic algorithms. Miller and I introduced the parallel program schema as a model of asynchronous parallel computation; in the course of this work we introduced vector addition systems and initiated the study of related decision problems. The most notorious of these was the reachability problem, which after many false tries was proved to be decidable through the efforts of several researchers, culminating in a 1982 paper by Rao Kosaraju.

A Zest for Teaching

I moved to Berkeley at the end of 1968 in order to lead a more rounded life than the somewhat isolated suburban environment of IBM could provide. One aspect of this was the desire to be more involved with students. My father was a junior high school mathematics teacher, and I have fond memories of visiting his classroom as a youngster. He was undoubtedly the role model responsible for my attraction to teaching. I have been involved in teaching throughout my career and have always enjoyed it. Recently Greg Sorkin, a former student, reminded me of some thoughts on teaching that I wrote up for the Berkeley students in 1988. I include them here.

Thoughts on Teaching

Preparation

Follow the Boy Scout motto: Be Prepared!

Never select material that doesn't interest you. Boredom is deadly and contagious. If the standard syllabus is boring, then disregard it and pick material you like.

Figure out your notation and terminology in advance. Know exactly where you're going, and plan in detail what you are going to write on the board.

Check out the trivial details. They're more likely to hang you up than the major points.

Make sure you understand the intuition behind the technical results you are presenting, and figure out how to convey that intuition.

Debug your assignments and exams. They're just as important as the lectures.

Don't teach straight out of a textbook or from old notes. Recreate the material afresh, even if you're giving the course for the tenth time.

Structuring the Material

Ideally, each lecture should be a cohesive unit, with a small number of clearly discernible major points.

In organizing your lecture, use the principles of structured design: top-down organization, modularity, information hiding, etc.

Make sure the students have a road map of the material that is coming up.

Conducting the Lecture

Take a few minutes before each class to get relaxed.

Start each lecture with a brief review.

Go through the material at a moderate but steady pace. Don't worry about covering enough material. It will happen automatically if you don't waste time.

Write lots on the board; it helps the students' comprehension and keeps you from going too fast. Print, even if your handwriting is very clear. Cultivate the skill of talking and writing at the same time.

Talk loud enough and write big enough.

Maintain eye contact with the class.

Develop a sense of how much intensity the students can take. Use humor for a change of pace when the intensity gets too high.

Be willing to share your own experiences and opinions with the students, but steer clear of ego trips.

Make it clear that questions are welcome, and treat them with respect. In answering questions, never obfuscate, mystify or evade in order to avoid showing your ignorance. It's very healthy for you and the students if they find out that you're fallible.

Be flexible, but don't lose control of the general direction of the lecture, and don't be afraid to cut off unproductive discussion. You're in charge; it's not a democracy.

If you sense from questions or class reaction that you're not getting through, back up and explain the material a different way. The better prepared you are, the better you will be able to improvise.

Try the scribe system, in which the students take turns writing up the lectures and typesetting the notes.

Start on time and end on time.

The Professorial Life

The move to Berkeley marked the end of my scientific apprenticeship. At IBM I had enjoyed the mentorship of Alan Hoffman and Michael Rabin, and the opportunity to work with a host of other experienced colleagues. At Berkeley I worked mainly with students and young visiting scientists, and I was expected to serve as their mentor. The move also caused a sudden leap in my professional visibility. From 1968 onward I have been steadily besieged with requests to write letters of reference, do editorial work and serve on committees. With the advent of e-mail, the flow of requests has become a deluge; I offer my sincere apologies to any readers to whom I have been e-brusque.

A professor's life is a juggling act. The responsibilities of teaching, research, advising, committee work, professional service and grantsmanship add up to a full agenda. Fortunately, I have always heeded the advice of my former colleague Beresford Parlett: "If it's not worth doing, it's not worth doing well."

Parlett's wise counsel has helped me resist administrative responsibilities, but there was one period when I could not avoid them. Berkeley in the 60's was a cauldron of political controversy, and the mood of dissent extended to computer science. The faculty were divided as to whether to maintain computer science in its traditional home within Electrical Engineering, or to establish it as a separate department in Letters and Science. In 1967 the administration decided to do both, and I was one of several faculty hired into the newly formed Computer Science Department. The two-department arrangement was awkward administratively and only exacerbated the tensions between the two groups. In 1972 the administration decreed that the two groups of computer science faculty should be combined into a Computer Science Division within the Department of Electrical Engineering and Computer Sciences. As the faculty member least tinged with partisanship I emerged as the compromise candidate to head the new unit, and I somewhat reluctantly took the job for two years. I expected a period of turmoil, but once the merger was a *fait accompli* the tensions dissipated and harmony reigned. Much of the credit should go to Tom Everhart, the department chair at the time and later the Chancellor of Caltech. Tom nurtured the new unit and respected its need for autonomy.

Once the political disputes had healed Berkeley was poised to become a great center for computer science. In theoretical computer science the merger created a strong group of faculty, anchored by Manuel Blum, Mike Harrison, Gene Lawler and myself. Berkeley became a mecca for outstanding graduate students, and has remained so to this day. It is one of the handful of places that have consistently had a thriving community of theory students, and there has always been a spirit of cooperation and enthusiasm among them. A major reason for the success of theory at Berkeley has been Manuel Blum, a deep researcher, a charismatic teacher, and the best research adviser in all of computer science.

Over the years at Berkeley I supervised thirty-five Ph.D. students. I have made it a rule never to assign a thesis problem, but to work together with each student to develop a direction that is significant and fits the student's abilities and interests. Each relationship with a thesis student is unique. Some students are highly independent and merely need an occasional sounding board. Others welcome collaboration, and in those cases the thesis may become a joint effort. Some students have an inborn sense of how to do research, while others learn the craft slowly, and only gradually develop confidence in their ability. My greatest satisfaction has come from working with these late bloomers, many of whom have gone on to successful research careers.

NP-Completeness

In 1971 I read Steve Cook's paper "The Complexity of Theorem-Proving Procedures," in which he proved that every set of strings accepted in polynomial time by a nondeterministic Turing machine is polynomial-time reducible to SAT (the propositional satisfiability problem). Cook stated his result in terms of polynomial-time Turing reducibility, but his proof demonstrates that the same result holds for polynomial-time many-one reducibility. It follows that $P = NP$ if and only if SAT lies in P. Cook also mentioned a few specific problems that were reducible to SAT as a consequence of this theorem. I was not following complexity theory very closely, but the paper caught my eye because it made a connection with real-world problems. I knew that there were people who really wanted to solve instances of SAT, and Cook's examples of problems reducible to SAT also had a real-world flavor. I realized that the class of problems reducible to SAT with respect to many-one polynomial-time reducibility (the class that we now call *NP*) included decision problems corresponding to all the seemingly intractable problems that I had met in my work on combinatorial optimization, as well as many of those that I had encountered in switching theory. Thus Cook's result implied that, if SAT can be solved in polynomial time, then virtually all the combinatorial optimization problems that crop up in operations research, computer engineering, economics, the natural sciences and mathematics can also be solved in polynomial time.

It was not by accident that I was struck by the significance of Cook's Theorem. My work in combinatorial optimization had made me familiar with the traveling-salesman problem, the maximum clique problem and other difficult combinatorial problems. Jack Edmonds had opened my eyes to the possibility that some of these problems might be intractable. I had read a paper by George Dantzig showing that several well-known problems could be represented as integer programming problems, and suggesting that integer programming might be a universal combinatorial optimization problem. My reading in recursion theory had made me aware of reducibilities as a tool for classifying computational problems.

It occurred to me that other problems might enjoy the same universal character as SAT. I called such problems *polynomial complete*, a term which was later supplanted by the more appropriate term *NP-complete.* I set about to construct reductions establishing the *NP*-completeness of many of the seemingly intractable problems that I had encountered in my work on combinatorial algorithms.

It was an exciting time because I had the clear conviction that I was doing work of great importance. Most of the reductions I was after came easily, but the *NP*-completeness of the Hamiltonian circuit problem eluded me, and the first proofs were given by Gene Lawler and Bob Tarjan, who were among the first to grasp the significance of what I was doing. The first opportunity to speak about *NP*-completeness came at a seminar at Don Knuth's home. In April, 1972 I presented my results before a large audience at a symposium at IBM, and in the following months I visited several universities to give talks about *NP*-complete problems. Most people grasped the significance of the work, but one eminent complexity theorist felt that I was merely giving a bunch of examples - and I suppose that, in a sense, he was right. A year or so later I learned that Leonid

Levin, in the Soviet Union, had independently been working along the same lines as Cook and myself, and had obtained similar results.

The early work on *NP*-completeness had the great advantage of putting computational complexity theory in touch with the real world by propagating to workers in many fields the fundamental idea that computational problems of interest to them may be intractable, and that the question of their intractability can be linked to central questions in complexity theory. Christos Papadimitriou has pointed out that in some disciplines the term *NP*-completeness has been used loosely as a synonym for computational difficulty. He mentions, for example, that Diffie and Hellman, in their seminal paper on public-key cryptography, cited *NP*-completeness as a motivation for positing the existence of one-way functions and trapdoor functions, even though it can be shown that, when translated into a decision problem, the problem of inverting a one-way function lies in $NP \cap co-NP$, and thus is unlikely to be *NP*-complete.

The study of *NP*-completeness is more or less independent of the details of the abstract machine that is used as a model of computation. In this respect it differs markedly from much of the earlier work in complexity theory, which had been concerned with special models such as one- or two-tape Turing machines and with low levels of complexity such as linear time or quadratic time. For better or for worse, from the birth of *NP*- completeness onward, complexity theory has been mainly concerned with properties that are invariant under reasonable changes in the abstract machine and distortions of the time complexity measure by polynomial factors. The concepts of reducibility and completeness have played a central role in the effort to characterize complexity classes.

After my 1972 paper I did little further work on NP-completeness proofs. Some colleagues have suggested that I had disdain for such results once the general direction had been established, but the real reason is that I am not particularly adept at proving refined NP-completeness results, and did not care to compete with the virtuosi of the subject who came along in the 70's.

Dealing with *NP*-Hard Problems

There is ample circumstantial evidence, but no absolute proof, that the worst-case running time of every algorithm for solving an NP-hard optimization problem must grow exponentially with the size of the instance. Since NP-hard problems arise frequently in a wide range of applications they cannot be ignored; some means must be found to deal with them.

The most fully developed theoretical approach to dealing with NP-hard problems is based on the concept of a polynomial-time approximation algorithm. An NP-hard minimization problem is said to be *r-approximable* if there is a polynomial-time algorithm which, on all instances, produces a feasible solution whose cost is at most r times the cost of an optimal solution. A similar definition holds for maximization problems.

NP-hard problems differ greatly in their degree of approximability. The minimum makespan problem in scheduling theory and the knapsack problem are

$(1+\epsilon)$- approximable for all positive ϵ. By a recent spectacular result due to Sanjeev Arora, the Euclidean traveling-salesman problem also enjoys this property, but the execution time of the approximation algorithm grows very steeply as a function of $\frac{1}{\epsilon}$. Many problems are r-approximable for certain values of r but, unless $P = NP$, are not r-approximable for every $r > 1$. Unless $P = NP$, the maximum clique problem and the minimum vertex coloring problem in graphs are not r-approximable for any r.

The theory of polynomial-time approximation algorithms is elegant, but for most problems the approximation ratios that can be proven are too high to be of much interest to a VLSI designer or a foreman in a factory who is seeking near-optimal solutions to specific problem instances. Even when a problem is $(1+\epsilon)$-approximable for an arbitrarily small ϵ, the time bound for the approximation algorithm may grow extremely rapidly as ϵ tends to zero.

In practice, many NP-hard problems can reliably be solved to near-optimality by fast heuristic algorithms whose performance in practice is far better than their worst-case performance. In order to explain this phenomenon it is necessary to depart from worst-case analysis and instead study the performance of fast heuristic algorithms on typical instances. The difficulty, of course, is that we rarely have a good understanding of the characteristics of typical instances.

In 1974 I decided to study the performance of heuristics from a probabilistic point of view. In this approach one assumes that the problem instances are drawn from a probability distribution, and tries to prove that a fast heuristic algorithm finds near-optimal solutions with high probability. Probabilistic analysis has been a major theme in my research. I have applied it to the traveling-salesman problem in the plane, the asymmetric traveling-salesman problem, set covering, the subset-sum problem and 1-dimensional and 2-dimensional bin-packing problems, as well as problems solvable in polynomial time, such as linear programming, network flow and graph connectivity. There is a significant school of researchers working along these lines, but the approach has certain technical limitations. To make the analysis tractable it is usually necessary to restrict attention to very simple heuristics and assume that the problem instances are drawn from very simple probability distributions, which may not reflect reality. The results are often asymptotic, and do not reveal what happens in the case of small problem instances. The results we are obtaining do shed some light on the performance of heuristics, but the reasons why heuristics work so well in so many cases remain a mystery.

Randomization and Derandomization

In the Fall of 1975 I presented a paper on probabilistic analysis at a symposium at Carnegie-Mellon University, giving some early results and a road map for future research. At the same symposium Michael Rabin presented a seminal paper on randomized algorithms. A randomized algorithm is one that receives, in addition to its input data, a stream of random bits that it can use for the purpose of making random choices. The study of randomized algorithms is necessarily probabilistic, but the probabilistic choices are internal to the algorithm, and no assumptions about the distribution of input data are required.

As I stated in a 1991 survey paper, " Randomization is an extremely important tool for the construction of algorithms. There are two principal types of advantages that randomized algorithms often have. First, often the execution time or space requirement of a randomized algorithm is smaller than that of the best deterministic algorithm that we know of for the same problem. But even more strikingly, if we look at the various randomized algorithms that have been invented, we find that invariably they are extremely simple to understand and to implement; often, the introduction of randomization suffices to convert a simple and naive deterministic algorithm with bad worst-case behavior into a randomized algorithm that performs well with high probability on every possible input."

Inspired by Rabin's paper and by the randomized primality test of Solovay and Strassen, I became a convert to the study of randomized algorithms. With various colleagues I have worked on randomized algorithms for reachability in graphs, enumeration and reliability problems, Monte Carlo estimation, pattern matching, construction of perfect matchings in graphs, and load balancing in parallel backtrack and branch-and-bound computations. We have also investigated randomized algorithms for a variety of on-line problems and have made a general investigation of the power of randomization in the setting of on-line algorithms. I take special pride in the fact that two of my former students, Rajeev Motwani and Prabhakar Raghavan, wrote the first textbook devoted to randomized algorithms.

In a 1982 paper Les Valiant suggested the problem of finding a maximal independent set of vertices in a graph as an example of a computationally trivial problem that appears hard to parallelize. Avi Wigderson and I showed that the problem can be parallelized, and in fact lies in the class NC of problems solvable deterministically in polylog time using a polynomial-bounded number of processors. It was fairly easy to construct a randomized parallel algorithm of this type, and the harder challenge was to convert the randomized algorithm to a deterministic one. We achieved this by a technique that uses balanced incomplete block designs to replace random sampling by deterministic sampling. This was one of the first examples of *derandomization* - the elimination of random choices from a randomized algorithm. Later, Mike Luby and Noga Alon found simpler ways, also based on derandomization, to place the problem in NC.

In 1985 Nick Pippenger, Mike Sipser and I gave a rather general method of reducing the failure probability of a randomized algorithm exponentially at the cost of a slight increase in its running time. Our original construction, which is based on expander graphs, has been refined by several researchers, and these refinements constitute an important way of reducing the number of random bits needed to ensure that a randomized algorithm achieves a specified probability of success.

The Complexity Year

Early in 1985 the mathematician Steve Smale asked me to join him in a proposal for a special year in computational complexity at the Mathematical Sciences Research Institute in Berkeley. I gladly agreed, and our proposal was accepted by the Advisory Committee of the Institute, which had already recognized the significance of complexity theory as a mathematical discipline, and had been hoping for just such a proposal. The National Science Foundation provided generous funding, and over the next several months I worked with Smale and Cal Moore, the Associate Director of the Institute, to arrange for an all-star cast of young computer scientists and mathematicians to work at the Institute for a year, and to attract a number of the leading senior scientists in the field.

In the informal atmosphere of the Institute we worked hard, both at the blackboard and on hikes in the Berkeley hills, on structural complexity, cryptography, computational number theory, randomized algorithms, parallel computation, on-line algorithms, computational geometry, graph algorithms, and numerical algorithms. Although we didn't crack the *P:NP* problem, the Complexity Year met its main goal of broadening the outlook of the young scientists by exposing them to a wide variety of areas within complexity theory. It also led to a marriage between two of the young scientists, whose romance flourished after I asked them to work together to organize the Institute's colloquium series.

Some theoretical computer scientists believe that precious research funds are better spent on individual investigators than on special programs such as the Complexity Year that create a concentration of researchers at a single location. Both kinds of funding are crucially important, but I believe that the special programs provide a breadth of exposure to new ideas that is extremely beneficial to young scientists, and could not be provided through individual grants. Special programs can attract large-scale funding that would not otherwise be available at all. Over the past decade I have served on the External Advisory Committee of DIMACS, the NSF Science and Technology Center for Discrete Mathematics and Theoretical Computer Science, and have seen how the special years there have strengthened the research communities in a number of emerging areas, including computational molecular biology, which has become one of my own main interests.

A Spirited Debate

In 1995 I agreed to chair a committee to provide advice on the directions NSF should take in the funding of theoretical computer science. The committee recognized that the theory community had a splendid record of achievement, but also felt that the community was not achieving the maximum possible impact on the rest of computer science and on the phenomenal developments in information processing that were transforming our society. We developed a set of recommendations for increasing the impact of theory by communicating our results to nonspecialists, restructuring computer science education to bring theory and applications closer together, and broadening the scope of theory research to connect it better with emerging applications.

In the Fall of 1995, just before the IEEE Symposium on Foundations of Computer Science, I had the pleasure of attending a program of lectures organized by some of my colleagues in honor of my sixtieth birthday. I thoroughly enjoyed the company of former students and other old friends, the interesting lectures, some of which pursued themes from my work, as well as the somewhat overstated praise that is customary at events of this sort.

Just a few days later, another memorable event occurred. I was asked to present an oral report at an evening session of the symposium, laying out the concerns of our committee. In the afterglow of my birthday celebration I put my report together hastily and made some fundamental errors. I spent very little time extolling the past achievements of theory, feeling that I was addressing an audience that didn't need to be reminded of them. Instead of stressing the advantages and opportunities that would come from reaching out to applications, I took a negative tone, criticizing the theory community for being ingrown, worshiping mathematical depth, working on artificial problems and making unsupported claims of applicability. As a result my positive message was almost completely lost, and I became the focus of a firestorm of criticism. I was accused of trying to prescribe what people should and shouldn't work on, failing to appreciate the achievements of theory, providing ammunition for the enemies of theory, and selling out to anti-theory forces in the funding agencies.

Upon reflection I realized that my criticisms had been excessively harsh, and could only detract from the effectiveness of our report. Over the ensuing months the committee produced a report with a more positive tone. My coauthors and I expected that, with the submission of our report, we had put the incident behind us, but upon arriving at the May, 1996 ACM Symposium on Theory of Computing I learned that this was not the case. Oded Goldreich and Avi Wigderson had circulated an extended critique of our report, and an *ad hoc* meeting was scheduled at which Wigderson and I were to present our viewpoints. The Goldreich-Wigderson critique stressed the fundamental importance of TOC as an independent discipline with deep scientific and philosophical consequences, and rejected the opinion that the prosperity of TOC depends on service to other disciplines and immediate applicability to the current technological development. In my reply I expressed my deep respect for the fundamental work that had been done in TOC, but continued to assert that the subject could gain intellectual stimulation not only by pursuing the deep questions that had originated within theory itself, but also by linking up with applications, and that the two approaches could be complementary rather than competitive.

In the end the debate surrounding our report was quite valuable for the TOC community. It provoked a lively and continuing dialogue on the directions our field should be taking, and stimulated many people in the community to do some soul-searching about their own research choices. For my part, I learned how tactfully and clearly one must communicate in order to contribute effectively to public debate on sensitive topics.

Computational Molecular Biology

In the second half of this century molecular biology has been one of the most rapidly developing fields of science. Fundamental discoveries in the 50's and 60's identified DNA as the carrier of the hereditary information that an organism passes on to its offspring, determined the double helical structure of DNA, and illuminated the processes of transcription and translation by which genes within the DNA direct the production of proteins, which mediate the chemical processes of the cell. The connections between these processes and digital computation are striking: the information within DNA molecules is encoded in discrete form as a long sequence of chemical subunits of four types, and the genes within these molecules can be thought of as programs which are activated under specific conditions. Technology for manipulating genes has led to many applications to agriculture and medicine, and nowadays one can hardly pick up a newspaper without reading about the isolation of a gene, the discovery of a new drug, the sequencing of yet another microbe, or new insights into the course of evolution.

In 1963 the Mathematical Sciences Department at IBM decided to look into the applications of mathematics to biology and medicine, and I visited the Cornell Medical Center in New York City and the M.D. Anderson Hospital in Houston, looking for a suitable research problem. Nothing came of this venture, except that I ran across the work of the geneticist Seymour Benzer, in which he invented the concept of an interval graph in connection with his studies of the arrangement of genes on chromosomes; this was one of the earliest connections between discrete mathematics and genetics.

Over the next decades I was an avid reader of popular literature about molecular biology and genetics, but it was not until 1991 that I began to think seriously about applying my knowledge of algorithms to those fields. By then the Human Genome Project had come into existence, and it was evident that combinatorial algorithms would play a central role in the daunting task of putting the three billion symbols in the human genome into their proper order. The databases of DNA and protein sequences, genetic maps and physical maps had begun to grow, and to be used as indispensable research tools. My friend and colleague Gene Lawler and my former student Dan Gusfield, as well as several Berkeley graduate students, were working closely with the genome group at the Lawrence Berkeley Laboratory up the hill from the Berkeley campus, and I began to attend their seminars, as well as Terry Speed's seminar on the statistical aspects of mapping and sequencing.

To get started in computational biology I decided to tackle the problem of physical mapping of DNA molecules. We can view a DNA molecule as a very long sequence of symbols from the alphabet $\{A, C, T, G\}$. Scattered along the molecule are features distinguished by the occurrence of particular short DNA sequences. The goal of physical mapping is to determine the locations of these features, which can then be used as reference points for locating the positions of genes and other interesting regions of the DNA. The map is inferred from the *fingerprints* of *clones*; a clone is a segment of the DNA molecule being mapped, and the fingerprint gives partial information about the presence or absence of features on the clone. The problem of determining the arrangement of the clones and the features along the DNA molecule is a challenging combinatorial puzzle, complicated by the fact that the fingerprint data may be noisy and incomplete.

Beginning around 1991 my students and I developed computer programs to solve a number of versions of the physical mapping problem, but at first we lacked the close connections with the Human Genome Project that would enable us to have a real impact. In 1994, through my friends Maria Klawe and Nick Pippenger at the University of British Columbia, I made contact with a group of computer scientists and biologists who were meeting from time to time at the University of Washington to discuss computational problems in genomics. In addition to Maria and Nick, the group included the computer scientists Larry Ruzzo and Martin Tompa , as well as the geneticist Maynard Olson and the computational biologist Phil Green. I found the meetings very useful, and realized that the University of Washington was a hotbed of activity in the application of computational methods to molecular biology and genetics.

During the early 90's the University of California had a rich pension fund but a lean operating budget. In order to solve its financial problems the University offered a series of attractive early retirement offers to its older and more expensive faculty. Although I knew that it would not be easy to leave Berkeley after twenty-five years, I succumbed to the third of these offers.

In 1988 I had been part of a group of Berkeley faculty who helped establish ICSI, an international computer science research institute at Berkeley. Mike Luby, Lenore Blum and I built up a theoretical computer science group at ICSI which attracted outstanding postdocs and visitors from around the world. For the first year of my 'retirement' I based myself at ICSI, but in 1995 I moved to the University of Washington. I was attracted by the congenial atmosphere and strong colleagues in the computer science department at UW, and by the strength and depth of the activity in molecular biotechnology, led by Lee Hood. Hood saw the sequencing of genomes as merely a first step towards the era of *functional genomics*, in which the complex regulatory networks that control the functioning of cells and systems such as the immune system would be understood through a combination of large-scale automated experimentation and subtle algorithms. I decided that nothing else I might work on could be more important than computational biology and its application to functional genomics.

How is it that liver cells, blood cells and skin cells function very differently even though they contain the same genes? Why do cancer cells behave differently from normal cells? Although each gene codes for a protein, complex regulatory networks within the cell determine which proteins are actually produced, and in what abundance. These networks control the rate at which each gene is transcribed into messenger RNA and the rate at which each species of messenger RNA is translated into protein. These rates depend on the environment of the cell, the abundance of different proteins within the cell and the presence of mutated genes within the cell. Newly developed technologies make it possible to take a detailed snapshot of a cell, showing the rates of transcription of thousands of genes and the levels of large numbers of proteins. In model organisms such as yeast we also have the ability to disrupt individual genes and observe how the effects of those disruptions propagate through the cell. The problem of characterizing the regulatory networks by performing strategically chosen disruption experiments and analyzing the resulting snapshots of the cell will be the focus of much of my future work. I expect to draw on the existing knowledge in statistical clustering theory and computational learning theory, and will need to advance the state of the art in these fields in order to succeed.

With the help of outstanding mentors I have enjoyed learning the rudiments of molecular biology. I have found that the basic logic of experimentation in molecular biology can, to some extent, be codified in abstract terms, and I have discovered that the task of inferring the structure of genomes and regulatory networks can lead to interesting combinatorial problems. With enough simplifying assumptions these problems can be made quite clean and elegant, but only at the cost of disregarding the inherent noisiness of experimental data, which is an essential aspect of the inference task. Combinatorial optimization is often useful, but unless the objective function is chosen carefully the optimal solution may not be the true one. Typically, the truth emerges in stages through an interplay between computation and experimentation, in which inconsistencies in experimental data are discovered through computation and corrected by further experimentation.

Conclusion

Being a professor at a research university is the best job in the world. It provides a degree of personal autonomy that no other profession can match, the opportunity to serve as a mentor and role model for talented students, and an environment that encourages and supports work at the frontiers of emerging areas of science and technology. I am fortunate to have come along at a time when such a career path has been available.

Solomon Marcus

Professor Solomon Marcus is affiliated to the Department of Mathematics, University of Bucharest, Romania, where he was successively, student 1945, instructor 1951, assistant professor 1955, associate professor 1964, professor 1966, and emeritus professor 1991. Research in set theory, real analysis, general topology, where he is quoted mainly for his results related to measure vs. Baire category, differentiation, Darboux property, Jensen convexity, quasicontinuity, determinant and stationary sets, symmetry of sets and functions, Riemann integrability in topological spaces, Hamel bases. In the late fifties and early sixties Professor Marcus became one of the initiators of mathematical and computational linguistics, proposing algebraic, logical and set-theoretic models for some fundamental linguistic categories in phonology, morphology and syntax. Later, he extended his interest to poetry, being one of the founders of mathematical poetics. Then, he became active in the semiotics of artificial languages (including also programming languages). Quoted by more than a thousand authors, he gave invited lectures and had temporary positions in most European countries, in U.S.A., Canada, Brazil, New Zealand and other countries. He is a member of the editorial boards of several journals of mathematics, computer science, linguistics, poetics and semiotics. Among his former students and pupils one can find many well known names in the fields of mathematics and computer science.

Bridging Linguistics and Computer Science, via Mathematics

From Poetry to Mathematical Analysis

When I was fifteen, I was fascinated by poetry. Mathematics was still a territory remaining to be discovered (school mathematics seems to be still today rather a failure). In the fall of 1944 I began a new life: I survived the second world war, I was classified the first among 156 candidates at the French type high-school final examination called "baccalaureat" in a city of a devastated Romania and I got the right to be accepted by any institute of higher education of the country with no further examination. I had a very short time to take a decision. I arrived with many complications (the war consequences were still visible) in Bucharest, the capital of Romania, and I began to visit various faculties and look at the programmes they offered. I was almost twenty. During the summer of 1944, I had my first contact with the true mathematics, reading a presentation of the non-Euclidean geometries. I was very impressed. I paid special attention to the announcements I saw at the Faculty of Sciences, Department of Mathematics. From the list of courses to be taught I remember: infinitesimal calculus, mathematical logic, topology, theory of functions, abstract algebra, analytic, projective and differential geometry, number theory. Although most of these words were unknown to me, I had a feeling that they refer to a universe responding to my intellectual needs. Like poetry some years earlier, words such as "infinitesimal" and "topology" gave me the impression that they refer to some aspects hidden to common, everyday observation. My assumption was that mathematics, again like poetry, refers to a second reality, to be discovered or invented. I became a student in mathematics, but poetry, my first love, remained for ever an essential component of my intellectual background. This belief was strengthen by my professor of "abstract algebra", Dan Barbilian, one of the greatest Romanian poets. The attribute "abstract" may seem today pleonastic when associated with "algebra", but at that time a rhetorical stress was necessary to call attention on the opposition with traditional, algorithmic and numerical algebra. Under the influence of my teachers Nicolescu, Stoilow, Froda and of the teacher of my teachers, Pompeiu, all with PhD at Sorbonne-Paris, guided by the great masters of French mathematical analysis, I chose real analysis, set theory and topology as my main field of research, attracted by the counter-intuitive, sometimes pathological phenomena such as the continuous functions nowhere differentiable. I published in this field of pure mathematics about a hundred articles, many of which are still used by the most active authors in the field (see the basic monographs devoted to it, as well as the specialized journal "Real Analysis Exchange"). Looking at them from today's perspective, one could consider them as a preliminary step to the "fractal geometry of nature" developed in the seventies by Benoit Mandelbrot.

Itineraries to Language

However, it is known that some of the most difficult problems considered in set theory, real analysis and topology in the first decades of our century proved to be related to some aspects of the foundations of mathematics and mathematical logic. On the other hand, the latter fields showed already from the beginning of our century (see Axel Thue's combinatorial systems and David Hilbert's formal systems) their strong connections with what we call today a structure of formal language. Chronologically however, my first step towards bridging mathematics and languages was a different one. Already in the late fifties, I learned of the emergence of the new field called "automatic translation", simultaneously developed in West and East. Some engineers in electronics and computers, sometimes associated with linguists, were optimistic in trying to build algorithms of translation from a natural language into another one, under the more or less explicit assumption that translation is mainly a lexical and syntactic activity that can be easily arranged in an algorithmic form and then transformed in a computer program. I was not attracted by such projects; it happened, however, that as an additional activity to the projects of automatic translation, various attempts of formalization of some basic aspects of the morphology and syntax of natural languages were developed. Among the authors of these formal models were A.N. Kolmogorov, V.A. Uspenskii, R.L. Dobrusin, O.S. Kulagina and A.V. Gladkii in the former Soviet Union, A. Sestier, Y. Lecerf, P. Ihm in Western Europe, David Hays in U.S.A., J. Lambek in Canada, Y. Bar-Hillel, H. Gaifman, M. Perles, E. Shamir in Israel, A. Trybulec in Poland, M. Novotny and L. Nebesky in Czechoslovakia, J. Kunze in Germany. Some of these authors adopted an approach based on set theory, free semigroups and algebra of binary relations, others were oriented towards ideas coming from the Polish school of logic (Ajdukiewicz, Lesniewski). A common denominator of all of them was their linguistic background. Each of them exploited some linguistic ideas, methods and/or concepts. Linguistics was for long-time dominated by various schools of structuralism. In the early fifties, two schools were dominant: American distributionalism (Zellig S. Harris, Charles Hockett) and Danish glossematics (Louis Hjelmslev), but the classical heritage of Ferdinand de Saussure and of Prague Linguistic School in Europe, of Leonard Bloomfield and Edwin Sapir in U.S.A. was still very important. I became interested in all these developments, trying to bridge linguistics and mathematics. Harris' view, exploiting the contextual behaviour of various linguistic units, was near to the spirit of the theory of free semigroups, while Hjelmslev conceived linguistics as a kind of algebra. I was also influenced by some trends in Romanian structural linguistics and I have to quote in this respect at least two names: Emanuel Vasiliu and Paula Diaconescu.

Continuous and Discrete Mathematics

It was not so easy for me to adapt to all these developments. My mathematical training and activity were oriented almost exclusively towards continuous mathematics, almost all of my research articles and my teaching at the University were devoted to it, while the mathematics of language was conceived, more or less explicitly, as a chapter of discrete mathematics. Algebra, combinatorics, mathematical logic were not so familiar to me and I had to change in some respects my habits of thinking, in order to adapt them to the study of language. Only in a next step I realized that these two kinds of mathematics, continuous and discrete, share an important common problem, the study of infinity by finite means, and they interact so much that we cannot separate them. My experience in mathematical analysis, including topology and set theory, proved to be a good source of inspiration in the field of language. Several (joint) papers I published in this respect have as their starting point some phenomena in continuous mathematics related, for instance, to symmetry or to convexity. The analogy between continuous and discrete phenomena remains a basic strategy, which still deserves to be used in the study of languages, with a great chance of interesting results. However, more important than its technicality, mathematics proved to be useful in the field of languages by its way of thinking. Despite its very controversial nature, the mathematical way of thinking is first of all characterized by its step by step procedure, each step being rigorously and explicitly based on the previous steps. This fact can be easily confirmed by a very simple experiment: take two books, one of mathematics, the other of geography, for instance: if we delete from each of them the first 20 pages, the remaining part in the former will be almost completely non-intelligible, while in the latter it will be to a large extent intelligible. The mathematical thinking is looking for explicitness and has the tendency to become marathonic; it clearly separates what is given from what is to be found; it proceeds from the simplest (apparently trivial) things and moves, step by step, to less simple things. Complexity is increasing gradually. At each step, we pack the already acquired concepts and results, by means of an adequate symbolism, in order to keep our language within reasonable limits of complexity.

Linguistic Structuralism and Automatic Translation through the Glasses of Mathematics

Having in front of me these two lines of development: linguistic structuralism and construction of algorithms for the automatic translation of languages, both dominated by the aim to formalize linguistic phenomena, I realized that the best help mathematics could give to make efficient their interaction is to read the basic ideas of linguistic structuralism through the glasses of what I described above as the mathematical way of thinking, by taking into account the mathematical models coming from the field of automatic translation. This project was realized in 1967 in two books, the first in French, the second in English; while in the former the main attention was directed towards linguistic structuralism, in the latter we tried to order in a uniform framework the European mathematical models inspired by automatic translation. To give some examples in this respect, we realized that the Trubetzkoy-Cantineau system of oppositions, Hjelmslev's system of relations and Harris' system of different types of distributions are isomorphic, so they can be captured in a unique framework, which in its turn permits to enlarge Dobrushin's model of elementary grammatical categories, in order to represent the phenomena of contextual ambiguity in their most general form (Marcus ed., 1981, 1983). Various types of syntactic projectivity introduced in Western Europe and in U.S.A. were captured in a unique framework and articulated with the graph-theoretic model of a grammatical proposition, due to M.I. Beleckii, V.M. Grigorjan and I.D. Zaslavskii. Two basic different approaches to the concept of phoneme, one based on binary distinctive features, the other on phonemic sequences, received their clear formal status, with the surprising fact that the relation between two sequences belonging to the same phoneme is not transitive. Different approaches to the concept of morpheme, the set-theoretic models of part of speech and of grammatical case were critically examined. One of the most interesting concept emerging from the preliminaries to automatic translation, the concept of syntactic configuration of various orders, received a uniform presentation, including all its variants (Kulagina, Gladkii, Novotny) and leading to some new ones (further ideas and systematization are due to Maria Semeniuk-Polkowska).

The Impact of some Political and Ideological Factors

Local conditions in Romania were difficult in the emergence period of information sciences, mainly in the fifties and in the sixties. Communication with scientists from Western countries were very reduced, while the necessary scientific journals and books available in Romanian universities were to a large extent those from Eastern Europe. We had to wait for the Russian translations of the most important Western books and articles. But in some respect we were lucky, because the policy of the communist party was to include the computer revolution as a basic component of the communist society. Romanian scholars such as the mathematician Gr. C. Moisil and the linguist Al. Rosetti knew how to take advantage of this opportunity. They articulated their efforts to make possible the early development of mathematical and computational linguistics at the University of Bucharest. In 1962, they founded the journal "Cahiers de Linguistique Théorique et Appliquée", specially devoted to this line of research. They also introduced regular courses on mathematical and computational linguistics at the University of Bucharest, which were chronologically among the first in the world in this respect. Professor Moisil, the founder of the Romanian school of mathematical logic, with fundamental contributions related to non-classical logics, was very efficient in attracting young mathematicians at the crossroad of logic, linguistics and computer science. He challenged us with new topics and problems, many of them very provocative. To give only one example in this respect, I will mention the following fact: Short time before his death, in May 1973, he told me, during a walk, that Gabriel Sudan, a former PhD student of Hilbert, is the author, before Ackermann, of an example of a recursive function which is not primitive recursive. He promised to give me more details with another occasion, which however never arrived. I told this fact to Cristian Calude, then one of my best students, and, after a long investigation, he succeeded to discover the corresponding result in an article by Sudan, whose title and introduction did not at all promise to hide this fact. Now, Ackermann and Sudan are both recognized as authors of a first example (Sudan's example is different from that proposed by Ackermann) of a recursive function which is not primitive recursive. But, as it usually happens, the search is richer than the object found. This exercise was enough for Calude to discover the pleasure of research in the field of recursiveness, which became one of his main fields of interest, as he confessed to me.

Chomsky Grammars. From Linguistics to Computer Science

In the late fifties, Chomskian linguistics emerged, immediately recognized as a revolution in the scientific humanities of the XXth century. The book "Syntactic Structures" published by Mouton (The Hague) in 1957 was initially received as a purely linguistic event, claiming a new, generative-transformational approach to language, as opposed to the analytical one, promoted by comparative-historical linguistics and by classical structural linguistics. Instead of accepting as given the well-formed strings of English and analysing their structure, Chomsky sees human language as the result of the activity of a hypothetical machine, called "generative-transformational grammar". Linguistics became for Chomsky a chapter of cognitive psychology, elaborating hypothetical-explanatory models of human linguistic competence. Despite the fact that all motivations brought by Chomsky were of linguistic nature, most of his pioneering articles were published in non-linguistic journals such as "IRE Transactions on Information Theory" and "Information and Control" and in a non-linguistic handbook such as "Handbook of Mathematical Psychology". How should we interpret this fact? Did Chomsky suggest in this way that his approach could be relevant to Shannon's information theory? This link remained so far rather weak. The only articles supporting this link are those in collaboration with G.A. Miller and they refer more or less explicitly to Markov processes. More convincing is the choice of a handbook of mathematical psychology (1963) for the publication of two other articles, because this choice was in conformity with his view of linguistics as a branch of cognitive psychology. Today, it appears very ironical the fact that Chomsky avoided to publish in journals of computer science (there are only two exceptions in this respect, one of them being his famous article in collaboration with M.P. Schutzenberger), despite the tremendous relevance of his grammars for the theory of programming languages and, generally, for theoretical computer science. Indeed, S. Ginsburg and H.G. Rice proved in 1961 that the Chomskian context-free grammars are equivalent to Backus normal forms defining the syntax of the programming language ALGOL 60, while R.W. Floyd has shown that, taking into account some semantic aspects of ALGOL 60, some context-sensitive rules which are not context-free are necessary too. Similar situations appear with virtually any programming language. The discovery of this fact made from Chomsky hierarchy of grammars and languages the basic tool to investigate the syntax and the semantics of programming languages; what is understood by the theory of programming languages is to a large extent the theory of formal languages, conceived in respect to Chomsky hierarchy. The standard presentation of this theory, in its form recognized today is due to Arto Salomaa (1973). The years 1960–1961 represent a turning moment in the evolution of Chomskian grammars; initially motivated by studies of natural languages, context-free and context-sensitive grammars prove their relevance to programming languages, as we pointed out above, while their role in linguistics is weak, being considered, to a large extent, inadequate for natural languages. In a second step, in the eighties, the question of non-context-freeness of natural languages was raised once more, due to the fact that the initial argument proposed by Chomsky, as well as some other arguments against the context-freeness of natural languages, are no longer accepted. In this way, context-free and context-sensitive grammars keep their importance to both natural and programming languages. However, this fact, generally accepted in theoretical computer science and particularly in computational linguistics, seems to be still ignored by many linguists.

Is Formal Linguistics a Pilot Science?

Going back to the period of the sixties, I remember that Chomsky's approach was popular among linguists mainly by the transformational component of his grammars, added to what was called a phrase structure grammar. However, in the field of computational linguistics a more balanced view prevailed, leaving room to both analytical and generative-transformational aspects. Beginning with the sixties, I tried to order all directions leading to a bridge between linguistics, computer science and mathematics and I identified one coming from linguistics (linguistic structuralism and various fields of applied linguistics), another from mathematics (free monoids, free semigroups, various combinatorial problems going from Gauss to Langford and the well known Thue combinatorial systems, number theory such as Conway's iterative reading of numbers), a third from logic (what we call today a "formal language" and a "formal grammar" can be associated with a Hilbert formal system as well as with any variant of modeling the idea of computation, such as Turing machine, Markov normal algorithm, recursive function etc.), another from automata theory (Rabin-Scott's article published in "IBM Journal of Res. and Dev.", in 1959 was the most accurate in this respect) and, related to it, from biology (mainly S.C. Kleene's "Representation of events in nerve nets and finite automata", 1956) and molecular genetics (Marcus, 1974). I remember that already in 1946 Claude Levi-Strauss proposed the slogan "linguistics as a pilot science" and I used it as a title for one of my articles (Marcus, 1969, 1974). My main idea was to consider not linguistics, but formal linguistics as a pilot science, involving also the association of the left hemisphere of the brain with sequential structures (mainly language and logic). I supposed that Roman Jakobson had a similar idea in mind when he stated that linguistics is the mathematics of social sciences. It happened that in the same year when I published my article "Linguistics as a pilot science" another article, with the same title, was published by Joseph Greenberg, the famous specialist in linguistic universals, arguing (implicitly polemical with Levi-Strauss and Jakobson) that the respective slogan is wrong. He had in view some failures of classical structuralism in its attempts to transfer from linguistics into other social sciences some concepts, methods and results. Examples were selected just in order to illustrate failures, while successes were ignored. My strategy was just the opposite: to point out the important successes (such as molecular genetics and programming languages); moreover, I argued that the function of a pilot science appears under the presupposition that linguistics increases its degree of formalization, a fact ignored by Greenberg and by other critics supportive of the mentioned slogan (for instance, Walter Koch).

Towards Contextual Grammars

My first monograph "Lingvistica matematică" was published in 1963. I realized only later that it was perhaps the first attempt to put in a systematic form the possibility to bridge linguistics and mathematics. Probably, this was the reason that it was required for translation in French, English, Russian and Czech, but I used this opportunity to publish in these languages completely new versions; the French version, "Introduction mathématique à la linguistique structurale" was published by Dunod, Paris, in 1967; the English version, "Algebraic Linguistics", appeared at Academic Press, New York in 1967; the Russian version, "Teoretiko mnojestvennye modeli jazykov", appeared in 1970 at Ed. Nauka, Moscow; the Czech version, "Algebraicke modely jazyka", appeared in 1969 at Ed. Academia, Prague. In 1964, I published in Romanian "Gramatici şi automate finite", devoted exclusively to different variants of regular grammars (type 3 in Chomsky hierarchy). A special section of this book was devoted to some characterizations of regular languages in terms of analytical models, but I did not push this study further. However, it became clear in the sixties that several authors were involved in bridging analytical and generative (Chomskian) models, A.V. Gladkii and M. Novotny being perhaps the most important of them. At the same time, some alternative generative models were developed with significant impact in computational linguistics (dependency grammars by David G. Hays), in logic (categorical grammars, by H. Gaifman, preceded by the syntactic calculus proposed by Joachim Lambek), in cellular biology (Lindenmayer systems, later with impact in other fields such as computer graphics). Under the influence of these facts, I tried to build a generative counterpart of contextual analysis and I proposed, at the International Conference on Computational Linguistics in Senga Saby (near Stockholm), in 1968, a new type of generative device I called "contextual grammars". The name was not appropriate, because it was confused with context-sensitive grammars. My intention was, in the spirit of that time, to connect the idea of contextual analysis (developed by Dobrusin, Sestier, Jurgen Kunze and myself) and that of a generative device; in other words, to transform contextual analysis in a generative machine. All concepts involved in this enterprise emerged from the structural analysis of natural languages. Contextual grammars were conceived just as one more way to bridge the analytical and the generative approach to the grammar of a natural language. However, in my article published in 1968 I avoided any explicit linguistic motivation of contextual grammars, giving directly their technical definition. Perhaps this is the reason why this formalism had no linguistic impact; in exchange, it became attractive for many theoretical computer scientists. Several surveys of this research were published, from time to time, by Gheorghe Păun, one of the main contributors in this field; his most recent survey "Marcus Contextual Grammars" has appeared at Kluwer Academic Publisher (Dordrecht, Holland, 1997). In the meantime, I devoted a special chapter to the linguistic aspects of contextual grammars in the second volume of the "Handbook of Formal Languages" (eds. G. Rozenberg, A. Salomaa), published by Springer in 1997. The link between contextual grammars and natural languages, for a long time ignored, is now a topic of increasing interest, as it can be seen from a series of articles considering various extensions of contextual grammars and the way they challenge other types of grammars, in respect to their linguistic relevance.

Texts, Contexts, Intertexts and Hypertexts

Contextual grammars also have an interesting mathematical aspect. Strings and contexts interact symmetrically, within the framework of a Galois connection. Despite this fact, in all already considered variants of contextual grammars the symmetry between strings and contexts is not respected. In order to fill this gap, I proposed some new variants, where the same grammar is able to generate both strings and contexts; moreover, I tried to articulate texts, contexts, intertexts and hypertexts, taking into account the development of text theory in both its linguistic-literary variant (for instance that given by Teun A. van Dijk) and its mathematical computer science variant (see the series of articles published by A. Ehrenfeucht and G. Rozenberg, some times in collaboration with P. Ten Pas and/or H.J. Hoogeboom). The intertext originated in M. Bakhtin's dialogic principle: each text is in a more or less explicit dialogue with other texts. Although the idea of intertext appeared in the framework of literary studies, it seems that the corresponding phenomenon is more and better visible in scientific texts (see the quotations and the bibliographic references). Hypertexts emerge from the need to transgress the sequentiality of usual texts, making possible all paradigmatic connections related to the elements of a text; this possibility is a consequence of recent computational capabilities developed in the field of hypermedia. Ultimately, this line of research leads to the need to develop suitable types of contextual grammars, where arbitrary strings are replaced by texts, while contexts are restricted to ordered pairs of texts. It is important to observe that "text" became in the last decades a real universal paradigm, leading to the vision of the world as a text. This fact brings a new argument in favor of the status of linguistics as a pilot science. As a matter of fact, already in the field of molecular genetics the idea of a text emerges from the need to consider higher levels of organization of DNAs and proteins. The natural trend of quantification of our field of knowledge leads to the need to find a basic alphabet of irreducible units and to use the sequential combinatorial capacities of these units—it seems that this process cannot avoid the fundamental pattern of our mother tongue.

Contextual Grammars and Chaitin-Kolmogorov Complexity. Another Perspective

Contextual grammars are naturally linked to reading processes and, through them, to Chaitin-Kolmogorov complexity. In this respect we are inspired by

Walter J. Savitch (1993), who was concerned mainly with Chomskian grammars. However, we claim that contextual grammars are more suitable for a natural and simple definition of algorithmic complexity, due to their intrinsic character (the absence of any auxiliary symbol). The intuitive idea in our approach is the remark that we read any (finite) text by extending it to an infinite one (this fact could be a way to interpret Umberto Eco's "opera aperta"). This happens because infinity is more structured than finiteness. Infinity can be understood only by an explicit rule and expliciteness requires, in its turn, simplicity. The most convincing example in this respect is the fact that we need n rules to generate the set $A(n)$ of the first n natural numbers, but we need only two rules to generate the set A of all natural numbers (this fact is true for both Chomskian grammars and contextual grammars). This means that the set $A(n)$ accepts a reading that is shorter than its size n (= cardinal of $A(n)$), as soon as we interpret the extension A of $A(n)$ as a reading of $A(n)$; indeed, the size of A is equal to 2. We adopted here the convention to define the size of a grammar as the number of its rules; it happens that for $A(n)$ this number is also the cardinal of $A(n)$. So, we are lead to the following general problem: given a finite language $L(1)$ on the alphabet V and a class H of generative devices on V (a device G in H consists of a finite set of rules by means of which we generate a language on V), we define a reading of $L(1)$ with respect to H to be any infinite language L generated by G in H, such that the finite language L_n of those strings in L whose length is not larger than the length n of the longest string in $L(1)$ is just $L(1)$. For instance, in the above example A is a reading of $A(n)$. Suppose we can define in a reasonable way the size of any device in H. In this case, the size of any language generated by a device in H is defined as the smallest possible size of such a device. The size of a finite language can be defined independently of the class H, either as its cardinal number or as the length of its longest string. This is the intrinsic size of the given finite language. The size is said to be H-consistent if any finite language can be generated by a device in H and its intrinsic size is identical to the size resulting from the definition above with respect to H. Now the interesting problem is to compare the intrinsic size of a given finite language $L(1)$ with the sizes of the possible readings of $L(1)$ in respect to H. Does there always exist a reading of $L(1)$ with respect to H? The following two conditions are sufficient for the existence of such a reading: a) any finite language on V has a grammar in H; b) the universal language on V has a grammar in H. These conditions are fulfilled, for instance, when H is any class of grammars in Chomsky hierarchy or when H is the class of simple contextual grammars. We will assume the existence in H of at least one reading of $L(1)$. However, the possibility of absence of any reading of $L(1)$ with respect to some choices of H deserves attention, as a possible approach to the lack of coherence or to the lack of cohesion of a text. Let us denote by size (reading of $L(1)$) the smallest possible size of a reading of $L(1)$. If it is larger than the size of $L(1)$, then we consider $L(1)$ as partially readable with respect to H. The degree of unreadability is given by the difference: size (reading of $L(1)$) - size ($L(1)$). Let us observe that this difference, if it is strictly positive, takes usually only small values; this is the case when the conditions a) and b) considered above are satisfied. In this case we can work with a grammar G in H obtained by adding to the rules used to generate $L(1)$ the necessary rules to generate the universal language on V, from which we have eliminated the strings whose lengths are not larger than the maximum length of strings in $L(1)$. If size (reading of $L(1)$) = size ($L(1)$), then $L(1)$ has the highest possible complexity with respect to H and it is close to the limit of readability (it corresponds, mutatis mutandis, to the status of a random string considered as the case of maximum algorithmic complexity in the Chaitin-Kolmogorov approach). However, in most situations size (reading of $L(1)$) is strictly smaller than size ($L(1)$) and the difference size ($L(1)$) - size (reading of $L(1)$) is a measure of the efficiency of the reading process of $L(1)$ with respect to H.

Reading via Simple Contextual Grammars

Now let us consider as H the class of simple contextual grammars on V. Such a grammar is composed of a finite set L_1 of strings and a finite set C_1 of contexts on V. Among various possibilities to define the size of such a grammar G, we choose size $G = \max(s, c)$, where s is the length of the longest string in L_1 and c is the length of the longest context in C_1 (the length of the context $< x, y >$ is the sum of the lengths of x and y). An alternative possibility is to take size of G to be $s + c$. Let us recall that the language generated by G is the smallest language L containing L_1 such that if $x \in L$ and $< u, v > \in C_1$, then $uxv \in L$. Let us take the example of the language $L_n = \{a_1, a_2, \ldots, a_n\}$ on $V = \{a\}$. It has the intrinsic size equal to n while the grammar $G = < V, \{a\}, \{< \lambda, a >\} >$ (where λ is the null string) has the size equal to 1 and it is a reading of L_n, because $L(G) = \{a_n \mid n = 1, 2, \ldots\}$; so, the efficiency of the reading process is equal to $n - 1$. If $L_1 = \{a_1, a_3\}$, then the intrinsic size of L_1 is 3, while the grammar $G = < \{a\}, L_1, \{< \lambda, \lambda >, < \lambda, a_2 >\} >$ has the size equal to 3 and leads to a reading of L_1 with size (G) = size (L_1) =3. However, there is a more efficient reading of L_1, given by the grammar $< \{a\}, \{a\}, \{< \lambda, \lambda >, < \lambda, a_2 >\} >$, whose size is equal to 2, while no reading of L_1 and of size 1 exists; so, the efficiency of the reading process is here equal to $3 - 2 = 1$. Combinatorial arguments may lead to the characterization of those finite languages on $\{a\}$ accepting an efficient reading, but the problem becomes more difficult when the cardinal of the alphabet takes values larger than one. Further steps are necessary to extend this approach to more general types of contextual grammars.

Back to Poetry

The situation we described as being at the limit of the possibility of a reading process brings back the case of poetry, having just the aim to realize maximum of semantic density by means of a minimal expression. In other words, in an ideal piece of poetry no abbreviation is possible, nothing can be eliminated, modified or added. The highest complexity associated with a piece of poetry goes in a happy marriage to its hidden simplicity; but the latter is obtained only as a result of the reading process. Talking about poetry, it is the right moment to recall the beginning of this story. We never forget our first love and its name was, for me, poetry. As a matter of fact, language and linguistics were a way to reach a deeper understanding of poetry for me. My first approach to the edification of a field that could deserve to be called "mathematical poetics" happened in the late sixties and the resulting monograph "Poetica matematică" published in Romanian in 1970, knew a German improved edition in 1973, at Athenaeum Verlag, Frankfurt/Main. Links between poetry and mathematics are known for a long time, but most investigations in this respect were related mainly to quantitative-statistical aspects. My aim was to build a theory of the poetic language seen as a continuous, topological structure, in contrast with the discrete, algebraic nature of the scientific (mathematical) language. My guide was the tradition represented by George D. Birkhoff (the father of ergodic theory), Matila C. Ghyka and Pius Servien, all three deserving to be considered as founders of the new field of mathematical aesthetics. Birkhoff, with his famous measure of the beauty of an artistic work, given by the ratio O/C, where O is the order and C is the complexity of the considered work, had to be reinterpreted in the light of today's information theory and still waits to be reconsidered in the perspective of Chaitin-Kolmogorov algorithmic complexity. Ghyka's approach via the common denominator of organic growth and artistic creativity through the glasses of the golden proportion and of Fibonacci sequences becomes now a particular case of the generative approach, via different types of grammars and machines (mainly the so-called picture grammars). Pius Servien idea to replace the traditional opposition between the poetic language and the ordinary one with the opposition poetic-scientific was a starting point. Our first approach to this opposition was reconsidered and improved in several steps (mainly in our book "Art and Science", in Romanian, Ed. Eminescu, Bucureşti, 1986), culminating with our plenary lecture at the Fifth Congress of the Hellenic Semiotic Society (Thessaloniki, May 1997), where we proposed a way to place all the oppositions between poetry and science within the framework of their similarities. The formal language point of view was adopted in the study of most creative processes, according to our slogan "formal linguistics as a pilot science". In a sequence of articles and books, I tried (in most cases in collaboration with some of my colleagues) to point out the hidden creative machine (usually, under the form of various types of grammars in Chomsky hierarchy) of a work of art, for instance, in the study of fairy-tales (S. Marcus, ed., "La semiotique formelle du folklore", Klincksieck, Paris, 1978), of theater (S. Marcus, ed., "The formal study of drama", special issues of the journal "Poetics", Amsterdam, 1976 and 1984), of visual arts (S. Marcus, ed., "Semiotica matematică a artelor vizuale", in Romanian, Ed. Ştiinţifică, Bucureşti, 1981), of music (see the sequence of articles in "Romanian Review", in the eighties). A special word deserves the mathematical linguistic approach to theater. The starting idea was to consider a Boolean matrix associated to a theatrical play; the matrix has m lines and n columns, where m is the number of the characters and n is the number of scenes in the play. At the intersection of line i and column j (i between 1 and m, j between 1 and n) we place the digit 1 if the ith character appears in the jth scene and the digit 0 in the contrary case. A very rich information is obtained by adequately processing this matrix. Due to its simplicity and its operational and intuitive nature, the proposed tool became very popular among researchers in the field of theatrical semiotics, being quoted, used and improved by many authors.

Transdisciplinarity. Between Risk and Chance

Trying to bridge such heterogeneous fields made my social life difficult enough. Mainly in the fifties and in the sixties, some of my colleagues mathematicians and computer scientists were reluctant towards my enterprise involving several fields usually considered far away from each other. When you try to connect two fields, you need to be in both of them, but you risk to be considered in none of them. Fortunately, in the sixties I was already the author of a large number of research articles in the field of mathematical analysis, well received by the experts, so my status as a mathematician was beyond doubt. Traditionally, mathematicians conceive their field in two variants: theoretical (pure) and applied, the latter being related mainly to mechanics, physics and chemistry. Pure and applied mathematics are considered in interaction and it is a great honor for a theorist to be able to obtain results with some relevance in physics. Gradually, biology too was accepted as a field where mathematics could have some impact. But, already in the XlXth century, mathematics showed its relevance for economics and to some extent, for psychology too. However, for the predominant mentality, humanities remain in a kind of incompatibility with exact sciences. My first attempt to bridge linguistics with mathematics under the label "mathematical linguistics" came as a shock for many people, good to be exploited in the newspapers and by other media, but not to be trusted. This happened despite the fact that structural linguistics, very popular among linguists, prepared the way to mathematization (structuralism was always a preliminary step to mathematical modeling). Paradoxically, it seems that this shock contributed to the social success of mathematical linguistics, which became in the meantime "computational linguistics". In the sixties, the skepticism was also nourished by the generally recognized failure of the huge projects of automatic translation. However, it happened that, beyond their initial motivation, the mathematical and computational models of languages proved to have an important linguistic impact (to give only one example, syntactic projectivity became a basic tool in the study of dependency grammars). On the other hand, as we have already pointed out, Chomskian models, initially conceived exclusively for linguistics, became a basic tool in the study of programming languages. Now we can add that the new field of computational biology is also using the experience of formal generative grammars. Such unexpected evolutions show that the reasons to connect linguistics and computer science (via mathematics) are multiple and very deep. Let us recall that important approaches in historical linguistics use the DNA metaphor, while the latter is a term of reference for non-conventional computation, and its understanding is via formal linguistic models. Computational linguistics is now a basic component of Artificial Intelligence. Journals such as "Linguistics and Philosophy", "Theoretical Linguistics", "Computational Linguistics", "Mathématiques, Informatique et Sciences Humaines" and many others contain regularly articles involving a mathematical and computational formalism. So, mathematics no longer needs some rhetorical means to reach the attention of specialists of language and of computer science. A similar process occurred in the field of poetics. Journals such as "Poetics", "Leonardo", "Computers and the Humanities", "Cahiers de Linguistique Théorique et Appliquée", "Symmetry Culture and Science" publish regularly articles bridging art, mathematics and computer science. The project of unification of knowledge, started already by the Vienna Circle, in the twenties and in the thirties of our century, is now strongly stimulated by the development of information sciences and particularly by theoretical computer science.

More References

N. Chomsky, G.A. Miller, 1958, Finite state languages, *Information and Control* 1, 91-112.

N. Chomsky, G.A. Miller, 1963, Finitary models of language users, in *Handbook of Mathematical Psychology* II, 419-491.

N. Chomsky, M.P. Schutzenberger, 1963, The algebraic theory of context-free languages, In P. Braffort, D. Hirschberg (eds.) *Computer Programming and Formal Systems*, North Holland, Amsterdam, 118-161.

S. Marcus, 1969, Lingvistica, ştiinţa pilot, *Studii şi Cercetări Lingvistice* 3, 235-245. (In Romanian)

S. Marcus, 1974, Linguistics as a pilot science. In Th. A. Sebeok (ed.) *Current Trends in Linguistics* 12, 2871-2887.

S. Marcus, 1974, Linguistic structures and generative devices in molecular genetics, *Cahiers de Linguistique Théorique et Appliquée* 11, 2, 77-104.

S. Marcus (ed.) 1981, 1983, *Contextual Ambiguities in natural and in Artificial Languages*, Communication and Cognition, Ghent, Belgium.

A. Salomaa, 1973, *Formal Languages* Academic Press, New York.

J. W. Savitch, 1993, Why it may pay to assume that languages are infinite, *Annals of Mathematics and Artificial Intelligence* 8, 17-25.

W. Kuich introducing
H. Maurer to A. Salomaa.

French connection is
always rational.
S. Eilenberg,
J.F. Perrot,
M. Nivat.

R. Book and A. Aho.

Quasi-uniform events. G Rozenberg.

L conferences attracted many people. S. Ulam.

Almost everybody smoked.
A. Salomaa,
E. Schmidt,
H. Maurer.

Ties of computability.
A. Salomaa,
F. Gécseg,
L. Kálmar.

The late M. Schützenberger,
Turku ICALP.

Early L meeting.
A. Salomaa, J. Opatrny, K. Culik, and A. Lindenmayer.

Complexity group, Copenhagen, 1996 (l–r):
J. Casti, G. Markowsky, G. Chaitin, C. Calude, and W. Meyerstein.

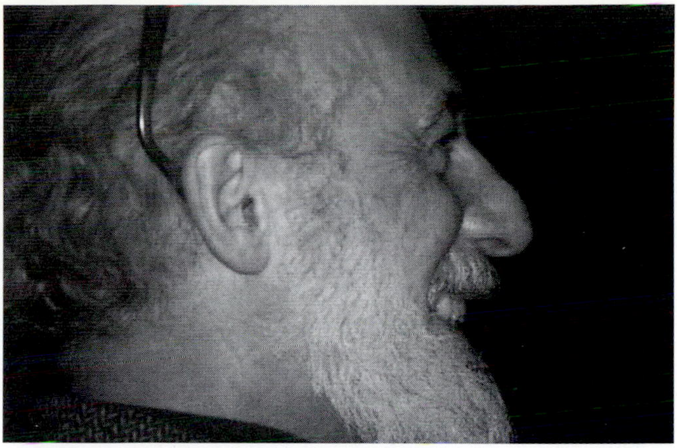

A. Lindenmayer.

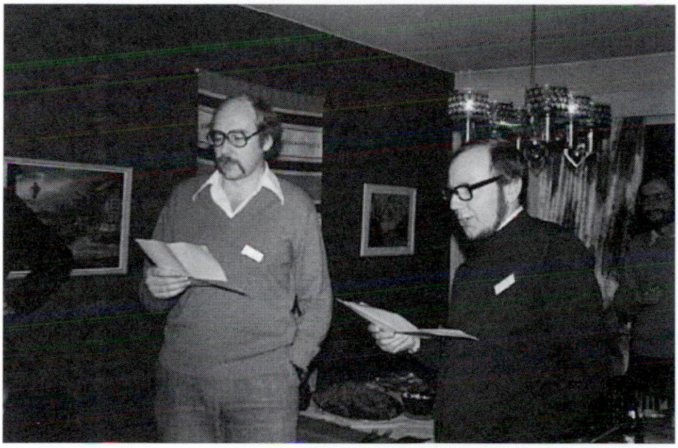

Sauna poetry.
D. Wood,
H. Maurer.

S. Eilenberg.

Gr.C. Moisil and
S. Marcus, 1973.

A.J. Greimas and
S. Marcus, 1983.

B.A. Trakhtenbrot and
A.V. Kuznetsov, 196?.

V.Yu. Sazonov,
J.M. Barzdins,
R.V. Freivalds,
A.Y. Dikovski,
B.A. Trakhtenbrot,
M.K. Valiev,
M.I. Kratko,
V.A. Nepomnyashchy,
Concurrency, 1992.

M.N. Sokolovski,
M.I. Dekhtyar,
M.K. Valiev,
B.A. Trakhtenbrot,
V.Yu. Sazonov,
I.A. Lomazova,
V.N. Agafonov,
Volga, 1992.

Row 1: Z. Galil, Yu. Gilderman, Mrs. B. Trakhtenbrot, B.A. Trakhtenbrot, A. Yehudai, R. Constable, A. Pnueli.
Row 2: K. Wagner, A. Avron, A. Rabinovich, G. Plotkin, A. Meyer, V. Pratt.
Row 3: ?, V. Tannen, ?, M. Broy, W. Reisig, ?, Tel-Aviv, 1991.

Concentration always helps.
W. Kuich and K. Salomaa.

Z. Manna and J. de Bakker.

Probabilistic automata.
A. Salomaa and A. Paz.

Yuri Matiyasevich

Professor Yuri Matiyasevich was born on March 2, 1947 in Leningrad, the USSR. In 1969 he graduated from *Department of Mathematics and Mechanics* of *Leningrad State University* and continued his study as post-graduate student at the *Steklov Institute of Mathematics, Leningrad Branch (LOMI)*. From 1970 till now he works in this Institute, currently as the Head of the Laboratory of Mathematical Logic.
The name of Yuri Matiyasevich became known worldwide in 1970 when he completed the last missing step in the "negative solution" of *Hilbert's tenth problem*. His book on this subject, originally published in Russian, was translated into English (The MIT Press, 1993) and French (Masson, 1995).
Yuri Matiyasevich is *Docteur Honoris Causa* de l'Université d'Auvergne, France (1996) and *Correspondent Member* of the Russian Academy of Sciences (1997).

Hilbert's Tenth Problem: A Two-way Bridge between Number Theory and Computer Science

Historical Background

Number Theory is one of the oldest branches of mathematics, which itself is one of the eldest among the sciences. In contrast, computer science is relatively young. However, it did not emerge on empty ground. Computer Science takes its stems from mathematics, more specifically, from mathematical logic. So if we want to trace the origin of computer science we need to look into the history of mathematics.

Elements of the future computer science can be found at very early stages of the development of mathematics. The word "algorithm" was derived from the name of great arab scientist Al-Khorezmi who lived in the 9th century A. C. However, the idea of an algorithm has a longer history. Every mathematician knows *Euclid's algorithm* for finding the greatest common divisor of two natural numbers. Euclid lived in the 4th century B. C. but the basic algorithms for performing arithmetical operations were invented much before Euclid.

Hilbert's Tenth Problem

Mathematics is a science to a great extent driven by problems. For some of them it took decades of years to be solved.

In 1900 scientists from many countries gathered together in Paris for the *Second International Congress of Mathematicians.* One of the invited lectures was to be delivered by the great German mathematician David Hilbert. It was the last year of the passing century, and he decided to survey the most important, in his opinion, open problems in mathematics which the pending 20th century was to inherit from its predecessor. Hilbert's famous paper *Mathematische Problemen* [18] lists 23 problems (in fact, most of them are collections of related problems).

One, and only one, of these 23 problems can be recognized today as a problem in computer science. This is the tenth problem. The section of Hilbert's paper devoted to this problem is so short that can be reproduced here entirely.

> **10. Entscheidung der Lösbarkeit einer diophantischen Gleichung.**
> Eine diophantische Gleichung mit irgendwelchen Unbekannten und mit ganzen rationalen Zahlkoefficienten sei vorgelegt: *man soll ein Verfahren angeben, nach welchen sich mittels einer endlichen Anzahl von Operationen entscheiden läßt, ob die Gleichung in ganzen rationalen Zahlen lösbar ist.*[1]

Of course, Hilbert and his contemporaries viewed this problem as a problem in number theory.

The range of unknowns. Solving equations was a major occupation for mathematicians from ancient times. The equations mentioned in the tenth problem were named after the Greek mathematician Diophantus who lived in the 3rd century A. C. They have the form

$$D(x_1, \ldots, x_m) = 0, \tag{1}$$

[1] **10. Determination of the Solvability of a Diophantine Equation.** Given a diophantine equation with any number of unknown quantities and with rational integral numerical coefficients: *To devise a process according to which it can be determined by a finite number of operations whether the equation is solvable in rational integers.*

where D is a polynomial with integer coefficients. The novelty which Diophantus introduced into the study of polynomial equations was as follows: he restricted the range of admissible values of unknowns to (positive) rational numbers. So for Diophantus the equation $x^2 = 2$ had no solution which contrasted with the previous tradition of solving equations in (positive) real numbers geometrically (represented by segments of the straight line).

In contrast with Diophantus, Hilbert—in the tenth problem—asks about solving Diophantine equations in *"rational integers"*. This terminology may sound strange and misleading to computer scientists. In fact, Hilbert had in mind nothing else but the familiar to everyone integers $0, \pm 1, \pm 2, \ldots$. He used the name *rational integers* because the term *integers* can be understood in a broader sense of *algebraic integers.* While the latter are very interesting and important objects, they are entirely outside of the scope of this paper, and we will use the word *integer* in its "elementary school sense" without risking a terminology confusion.

Very often the range of unknowns is further restricted to *natural numbers.* By the latter one understands either positive integers 1,2,3,... (tradition of number theory) or non-negative integers 0,1,2,... (tradition of mathematical logic). In this paper, we will adhere to the number-theoretical understanding of natural numbers.

Considering Diophantine equations only with unknowns in natural numbers in fact does not restrict the generality of results. It is easy to see that solving the equation

$$D(z_1, \ldots, z_m) = 0 \tag{2}$$

in integers is equivalent to solving equation

$$D(x_1 - y_1, \ldots, x_m - y_m) = 0 \tag{3}$$

in natural numbers. Reduction in the other direction is a bit less evident: solving an equation

$$D(x_1, \ldots, x_m) = 0 \tag{4}$$

in natural numbers is equivalent to solving the equation

$$D(s_1^2 + t_1^2 + u_1^2 + v_1^2 + 1, \ldots, s_m^2 + t_m^2 + u_m^2 + v_m^2 + 1) = 0 \tag{5}$$

in integers because every positive integer is the sum of the squares of four integers. Technically, it is often more convenient to work with natural numbers only, and from now on we restrict the range of unknowns to positive integers.

Sometimes the range of unknowns is considered as the only distinctive feature of Diophantine equations and then one calls "Diophantine" an equation of arbitrary form as soon as only integer or natural number solutions of it are of interest. However, we will always understand by a "Diophantine equation" an equation of the form (1).

The tenth problem as a decision problem. Why Hilbert considered solving Diophantine equations as an open problem? In fact, since Diophantus time number-theorists have found solutions for plenty of Diophantine equations and also have proved the unsolvability of a large number of other equations. Unfortunately, for different classes of equations, or even for different individual equations, one had to invent different specific methods. In the 10th problem Hilbert asked for a *universal* method for recognizing the solvability of Diophantine equations.

In today's terminology we consider this problem as a *decision problem.* This means that the problem consists of infinitely many subproblems (specified by particular equations) each of which requires an answer "YES" or "NO" ("there is" or "there is no" solution). An expected solution to the problem should be an algorithm applicable to an arbitrary equation and producing the correct answer.

A remark is appropriate here. Hilbert did *not* ask whether such an algorithm exists. On the one hand, he seemed to be an optimist in mathematics and, most likely, he was sure of the existence of such an algorithm (he posed his problems much before the revolutionary works of K. Gödel). On the other hand, Hilbert even did not use the word "algorithm" in the statement of the problem. Instead, he used a rather vague wording "*a process according to which it can be determined by a finite number of operations ...*". He could have used the *word* "algorithm" but it would have not helped much in clarifying the statement, because at that time there was no rigorous *general notion* of an algorithm. What was known were different examples of particular mathematical algorithms like Euclid's algorithm for finding GCD.

The absence of a general definition of an algorithm was not by itself an obstacle for a positive solution of Hilbert's tenth problem. If somebody would have invented the required "*process*" it should have been clear that in fact the process does the job.

The situation is essentially different if there is no required algorithm as it turned out to be the case with Hilbert's 10th problem. To prove this fact, or even to state it rigorously, one need a definition of an algorithm.

So this was a point where computer science could have come for help to number theory in clarifying difficulties arose in solving Diophantine equations. However, there was no computer science at that time, and even the mathematical logic, a connecting link between mathematics and computer science, was yet to mature.

It is interesting to see that Hilbert foresaw the possibility of the future development of the 10th problem. He wrote [18]:

> Occasionally it happens that we seek the solution under insufficient hypotheses or in an incorrect sense, and for this reason do not succeed. The problem then arises: to show the impossibility of the solution under the given hypotheses, or in the sense contemplated. Such proofs of impossibility were effected by the ancients, for instance when they showed that the ratio of the hypotenuse to the side of an isosceles triangle is irrational. In later mathematics, the question as to the impossibility of certain solutions plays a preeminent part, and we perceive in this way that old and difficult problems, such as the proof of the axiom of parallels, the squaring of circle, or the solution of equations of the fifth degree

by radicals have finally found fully satisfactory and rigorous solutions, although in another sense than that originally intended. It is probably this important fact along with other philosophical reasons that gives rise to conviction (which every mathematician shares, but which no one has as yet supported by a proof) that every definite mathematical problem must necessary be susceptible of an exact settlement, either in the form of an actual answer to the question asked, or by the proof of the impossibility of its solution and therewith the necessary failure of all attempts.

It was possible to prove the unsolvability of the above mentioned classical problems because they contained an explicit description of admissible tools: a fixed axiomatic system, constructions with ruler and compass or expressions in radicals. However, in Hilbert's tenth problem the admissible tools were not fixed. So the failure in finding a universal method for Diophantine equations has been a stimulus for developing a general notion of an algorithm.

Such a definition emerged much later, only in the 30's of this century in the works of Kurt Gödel, Alan Turing, Emil Post, Alonzo Church and other logicians. However, I cannot trace any *direct* influence of Hilbert's tenth problem on this process.

Soon after the appearance of the rigorous notion of algorithm and *Church Thesis*, the first algorithmically undecidable problems were found, first in mathematical logic itself. In 1947 another breakthrough took place. Andrei Markov [33] and Emil Post [49] proved that so called *word problem* for semigroups was undecidable. This problem was stated by Axel Thue [58] in 1914 and is known also as *Thue's problem.* The importance of this result comes from the fact that it was the first decision problem which arose naturally in mathematics and was finally shown undecidable.

Davis' conjecture

After the success in the proof of the undecidability of the Thue problem and all failures to find a decision procedure for Diophantine equations it was quite natural to suspect that Hilbert's tenth problem is undecidable as well. The American mathematician Martin Davis [6, 7] stated at the beginning of 50's a much stronger hypothesis.

Diophantine sets. In order to be able to present Davis's Conjecture we need more terminology. Besides single Diophantine equations, number-theorist consider also *families of Diophantine equations.* Such a family is an equations of the form

$$D(a_1, \dots, a_n, x_1, \dots, x_m) = 0, \tag{6}$$

where D is again a polynomial with integer coefficients, the variables of which are split into two groups: the *parameters* $a_1, \dots, a_n$ and the *unknowns* $x_1, \dots, x_m$. We will suppose that parameters can, similarly to unknowns, assume positive integer values only. For some choice of the values of the parameters $a_1, \dots, a_n$ the equation (6) can have a solution in the unknowns $x_1, \dots, x_m$, for other choices of the values of the parameters it can have no solution. We can consider the set $\mathcal{M}$ of all n-tuples $\langle a_1, \ldots, a_n \rangle$ for which our parametric equation has a solution, that is

$$\langle a_1, \ldots, a_n \rangle \in \mathcal{M} \iff \exists x_1 \ldots x_m \{D(a_1, \ldots, a_n, x_1, \ldots, x_m) = 0\}. \qquad (7)$$

Sets having such representations are called *Diophantine.* An equivalence of the form (7) is called a *Diophantine representation* of the set $\mathcal{M}$. With an abuse of language, one can say that the equation (6) itself is a representation of the set.

Similarly, one says that a relation among natural numbers is Diophantine if so is the set of all tuple of natural numbers satisfying this relation.

Easy examples of Diophantine sets are the following:

- *the set of all squares* represented by equation
$$a - x^2 = 0; \qquad (8)$$
- *the set of all composite numbers* represented by equation
$$a - (x_1 + 1)(x_2 + 1) = 0; \qquad (9)$$
- *the set of all positive integers which are not powers of* 2 represented by equation
$$a - (2x_1 + 1)x_2 = 0. \qquad (10)$$

It is easy to prove that the union and the intersection of two Diophantine sets are also Diophantine. However, the complement (to the set of all n-tuples of corresponding size) of a Diophantine can be non-Diophantine. The latter fact is not trivial at all. It was proved by Martin Davis [7].

In the case of our first example (8) of a Diophantine set, it can be seen that its complement, i.e. *the set of all numbers which are* not *squares*, is Diophantine; it is represented by equation

$$(a - (z-1)^2 - x)^2 + (z^2 - a - y)^2 = 0. \qquad (11)$$

However, if we ask about the complements of the other two sets, the answers are not evident at all.

- Is *the set of all prime numbers* Diophantine?
- Is *the set of all powers of 2* Diophantine?

It seems that questions like these have never been posed in number theory. There an equation was a primary object of study, and the typical problem was *to describe the set of all solutions of a given Diophantine equation.* The inverse problem of *constructing a Diophantine representation for a given set* was stranger to number theory.

Diophantine sets from a computational point of view. Every Diophantine set is clearly *listable*, or, in another terminology, *recursively enumerable* (r.e.). (So if the complement of every Diophantine set were Diophantine too, then every set described by an arithmetical formula with both existential and universal quantifiers would be listable too, which is not the case. This was Davis's proof of the existence of a Diophantine set with non-Diophantine complement.)

Martin Davis conjectured that *to be listable* is not only necessary but also a sufficient condition for a set of n-tuples of natural numbers *to be Diophantine.*

M. Davis's conjecture. The notions of Diophantine set and recursively enumerable set coincides, i.e. a set is Diophantine if and only if it is recursively enumerable.

The notion of an r.e. set is essentially equivalent to the notion of an algorithm and so it is a fundamental notion of computability theory (see, for example, the book of Martin-Löf [34] where computability theory is developed in terms of listable sets rather than in terms of algorithms). Davis's conjecture states that the general notion of recursively enumerable set is equivalent to the seemingly rather narrow notion of a Diophantine set.

From conjecture to theorem

It took two decades before Davis's conjecture became a theorem, i.e. before it was proved.

Davis' normal form. In the early 50's Martin Davis made the first step towards the proof of his conjecture. Namely he showed that every r.e. set $\mathcal{M}$ of n-tuples of natural numbers has an "almost Diophantine" representation of the form

$$\langle a_1, \ldots, a_n \rangle \in \mathcal{M} \Longleftrightarrow$$
$$\exists z (\forall y \leq z) \exists x_1 \ldots x_m \{ D(a_1, \ldots, a_n, x_1, \ldots, x_m, y, z) = 0 \}. \qquad (12)$$

Note that the universal quantifier in (12) is bounded, so every formula of this kind determines an r.e. set. Representations of the type (12) became known as *Davis normal form.* They were a quantitative improvement over the classical result of Kurt Gödel [16] who demonstrated the existence of similar arithmetical representations with arbitrary number of (bounded) universal quantifiers.

Back to equations. The single remaining universal quantifier was finally eliminated from Davis normal form. At first this was done conditionally and at the cost of extending the set of admissible equations to the so called *exponential Diophantine equations.* They are equations of the form

$$E_{\mathrm{L}}(a_1, \ldots, a_n, x_1, x_2, \ldots, x_m) \;=\; E_{\mathrm{R}}(a_1, \ldots, a_n, x_1, x_2, \ldots, x_m) \qquad (13)$$

where E_{L} and E_{R} are *exponential polynomials*, i.e. expression constructed by traditional rules from the variables and particular natural numbers by addition, multiplication and exponentiations. Namely, Martin Davis and Hilary Putnam [12] proved in 1960 that every r.e. set $\mathcal{M}$ has a purely existential *exponential Diophantine representation*

$$\langle a_1, \ldots, a_n \rangle \in \mathcal{M} \Longleftrightarrow \qquad (14)$$

$$\exists x_1 \ldots x_m \{E_{\mathrm{L}}(a_1, \ldots, a_n, x_1, x_2, \ldots, x_m) = E_{\mathrm{R}}(a_1, \ldots, a_n, x_1, x_2, \ldots, x_m)\}$$

under the assumption that *there are arbitrary long arithmetical progressions consisting entirely of prime numbers.*

(The report [12] is not easy to find. When by my request Martin Davis presented a copy of it to me, he said: "Do not read it!")

Thus, the existence of arbitrary many primes in arithmetical progressions would imply an algorithmical result: the undecidability of the weaker form of Hilbert's tenth problem corresponding to a broader class of exponential Diophantine equations. This implication motivated the search of long arithmetical progressions of primes, but why should number-theorist consider this problem? There was an opinion (expressed, as I remember, by physicists not mathematicians) that "prime numbers were born to be multiplied" (for generating all natural numbers) and respectively "additive problems about prime numbers are artificial and need not be considered at all". Nevertheless the problem of how long an arithmetical progressions of primes could be was considered in number theory much before the work of Davis and Putnam. However, "being born for multiplication", prime numbers are very stubborn when one tries to add or subtract them, and we still (1998) do not know whether they form arithmetical progressions of arbitrary large length.

Luckily, Julia Robinson [52] was able to avoid the assumption about primes in progressions, and in 1961 Martin Davis, Julia Robinson and Hilary Putnam published a famous joint paper [11] with an unconditional proof of the existence of an exponential Diophantine representation for every r.e. set.

(The need of prime numbers was caused by the use of Gödel's [16] method of representing n-tuples of natural numbers of arbitrary length based on *Chinese Remainder Theorem.* It is interesting to note that much later I [36] was able to show that primality is not essential at all and prime numbers can be completely avoided.)

Exorcizing exponentiation. With the work of Davis, Putnam and Robinson the hard algorithmical part of the job was done: an analog of Hilbert's tenth problem for exponential Diophantine equation was proved undecidable. All what remained in order to prove the undecidability of the original Hilbert's problem was to learn how to transform an arbitrary exponential Diophantine equation (13) into an equivalent Diophantine equation, i.e. an equation of the form (6) having solutions for the same values of the parameters. Moreover, in order to prove Davis's conjecture in its full form it was sufficient to prove a very particular case of it, namely, to prove that the ternary relation of exponentiation $a = b^c$ was Diophantine. As soon as we have a Diophantine representation for this relation,

$$\langle a, b, c\rangle \in \mathcal{M} \Leftrightarrow a^b = c \Leftrightarrow \exists z_1 \ldots z_w \{A(a, b, c, z_1, \ldots, z_m) = 0\} \qquad (15)$$

we can easily use the polynomial A for transforming exponential Diophantine equations into Diophantine.

However, as it was mentioned above, the inverse problem of finding an equation for a given set was not popular in number theory, and so in 1961 the existence of (15) was open. Luckily, this problem has already been attacked in mathematical logic. The starting point of this investigation was quite opposite to Davis's conjecture: Alfred Tarski suspected at the end of 40's that even the set of all powers of 2 is *not* Diophantine. Julia Robinson spent some time trying to prove it but then she switched to proving that the relation $a = b^c$ was Diophantine. She failed to do it but found some sufficient conditions. Namely, Julia Robinson proved that the relation of exponentiation is Diophantine provided that there is a Diophantine relation

$$J(u,v) \Leftrightarrow \exists z_1 \ldots z_m \{B(u,v,z_1,\ldots,z_m) = 0\} \tag{16}$$

such that

1. if $J(u,v)$ holds then $u < v^v$;
2. for every k there are u and v such that $J(u,v)$ holds and $u > v^k$.

Julia Robinson called the relations satisfying the above two conditions *relations of exponential growth*; they became known in the literature as *Julia Robinson relations*.

All it remained to prove Davis conjecture was to find a single relation of exponential growth defined by a Diophantine equation. Surprisingly, among numerous two-parameter equations studied in number theory since Diophantus up to the middle of 20th century there was no Diophantine equation defining a relation of exponential growth.

A few words about words. I began to be involved in investigations on Hilbert's tenth problem at the end of December 1965 when I was a second year student at the department of mathematics and mechanics of Leningrad State University. This subject was suggested to me by my scientific adviser Serguei Maslov (see [13]). However, I studied the works of Martin Davis, Hilary Putnam and Julia Robinson sometime later. Maslov believed that their approach cannot lead to solution because it had not led already. Instead Maslov suggested to try another approach initiated by A. A. Markov.

The idea was as follows. A universal computer science tool for representing information uses words rather than numbers. However, there are many ways to represent words by numbers. One of such methods is naturally related to Diophantine equations. Namely, it is not difficult to show that every 2×2 matrix

$$\begin{pmatrix} m_{11} & m_{12} \\ m_{21} & m_{22} \end{pmatrix} \tag{17}$$

with the m's being non-negative integers and the determinant equal to 1 can be represented, in an unique way, as a product of matrices

$$M_0 = \begin{pmatrix} 1 & 1 \\ 0 & 1 \end{pmatrix} \qquad \text{and} \qquad M_1 = \begin{pmatrix} 1 & 0 \\ 1 & 1 \end{pmatrix}. \tag{18}$$

It is evident that any product of such matrices has non-negative integer elements and the determinant equals 1. This implies that we can uniquely represent a word

$a_{i_1} \dots a_{i_m}$ in a two-letter alphabet by the four-tuple $\langle m_{11}, m_{12}, m_{21}, m_{22} \rangle$ such that

$$\begin{pmatrix} m_{11} & m_{12} \\ m_{21} & m_{22} \end{pmatrix} = M_{i_1} \dots M_{i_m}; \tag{19}$$

the numbers evidently satisfy the Diophantine equation

$$m_{11}m_{22} - m_{21}m_{12} = 1. \tag{20}$$

Under this representation of words by matrices, the concatenation of words corresponds to matrix multiplication and thus can be easily expressed as a system of Diophantine equations. This opens a way to transform an arbitrary system of *word equations* into equivalent Diophantine equations. Many decision problems about words had been shown undecidable, so it was quite natural to try to attack Hilbert's tenth problem by proving the undecidability of systems of word equations.

I spent some time trying to show the undecidability of word equations. Much later it became known that this approach was fruitless as G. S. Makanin [29] found an algorithm for word equations. Luckily, I soon abandoned this approach.

My next attempt was to consider a broader class of word equations with additional predicates. Since the ultimate goal was always Hilbert's tenth problem, I could consider only such predicates which (under suitable coding) would be represented by Diophantine equations. In this way I came to what I have called *equations in words and lengths*. Reduction of such equations to Diophantine equations was based on celebrated Fibonacci numbers. It is well-known that every natural number can be represented, in an almost unique way, as the sum of different Fibonacci numbers, none of which are consecutive (so called *Zeckendorf's representation*). Thus we can look at natural numbers as words in two-letter alphabet {0,1} with the additional constraint that there cannot be two consecutive 1's. I [35] managed to show that under this representation of words by numbers both the concatenation of words and the equality of the lengths of two words can be expressed by Diophantine equations.

However, I was unable to show the undecidability of equations in words and length (and this still remains an open problem).

Rabbits strike again. Finally, I switched to Davis-Putnam-Robinson approach and tried to construct a Diophantine relation of exponential growth. Due to my previous work, I realized the importance of Fibonacci numbers for Hilbert's tenth problem. That is why during the summer of 1969 I was reading with great interest the third augmented edition [61] of a popular book on Fibonacci numbers written by N. N. Vorob'ev from Leningrad. It seems incredible that in the 20th century one can still find something new about the numbers introduced by Fibonacci in the 13th century in connection with multiplying rabbits. However, the new edition of the book contained, besides traditional stuff, some original results of the author. In fact, Vorob'ev had obtained them a quarter of a century earlier but he never published anything before. His results attracted my attention at once but I was not able to use them immediately for constructing a Diophantine representation of a relation of exponential growth.

After the summer I had to pass the entrance examinations in order to continue my postgraduate studies at Steklov Institute of Mathematics (Leningrad Branch). Three years are given for the preparation of a Ph. D. thesis; at that time I had already spent 4 years in vain on Hilbert's tenth problem, and there was no real hope to solve it in another 3 years. So I switched to a quite different subject, automatic theorem proving, which was the main area of investigations of Maslov.

My previous work on Hilbert's tenth problem unexpectedly paid in the end. Thanks to it I was considered an expert on the problem, and when Julia Robinson published her new paper [53], it was sent to me for a review for the soviet counterpart of *Mathematical Reviews.* The period when I was not thinking about Diophantine equations, Vorob'ev's theorem and new ideas of Julia Robinson led me to the negative solution of Hilbert's tenth problem. On January 29, 1970 I gave at my institute the first talk on a Diophantine relation of exponential growth. It was the relation *u is the $2v$th Fibonacci number.*

Surprisingly, in order to construct a Diophantine representation for this relation I needed to proof a yet new purely number-theoretical result about Fibonacci numbers, namely, that *kth Fibonacci number is divisible by the square of the lth Fibonacci number if and only if k itself is divisible by the lth Fibonacci number.* This property is not difficult to prove; what is striking is that this beautiful fact has not been discovered, even empirically, since Fibonacci times.

With my example of a Diophantine relation of an exponential growth Davis's conjecture became a theorem which is often referred to as DMPR theorem after Davis-Matiyasevich-Putnam-Robinson. Nowadays detailed and simplified proofs of this theorem can be found in many publications, in particular, in [1, 5, 9, 10, 23, 31, 32, 36, 41, 57].

From Computer Science to Number Theory

DMPR theorem, establishing the equivalence of a number-theoretical notion of Diophantine set and the notion of r.e. set, can serve as a bridge for transporting ideas and methods from computer science to number theory and backward. First we shall survey what number theory gained from DMPR theorem.

The undecidability of Hilbert's Tenth Problem

Of course, the first evident gain is the proof of the undecidability of Hilbert's tenth problem. With it number-theorists have got a "moral right" to abandon further attempts to find a universal method for Diophantine equations and use *ad hoc* tools for particular equations.

DMPR theorem implies an undecidability result stronger than what is required just to "close" Hilbert's tenth problem. Hilbert asked for a universal method suitable for an arbitrary Diophantine equation. Now we can take a particular r.e. but undecidable set $\mathcal{M}$ of natural numbers and construct its Diophantine representation

$$a \in \mathcal{M} \iff \exists x_1 \dots x_m \{M(a, x_1, \dots, x_m) = 0\}. \qquad (21)$$

The undecidability of $\mathcal{M}$ implies that there is no universal method to decide, for a given a, whether the equation

$$M(a, x_1, \ldots, x_m) = 0 \tag{22}$$

has a solution in $x_1, \ldots, x_m$. That is, to get the undecidability we need not consider all Diophantine equations, in any number of unknowns and of arbitrary large degree.

The undecidability of one-parameter Diophantine equation (22) means the following. Suppose that $\mathcal{A}$ is some algorithm hypothetically capable to tell for a given a whether the equation (22) has a solution or not. Now we know that the algorithm $\mathcal{A}$ should fail for some particular number $a_{\mathcal{A}}$, that is, either $\mathcal{A}$ never stops on input $a_{\mathcal{A}}$ or its output, if any, is wrong. The set $\mathcal{M}$ can be chosen in such a way that this counterexample $a_{\mathcal{A}}$ could be effectively found for every algorithm $\mathcal{A}$.

Speeding-up Diophantine equations

The undecidability of Hilbert's tenth problem, being very important for number theory, still cannot be considered as a purely number-theoretical result because its formulation involves notions from computability theory. However, DMPR theorem allows us to obtain many new results about Diophantine equations which are, at least in the form, number-theoretical. We can just take any theorem about r.e. sets and replace words "recursively enumerable" in its statement by "Diophantine". Furthermore, we can use the definition of a Diophantine set and obtain a theorem about Diophantine equations.

Such reformulations of theorems from computer science in terms of Diophantine equations usually give results which are not typical for number theory. As an example, let us consider a Diophantine version of M. Blum's [3] *speed-up theorem* obtained by M. Davis [8]:

For every general recursive function $\alpha(a, w)$ there are Diophantine equations

$$B(a, x_1, \ldots, x_n) = 0, \tag{23}$$

$$C(a, y_1, \ldots, y_m) = 0 \tag{24}$$

such that:

1. *for every value of a, one and only one of these two equations has a solution;*
2. *if the equations*

$$B'(a, x'_1, \ldots, x'_{n'}) = 0, \tag{25}$$

$$C'(a, y'_1, \ldots, y'_{m'}) = 0 \tag{26}$$

are solvable exactly for the same values of the parameter a, as equations (23) *and* (24)*, respectively, then there is a third pair of equations*

$$B''(a, x''_1, \ldots, x''_{n''}) = 0, \tag{27}$$

$$C''(a, y''_1, \ldots, y''_{m''}) = 0 \tag{28}$$

such that:

- *these equations are also solvable exactly for the same values of the parameter a, as equations* (23) *and* (24), *respectively;*
- *for almost all a, for every solution of equation* (25) (*equation* (26)) *there is a solution of equation* (27) (*respectively, equation* (28)) *such that*

$$x'_1 + \ldots + x'_{m'} > \alpha(a, x''_1 + \ldots + x''_{m''}) \tag{29}$$

(*or*

$$y'_1 + \ldots + y'_{n'} > \alpha(a, y''_1 + \ldots + y''_{n''}) \tag{30}$$

respectively).

This theorem, in its full generality, is about an arbitrary general recursive function; replacing it by particular (fast growing) functions we obtain theorems which are purely number-theoretical but quite non-standard for number theory.

Universal Diophantine equations

Now we consider a transfer of another idea from computer science to number theory which is of more interest for our considerations.

Jack of all trades. Universal objects like *universal Turing machines*, *universal r.e. sets* and so on are quite ordinary objects in computer science. Now with DMPR theorem we can construct their counterparts in number theory. Namely, for every fixed n, we can construct in an effective way a list

$$D_0, D_1, \ldots \tag{31}$$

consisting of all polynomials with integer coefficients having n parameters and arbitrary number of unknowns. The set $\mathcal{U}_n$ of $(n+1)$-tuples defined by

$$\langle k, a_1, \ldots, a_n \rangle \in \mathcal{U}_n \iff \exists x_1 x_2 \ldots \{D_k(a_1, \ldots, a_n, x_1, x_2, \ldots) = 0\}, \tag{32}$$

being recursively enumerable is Diophantine, so we can construct a single polynomial U_n with $n+1$ parameter such that

$$\langle k, a_1, \ldots, a_n \rangle \in \mathcal{U}_n \iff \exists x_1 \ldots x_m \{U_n(k, a_1, \ldots, a_n, x_1, x_2, \ldots, x_m) = 0\}. \tag{33}$$

The corresponding equation

$$U_n(k, a_1, \ldots, a_n, x_1, x_2, \ldots, x_m) = 0 \tag{34}$$

is *universal* in the following sense: solving an arbitrary Diophantine equation with n parameters

$$D(a_1, \ldots, a_n, x_1, x_2, \ldots) = 0 \tag{35}$$

is equivalent to solving the equation

$$U_n(k_D, a_1, \ldots, a_n, x_1, x_2, \ldots, x_m) = 0 \tag{36}$$

resulting from the single equation (34) by choosing a particular value k_D for the first parameter (this value is, of course, the number of the equation (35) in the list (31)).

The degree and the number of unknowns of the equation (36) is fixed while the equation (35) can have any number of unknowns and can be of arbitrary large degree. As an example let us put $n = 1$ and let s be the number of arithmetical operations required to calculate the value of the *universal polynomial* U_1. Let us take for (35) the equation

$$a = (x_1 + 1) \dots (x_{2^s} + 1). \tag{37}$$

We see that in order to verify that a number a has at least 2^s prime factors it is sufficient to perform only s arithmetical operations!

Is it Number Theory?. To what extent the existence of universal Diophantine equations belongs to computer science and to what extent to number theory? First of all, number-theorists never anticipated such a possibility. It was a surprise even to some logicians familiar with other universal objects. For example, when in 1961 Martin Davis, Hilary Putnam and Julia Robinson proved the existence of exponential Diophantine representations, they immediately got as a corollary the existence of universal exponential Diophantine equations. However, the following paragraph appears in George Kreisel review [26] (for *Mathematical Reviews*) of the celebrated paper by Davis, Putnam and Robinson [11]:

> These results are superficially related to Hilbert's tenth Problem on (ordinary, i.e., non-exponential) Diophantine equations. The proof of the authors' results, though very elegant, does not use recondite facts in the theory of numbers nor in the theory of r.e. sets, and so it is likely that the present result is not closely connected with Hilbert's tenth Problem. Also it is not altogether plausible that all (ordinary) Diophantine problems are uniformly reducible to those in a fixed number of variables of fixed degree, which would be the case if all r.e. sets were Diophantine.

We can look at the existence of universal Diophantine equations as a number-theoretical result *inspired* by computer science. The following question naturally arises: *can the existence of universal Diophantine equations be proved by purely number-theoretical tools*, i.e. without any reference to the notion of r.e. set and constructions of universal r.e. sets? In my book [41] I managed to give what I believe to be such a number-theoretical construction of universal Diophantine equations.

Collapse of Diophantine hierarchy. The possibility to bound both the degree and the number of unknowns in Diophantine representations means that traditional number-theoretical classifications of Diophantine equations as equations in 1, 2, ... unknowns and as equations of degree 1, 2, ... collapse. Of course, it is of considerable interest to find out where exactly this happens.

Let us call a pair $\langle \nu, \mu \rangle$ of natural numbers *a universal Diophantine complexity bound* if every r.e. set can be defined by a Diophantine equation of degree ν with respect to its μ unknowns.

As usual, there is a natural trade-off between ν and μ. The best known today (1998) value of ν is 4. This follows from the existence of universal equations and an old result of Thoralf Skolem [56] who showed that solving an arbitrary Diophantine equation can be easily reduced to solving another Diophantine equation of degree 4 at the cost of introduction of many additional unknowns.

The best known today (1998) value of μ is 9. I obtained this result in [37] but for various reasons (see [41]) I have never published a detailed proof. This was finally done by James P. Jones [21]. He also calculated a value of ν corresponding to $\mu = 9$ and found a number of "intermediate values": the following pairs are all universal Diophantine complexity bounds:

$$\langle 4,58\rangle,\ \langle 8,38\rangle,\ \langle 12,32\rangle,\ \langle 16,29\rangle, \langle 20,28\rangle,\ \langle 24,26\rangle,\ \langle 28,25\rangle,\ \langle 36,24\rangle,$$
$$\langle 96,21\rangle,\ \langle 2668,19\rangle,\ \langle 2\times 10^5,14\rangle, \langle 6.6\times 10^{43},\ 13\rangle,\ \langle 1.3\times 10^{44},12\rangle,$$
$$\langle 4.6\times 10^{44},11\rangle,\ \langle 8.6\times 10^{44},10\rangle,\ \langle 1.6\times 10^{45},9\rangle.$$

It is known that universal Turing machines can be rather small. The above universal bound might produce the impression that writing down a universal Diophantine equation would require hundreds of pages. This is not the case. While the degree of a universal polynomial could be very large, the majority of coefficients can be equal to 0. Here is a universal system of Diophantine equations also constructed in [21]:

$$elg^2 + \alpha = (b - xy)q^2, \quad q = b^{5^{60}}, \quad \lambda + q^4 = 1 + \lambda b^5, \quad \theta + 2z = b^5,$$
$$l = u + t\theta, \quad e = y + m\theta, \quad n = q^{16},$$
$$r = \Big[g + eq^3 + lq^5 + \big(2(e - z\lambda)(1 + xb^5 + g)^4 + \lambda b^5 + \lambda b^5 q^4\big)q^4\Big][n^2 - n]$$
$$+ [q^3 - bl + l + \theta\lambda q^3 + (b^5 - 2)q^5][n^2 - 1],$$
$$p = 2ws^2r^2n^2, \quad p^2k^2 - k^2 = 1 = \tau^2, \quad 4(c - ksn^2)^2 + \eta = k^2,$$
$$k = r + 1 + hp - h, \quad a = (wn^2 + 1)rsn^2, \quad c = 2r + 1 + \phi,$$
$$d = bw + ca - 2c + 4a\gamma - 5\gamma, \quad d^2 = (a^2 - 1)c^2 + 1,$$
$$f^2 = (a^2 - 1)i^2c^4 + 1,$$
$$(d + of)^2 = \Big(\big(a + f^2(d^2 - a)\big)^2 - 1\Big)(2r + 1 + jc)^2 + 1$$

(clearly, a system of Diophantine equations can be easily combined into a single equation by squaring).

Most likely, 9 unknowns is rather far from the least possible values which could be as low as 3 (equations in a single unknown are evidently decidable and great progress for equations in 2 unknowns gives hope to number-theorists to find a decision procedure for such equations). For exponential Diophantine equations I was able to find a much better bound (see [40]): a universal exponential Diophantine equation can have only 3 unknowns.

Prime numbers are of main interest in number theory

With DMPR theorem the inverse problem, i.e. constructing Diophantine equations solvable for prescribed values of the parameters, has got a complete solution. The proof of DMPR theorem was constructive so for every r.e. set we can effectively write down its Diophantine representation.

Of course, the most interesting set in number theory is the set of all prime numbers. According to DMPR theorem, we can find a particular Diophantine equation

$$P(a, x_1, \ldots, x_m) = 0 \tag{38}$$

solvable in $x_1, \ldots, x_m$ if and only if the parameter a is a prime.

Consider now the equation

$$x_0(1 - P^2(x_0, x_1, \ldots, x_m)) = a. \tag{39}$$

Clearly, every solution of equation (38) can be extended to a solution of equation (39) just by putting

$$x_0 = a. \tag{40}$$

On the other hand, for every positive a in every solution (in positive integers) of equation (39) the expression in brackets should be positive too. This is possible only when

$$P(x_0, x_1, \ldots, x_m) = 0. \tag{41}$$

But (39) implies (40) and hence (38).

We see that equation (39) is also a representation of the set of all primes. This fact can be stated a bit differently thanks to the special form of (39): *the set of all prime numbers is identical with the set of all positive values assumed by the polynomial* $x_0(1 - P^2(x_0, x_1, \ldots, x_m))$, *for positive values of its variables.*

The above trick of passing from (38) to (39) was invented by Hilary Putnam [50] and published in 1960; at that time the possible existence of prime representing polynomials was considered as an informal argument against Davis' conjecture.

The discovery of prime number representing polynomials was quite a surprise for number-theorists. When prominent Soviet mathematician Yu. V. Linnik was told about it, he reacted: "That's wonderful, most likely we shall soon learn a lot of new things about primes. " But then it was explained to him that this was just one corollary of a much more general result about the existence of Diophantine representations for every r.e. set. "It's a pity," Linnik then said. "Most likely, we shall not learn anything new about the primes. "

The generality of DMPR theorem is both its power and the weak point. Hardly we can prove anything new looking at the equation (36) with k_D corresponding to the set of prime numbers. Luckily, for constructing a Diophantine representation of primes we need only part of the machinery developed for the case of arbitrary r.e. set.

Here is a particular prime representing polynomial taken from [24]:

$$\begin{aligned}(k+2)\{\ 1 &-[wz+h+j-q]^2 \\ &-[(gk+2g+k+1)(h+j)+h-z]^2 \\ &-[2n+p+q+z-e]^2\end{aligned}$$

$$
\begin{aligned}
&- \left[16(k+1)^3(k+2)(n+1)^2+1-f^2\right]^2 \\
&- \left[e^3(e+2)(a+1)^2+1-o^2\right]^2 \\
&- \left[(a^2-1)y^2+1-x^2\right]^2 \\
&- \left[16r^2y^4(a^2-1)+1-u^2\right]^2 \\
&- [n+l+v-y]^2 \\
&- \left[\left((a+u^2(u^2-a))^2-1\right)(n+4dy)^2+1-(x+cu)^2\right]^2 \\
&- \left[(a^2-1)l^2+1-m^2\right]^2 \\
&- \left[q+y(a-p-1)+s(2ap+2a-p^2-2p-2)-x\right]^2 \\
&- \left[z+pl(a-p)+t(2ap-p^2-1)-pm\right]^2 \\
&- [ai+k+1-l-i]^2 \\
&- \left[p+l(a-n-1)+b(2an+2a-n^2-2n-2)-m\right]^2 \}.
\end{aligned}
$$

The above polynomial for primes has 26 variables $a,\ldots,z$ (all letters of the alphabet, including o!) ranging over non-negative integers.

After one of his lectures about this polynomial, J. P. Jones got a telegram saying that this particular polynomial certainly was not representing the primes because itself was a product of two polynomials. J. P. Jones was asked to send as soon as possible a "correct polynomial". But nothing was wrong, it was just Putnam's trick!

For a long time the current (at that moment) known least number of variables in primes representations was smaller than the current (at the same moment) least known number of unknowns required for the representation of an arbitrary r.e. set. Today both records (i.e. for primes and for a general r.e. set) are the same—10 variables. Of course, polynomial constructed especially for primes [38] has much smaller degree than the universal polynomial.

Being unable to construct a Diophantine representation with a smaller number of unknowns for the whole set of prime numbers, we can look for Diophantine representations of its infinite subsets. Such a broader problem is, of course, of less interest for traditional number theory; however, it has direct connections with possible proofs of DMPR theorem. Namely, just before the theorem was proved, Julia Robinson [53] established the following conditional result: *every r.e. set is Diophantine as soon as at least one infinite set of prime numbers is Diophantine.*

J. P. Jones [25] proved that 7 unknowns are sufficient for Diophantine representations of the set of all *Mersenne primes* (i.e. primes of the form $2^n - 1$) and for the set of all *Fermat primes* (i.e. odd primes of the form $2^n + 1$). Up to now, only 5 Fermat primes have been found, and it is considered most likely that there are no others. As for Mersenne primes, (non-rigorous) probabilistic arguments suggest that there are infinitely many such primes but so far nobody was able to give a real proof.

Jones's success in reducing the number of unknowns in the cases of Mersenne and Fermat primes was due to the fact that for numbers of the forms $2^n \pm 1$ there are very efficient tests for primality. I suspected that other classes of primes which can be tested for primality quickly, could also be defined by a Diophantine equation with smaller number of unknowns.

My post-graduate student Maxim Vsemirnov confirmed my supposition. Large prime numbers are vital for modern cryptography, and efficient methods for generating them were developed. In particular, J. Pintz, W. L. Steger, and E. Szemeredi [48] described a (provably) infinite set of prime numbers of special form admitting fast primality test. M. Vsemirnov [62] showed that similar infinite set of primes admits a Diophantine representation with 8 unknowns only.

Non-effective estimates can be non-effectivizable

What are the "duties" of computer science with respect to number theory? One role which could/should be played by computer science is to explain difficulties arising in number theory. Computer science managed to help number theory by proving that Hilbert's tenth problem was undecidable. However, there are other situations in number theory where all attempts fail, and, possibly, because of some deep algorithmical reasons.

Suppose that we have a Diophantine equation

$$D(a, x_1, \ldots, x_m) = 0, \tag{42}$$

such that for every value of the parameter a has at most finitely many solutions in $x_1, \ldots, x_m$. This fact can be expressed in two form:

1. the equation (42) has at most $\nu(a)$ solutions;
2. in every solution of (42) $x_1 < \sigma(a)$, …, $x_m < \sigma(a)$

for suitable functions ν and σ.

From a mathematical point of view these two statements are equivalent. However, they are rather different from a computational point of view. Having $\sigma(a)$ we can find $\nu(a)$, but not *vice versa*. Number-theorists have found many classes of Diophantine equations with computable $\nu(a)$ for which they fail to compute $\sigma(a)$. In such cases number-theorists say that "the estimate of the size of solutions is *non-effective*".

I was able to show that at least in the theory of exponential Diophantine equations there are estimates which cannot be effective in principle. Namely, we can construct an exponential Diophantine equation

$$E_{\mathrm{L}}(a, x_1, x_2, \ldots, x_m) = E_{\mathrm{R}}(a, x_1, x_2, \ldots, x_m) \tag{43}$$

with the following properties:

1. for every value of the parameter a, the equation (43) has at most one solution in $x_1, \ldots, x_m$;

2. for every general recursive function σ there is a value of a for which the equation (43) has a solution $x_1, \ldots, x_m$ such that $x_1 > \sigma(a)$.

Improving this result to the case of Diophantine equations remains a challenge for computer science.

From Number Theory to Computer Science

In this section I will show that computer science has also gained something from the collaboration with number theory on Hilbert's tenth problem.

Diophantine equations as computing devices

It was already mentioned above that the notion of r.e. set is as fundamental as the general notion of algorithm. Respectively, we can treat Diophantine equations as computing devices. This was done in a picturesque form by Leonard Adleman and Kenneth Manders in [2, 30]. There they introduced the notion of *Non-Deterministic Diophantine Machine*, NDDM for short.

A NDDM is specified by a Diophantine equation

$$D(a_1, \ldots, a_n, x_1, \ldots, x_m) = 0, \tag{44}$$

and works as follows: on input $a_1, \ldots, a_n$ it guesses the numbers $x_1, \ldots, x_m$ and checks (44); if the equality holds, the n-tuple $\langle a_1, \ldots, a_n \rangle$ is accepted.

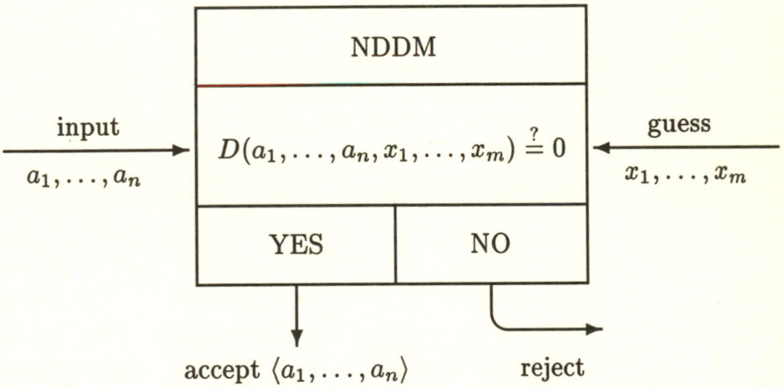

The idea behind the introduction of a new computing device was as follows: in NDDM we have full separation of guessing and deterministic computation, and the latter is very simple—just the calculation of the value of a polynomial.

Now the DMPR theorem states that NDDMs are as powerful as, say, Turing machines: every set recognizable by a Turing machine is recognized by some NDDM, and, of course, *vice versa*.

The original proof of the DMPR theorem was quite constructive, that is, given a Turing machine, we can effectively write down the corresponding Diophantine equation. Unfortunately, such a construction was rather roundabout. Namely, one had:

1. to construct an arithmetical formula with many bounded universal quantifiers describing the work of the Turing machine by the technique introduced by K. Gödel [16];
2. to bring this formula into Davis normal form (12) by gluing bounded universal quantifiers as it is described in Davis' paper [7];
3. to eliminate the last remaining bounded universal quantifier at the cost of passing to exponential Diophantine equations;
4. to eliminate the exponentiation making use of its Diophantine representation (15).

According to a footnote in Davis' paper [7], the idea of obtaining the representation (12) by combining universal quantifiers from a general arithmetic representation was due to the (anonymous) referee of the paper. The original proof of Davis (outlined in [6] and given with details in [10]) was quite different. Technically, it was more involved but in some respects the original construction was very appealing. Namely, Davis managed to arithmetize *Post normal systems* using only one universal quantifier. Now that we know that in fact no universal quantifier is necessary at all, would not it be more natural to try to go further in this direction and obtain, by an arithmetization of a Turing machine, directly a purely existential exponential Diophantine representation (14)?

This idea was very attractive for me and at last I was able to implement it in my paper [37]. There I arithmetized by purely existential formulas the work of a Turing machine. Later it turned out that another kind of computing devices, so called *register machines*, are even more suitable for such existential arithmetization. Register machines were introduced almost simultaneously by several authors: J. Lambek [28], Z. A. Melzak [43], M. L. Minsky [44, 45], and J. C. Shepherdson and H. E. Sturgis [54]. Like Turing machines, register machines have very primitive instructions but, in addition, they deal directly with numbers rather than with words. This leads to a "visual proof" of simulation of register machines by exponential Diophantine equations (see my joint papers with J. P. Jones [22, 23]). Such a proof of DPMR theorem is more suitable for a course in computability theory because its prerequisites in number theory are minimal.

I had a large choice of techniques for proving DMPR theorem during the writing of my book [41]. I decided to simulate Turing machines because they are the classical computing devices. The proof presented in the book differs very much from the first proof via Turing machines given in [37].

Later I found yet another direct way to prove DMPR theorem using induction on the construction of partial recursive functions (see [42]).

I have given talks about Hilbert's tenth problems in many universities, both at departments of mathematics and at departments of computer science. Mathematicians often complained that their salaries are less than salaries of computer scientists. So I often explained to mathematicians that if they are solving Diophantine equations, they are doing some kind of computer science and so they can demand an increase in payment.

The magic of old number theory

The pure number-theoretical proof of the existence of universal equations, the reductions to 9 and 3 unknowns, "visual" simulations of Turing and register machines and many other interesting new results were obtained thanks to a rather old result of always young number theory.

According to the Main Theorem of Arithmetic, the binomial coefficient $\binom{a+b}{b}$, as any natural number, can be represented as the product of powers of prime numbers:

$$\binom{a+b}{a} = 2^{\alpha(2)}3^{\alpha(3)}5^{\alpha(5)}\dots \tag{45}$$

What are these exponents $\alpha(1)$, $\alpha(2)$, $\alpha(3)$, ...? A surprising answer was found by Ernst Kummer [27]: *in order to find $\alpha(p)$ one can first write down both numbers a and b in p-base notation and then add them according to school rules; during this addition carries from digit to digit can occur; the number of this carries is exactly $\alpha(p)$.*

The power of Kummer's theorem in constructing Diophantine representations can be informally explained as follows: it connects divisibility properties of numbers with their properties as strings of digits. We can look at numbers as, say, words in the two-letter alphabet $\{0, 1\}$. Binary notation does not forbid two consecutive 1's as it was in above mentioned Zeckendorf representation. We can easily express concatenation by an (exponential) Diophantine equation. Kummer's theorem allows us to express in a Diophantine way many relations among numbers formulated in terms of their binary digits. For example, the property *every binary digit of the number a is less or equal to corresponding binary digit of number b* is equivalent, according to Kummer's theorem, to the congruence

$$\binom{b}{a} \equiv 1 \pmod 2 \tag{46}$$

which can be easily rewritten, with the aid of the binomial theorem, as an exponential Diophantine equation.

Kummer published this beautiful theorem in 1852. At that time this fact did not find many applications in number theory and the theorem was forgotten; several authors rediscovered Kummer's result in this century, and now it is fruitfully used for constructing Diophantine representations of arbitrary r.e. sets.

Diophantine complexity

For Turing machines there are two natural complexity measures: TIME and SPACE. For NDDMs there is only one natural complexity measure which plays the role of both TIME and SPACE. This measure is SIZE, which is the size (in bits) of the smallest solution of the equation. It is not essential whether we define this solution as the one with the smallest possible value of $\max\{x_1, \ldots, x_m\}$, or of $x_1 + \ldots + x_m$.

Naturally, SIZE can be used for defining complexity classes. Kosovskii and Vinogradov [60] showed that bounding SIZE by suitable function we can define Grzegorchyk's hierarchy starting from $\mathcal{E}_3$.

We know from the DMPR theorem that NDDMs are as powerful as Turing machines. The main intriguing question is how efficient are the former. Adleman and Manders supposed that NDDN is as efficient as Turing machine. They obtained in [2, 30] the first results in this direction by estimating the SIZE of a NDDM simulating a Turing machine with TIME in special ranges.

They also introduced the class D of all sets $\mathcal{M}$ having representations of the form

$$\langle a_1, \ldots, a_n \rangle \in \mathcal{M} \iff \exists x_1 \ldots x_m \{D(a_1, \ldots, a_n, x_1, \ldots, x_m) = 0 \\ \& x_1 + \ldots + x_m \le 2^{|a_1 + \ldots + a_n|^k}\},$$

where $|a|$ denotes, as usual, the (binary) length of a. (Note that in this definition it is not supposed that the polynomial D supplies a Diophantine representation for $\mathcal{M}$, i.e. the equivalence

$$\langle a_1, \ldots, a_n \rangle \in \mathcal{M} \iff \exists x_1 \ldots x_m \{D(a_1, \ldots, a_n, x_1, \ldots, x_m) = 0\}$$

needs not to hold.) It is easy to see that $\mathbf{D} \subset \mathbf{NP}$. The class $\mathbf{D}$ is known to contain **NP**-complete problems and Adleman and Manders asked whether in fact $\mathbf{D} = \mathbf{NP}$.

As a partial progress towards proving this equality we can consider the papers [63, 64, 19] where the class **NP** and other classes have been defined via an analog of Davis' normal form (with proper bounds on the universal and existential quantifiers). Historically, Davis' normal form was the first step towards the proof of DMPR theorem. Unfortunately, all known methods of eliminating the universal quantifier are too costly in terms of the size of solutions of resulting equations.

I will mention another result of Adleman and Manders which is not technically connected with the works on Hilbert's tenth problem but is formally very close. Namely, they [30] proved that deciding whether a Diophantine equation of the form $ax^2 + by = c$ is solvable in positive integers x and y is **NP**-complete.

Reductions of Hilbert's tenth problem to other decision problems

It was difficult to establish the undecidability of Diophantine equations, and to a great extent this was due to their simple form. This pays off: with the proof of the undecidability of Hilbert's tenth problem computer science got a powerful tool for proving undecidability of other decision problems. Here the simplicity of Diophantine equations is often of great help in reducing Hilbert's tenth problem to other problems.

In this way new, simpler proofs were given for some decision problems already known to be undecidable; also, a number of decision problems were originally shown undecidable by reduction of Hilbert's tenth problem to these problems. Below I list only several such examples just to show the great variety of situations in which we can encounter Diophantine equations, sometimes in disguised form; more than a hundred relevant references are given in my book [41] (the bibliography from the book is available on the Internet [65]).

Straightline programs. It is well-known that the *program equivalence problem* is undecidable: *it is impossible to determine, for arbitrary two computer programs, whether they are equivalent or not.* The undecidability of Hilbert's tenth problem implies that we can restrict the class of programs to very simple ones. For example, in order to get undecidability it is sufficient to compare two programs like

```
if D(x1,x2,x3,x4,x5,x6,x7,x8,x9)=0
   then return 0
   else return 1
```

and

```
return 1
```

(as it was mentioned above, 9 is the best known today (1998) bound for the number of unknowns sufficient for a Diophantine representation of every r.e. set). This trivial corollary can be improved in two directions. First, there is no need for conditional operators or loops, the equivalence problem is undecidable already for straightline programs (see [20]). Second, the required number of variables can be smaller than the number of unknowns required for the undecidability of Diophantine equations. The following is one of several theorems of this type proved by my former student D. V. Shiryaev [55]: *it is impossible to decide, for a given straightline program consisting of operators of the forms* `x ← 1`, `x ← x+y`, `x ← x div y` *whether it always returns 1, even if we restrict ourselves to programs with 3 input and 4 local variables.* For such programs the *problem of simplification* is also undecidable.

Word problem for groups. It was mentioned above that the first decision problem which arose in mathematics and was shown undecidable, was Thue's problem for semigroups. Its counterpart, the *word problem for groups*, was stated a few years earlier by M. Dehn [15], but it turned out much more difficult. The first proofs of its undecidability given by P. S. Novikov [46,47], W. W. Boone [4] were based on the undecidability of Thue's problem. M. K. Valiev [59] used the undecidability of Hilbert's tenth problem for giving a much simpler proof of the undecidability of Dehn's problem.

Petri nets. One of the important notions in computer science is concurrency. Different tools were proposed for describing and analyzing parallel computations, in particular, *Petri nets.* M. Rabin used the undecidability of (exponential) Diophantine equations in order to prove that the *inclusion problem for Petri nets* is undecidable. He never published this result. A proof can be found in the paper [17] by M. Hack who also proved a stronger result about the undecidability of the *equality problem for Petri nets.*

Calculi. When solving Diophantine equations we seek for solutions in the discrete set of integers. Nevertheless, Hilbert's tenth problem was used for establishing the undecidability of many problems in calculi. Such results put intrinsic limitations on capabilities of systems of computer algebra. For example, it is impossible to determining, for an arbitrary system of polynomial *differential* equations, whether it has a solution on a given interval or whether its solution is unique or not.

Unification problems. The basic idea behind the use of Hilbert's tenth problem for establishing the undecidability of some other decision problems is always essentially the same: we need to find in the latter problem some objects which can represent natural numbers, and express addition and multiplication in the language of the problem which we want to show to be undecidable. If the problem is far from arithmetic, such a reduction might be technically not easy. Sometimes it is difficult to anticipate that Hilbert's tenth problem can be reduced to another problem looking quite differently.

In the winter 1994/1995 I was visiting the Computing (sic!) Science Department of Uppsala University on an invitation of A. Voronkov. At that time he was working on the so called *simultaneous rigid E-unification*, SREU for short. The need for such unification arises in some approaches to automatic deduction of mathematical theorems in first order theories with equality. Several papers were published about SREU proving that this problem is NP-complete, EXPTIME-complete and NEXPTIME-complete; each new paper stated that the previous one was wrong, but was wrong itself too.

Voronkov suggested to search for the undecidability of SREU. It was quite natural for me to try to reduce Hilbert's tenth problem to SREU. However, there was no evident way to simulate multiplication by rigid equations because they deal with arbitrary terms rather than with numbers. I failed, and soon left Uppsala.

A few month later A. Voronkov, in collaboration with A. Degtyarev, proved the undecidability of SREU. Their proof was by reduction of the so called *monadic semi-unification problem* to SREU. The former problem was previously shown to be undecidable by M. Baaz using a reduction of the *second order unification problem*. In its turn, that problem was shown undecidable by W. D. Goldfarb, and that was ultimately done by a reduction of Hilbert's tenth problem! Later, A. Degtyarev and A. Voronkov [14] gave a more direct proof of the undecidability of SREU by reducing Diophantine equations to systems of rigid E-equations.

References

1. Adamowicz Zofia, Zbierski Paweł. *Logic of Mathematics*. John Wiley & Sons, New York a. o., 1997.
2. L. Adleman, K. Manders. Diophantine complexity. In: *17th Annual Symposium on Foundations of Computer Science*, pages 81–88, Houston, Texas, 25-26 October 1976. IEEE.

3. M. Blum. A machine-independent theory of the complexity of recursive functions. *Journal of the ACM*, 14(2):322–336, 1967.
4. W. W. Boone. The word problem. *Ann. Math.* , 70(2):207–265, 1959.
5. E. Börger. *Computability, Complexity, Logic.* North-Holland, Amsterdam, 1989.
6. M. Davis. Arithmetical problems and recursively enumerable predicates (abstract). *J. Symbolic Logic*, 15(1):77–78, 1950.
7. Davis M. Arithmetical problems and recursively enumerable predicates. *Journal of Symbolic Logic*, 18(1):33–41, 1953.
8. M. Davis. Speed-up theorems and Diophantine equations. In Randall Rustin, editor, *Courant Computer Science Symposium 7: Computational Complexity*, pages 87–95. Algorithmics Press, New York.
9. M. Davis. Hilbert's tenth problem is unsolvable. *Amer. Math. Monthly*, 80(3):233–269, 1973 (reprinted in [10]).
10. M. Davis. *Computability and Unsolvability.* Dover Publications, New York, 1982.
11. M. Davis, H. Putnam, and J. Robinson. The decision problem for exponential Diophantine equations. *Ann. Math.* , 74(3):425–436, 1961.
12. M. Davis and H. Putnam. A computational proof procedure; Axioms for number theory; Research on Hilbert's Tenth Problem. O. S. R. Report AFOSR TR59-124, U. S. Air Force, October, 1959.
13. G. V. Davydov, Yu. V. Matijasevich, G. E. Mints, V. P. Orevkov, A. O. Slissenko, A. V. Sochilina, and N. A. Shanin, "Sergei Yur'evich Maslov" (obituary). *Russian Math. Surveys* 39(2) (1984), 133-135. [translated from *Uspekhi Matem. Nauk* 39(236) (1984), 129-131.]
14. A. Degtyarev and A. Voronkov. Simultaneous Rigid *E*-Unification is Undecidable. CSL'95, Computer Science Logic, 9th International Workshop, Paderborn, Germany, September 1995. (Kleine H. Büuming, editor) *Lecture Notes in Computer Science* 1092, pp. 178-190, 1996.
15. M. Dehn. Über die Topologie des dreidimensionalen Raumes. *Math. Ann.* , 69:137–168, 1910.
16. K. Gödel. Über formal unentscheidbare Sätze der Principia Mathematica und verwandter Systeme. I. *Monatsh. Math. und Phys.* , 38(1):173–198, 1931.
17. M. Hack. The equality problem for vector addition systems is undecidable. *Theoretical Computer Science*, 2(1):77–95, 1976.
18. David Hilbert, Mathematische Probleme. Vortrag, gehalten auf dem internationalen Mathematiker Kongress zu Paris 1900, *Nachr. K. Ges. Wiss., Göttingen, Math.-Phys. Kl.* (1900), 253-297. See also *Arch. Math. Phys.* (1901) 44-63, 213-237. See also David Hilbert, *Gesammelte Abhandlungen*, Berlin : Springer, vol. 3 (1935), 310 (Reprinted: New York : Chelsea (1965)). French translation with corrections and additions : *Compte rendu du Deuxième Congrès International des Mathématiciens tenu à Paris du 6 au 12 août 1900*, Gauthier-Villars, 1902, pp. 58-114 (réédition : Editions Gabay, Paris 1992). English translation : *Bull. Amer. Math. Soc.* (1901-1902) 437-479. Reprinted in: *Mathematical Developments arising from Hilbert problems*, Proceedings of symposia in pure mathematics, vol. 28, American Mathematical Society, Browder Ed., 1976, pp. 1-34.
19. Bernard R. Hodgson and Clement F. Kent. A normal form for arithmetical representation of $\mathcal{NP}$-sets. *Journal of Computer and System Sciences*, 27(3):378–388, 1983.
20. O. H. Ibarra and B. S. Leininger. On the simplification and equivalence problems for straight-line programs. *Journal of the ACM*, 30(3):641–656, 1983.
21. James P. Jones. Universal Diophantine equation, *J. Symbolic Logic* 47 (1982), 549-571.

22. J. P. Jones and Y. V. Matijasevič. Register machine proof of the theorem on exponential Diophantine representation of enumerable sets. *J. Symbolic Logic*, 49(3):818–829, 1984.
23. J. P. Jones, Y. V. Matijasevič. Proof of recursive unsolvability of Hilbert's tenth problem. *Amer. Math. Monthly* 98(8):689–709, 1991.
24. James P. Jones, Daihachiro Sato, Hideo Wada, and Douglas Wiens, Diophantine representation of the set of prime numbers. *The American Mathematical Monthly*, 83(6):449–464, 1976.
25. J. P. Jones Diophantine representation of Mersenne and Fermat primes. *Acta Arithmetica*, 35(3):209–221, 1979.
26. G. Kreisel, A3061: Davis, Martin; Putnam, Hilary; Robinson, Julia. The decision problem for exponential Diophantine equations. *Mathematical Reviews*, 24A (6A):573, 1962.
27. E. E. Kummer. Über die Ergänzungssätze zu den allgemeinen Reciprocitätsgesetzen. Journal für die Reine und Angewandte Mathematik, 44:93–146, 1852.
28. J. Lambek. How to program an infinite abacus. *Canad. Math. Bull.* , 4:295–302, 1961.
29. G. S. Makanin. The problem of solvability of equations in a free semigroup. *Math. USSR Sbornik*, 32(2):129–198, 1977.
30. K. L. Manders and L. Adleman. *NP*-complete decision problems for binary quadratics. *J. Comput. System Sci.* , 16(2):168–184, 1978.
31. Yu. I. Manin. *A Course in Mathematical Logic*. Springer, New York; Heidelberg; Berlin, 1977.
32. M. Margenstern. Le théorèm de Matiyassévitch et résultats connexes. In C. Berline, K. McAloon, and J.-P. Ressayre, editors, *Model Theory and Arithmetic*, volume 890 of *Lecture Notes in Mathematics*, pages 198–241. Springer-Verlag, 1981.
33. A. A. Markoff. Impossibility of certain algorithms in the theory of associative systems (in Russian). *Dokl. Akad. Nauk SSSR*, 55(7):587–590, 1947.
34. Per Martin-Löf. *Notes on Constructive Mathematics*. Almqvist and Wiksell, Stockholm, 1970.
35. Yu. Matiyasevich. Svyaz' sistem uravneniĭ v slovakh i dlinakh s 10-ĭ problemoĭ Gil'berta, *Zap. nauch. Seminar. Leningr. otd. Mat. in-ta AN SSSR*, 8:132–144, 1968. English translation: The connection between Hilbert's Tenth Problem and systems of equations between words and lengths. *Seminars in Mathematics, V. A. Steklov Mathematical Institute*, 8:61–67, 1970.
36. Yu. Matiyasevich. Diofantovy mnozhestva. *Uspekhi Mat. Nauk*, 27:5(167),185–222,1972. Translated in: *Russian Mathematical Surveys*, 27(5):124–164, 1972.
37. Yu. V. Matiyasevich. Novoe dokazatel'stvo teoremy ob èksponentsial'no diofantovom predstavlenii perechislimykh predikatov. *Zap. nauchn. seminar. Leningr. otd. Mat. in-ta AN SSSR*, 60:75–92, 1976. Translated in: *J. Soviet Math.* , 14(5):1475-1486, 1980.
38. Yu. Matiyasevich. Prostye chisla perechislyayutsya polinomom ot 10 peremennykh. *Zapiski Nauchnykh Seminarov Leningradskogo Otdeleniya Matematicheskogo Instituta im. V. A. Steklova AN SSSR* , 68:62–82. (Translated as Yu. V. Matijasevič. Primes are nonnegative values of a polynomial in 10 variables. *Journal of Soviet Mathematics*, 15(1):33–44, 1981.)
39. Yu. Matijasevich. Some purely mathematical results inspired by mathematical logic, *Proceedings of Fifth International Congress on Logic, Methodology and Philosophy of Science, London, Ontario, 1975*, Dordrecht : Reidel (1977), 121-127.

40. Yu. V. Matiyasevich. Algorifmicheskaya nerazreshimost' èksponentsial'no diofantovykh uravneniĭ s tremya neizvestnymi. *Issledovaniya po teorii algorifmov i matematicheskoĭ logike*, A. A. Markov and V. I. Homič, Editors, Akademiya Nauk SSSR, Moscow 3:69–78,1979. Translated in: *Selecta Mathematica Sovietica*, 3(3):223–232, 1983/84.
41. Yu. Matiyasevich. *Desyataya Problema Gilberta.* Moscow, Fizmatlit, 1993. English translation: Hilbert's tenth problem. MIT Press, 1993. French translation: Le dixième problème de Hilbert, Masson, 1995
42. Yu. Matiysevich. A direct method for simulating partial recursive functions by Diophantine equations. Annals Pure Appl. Logic, 67, 325–348, 1994.
43. Z. A. Melzak. An informal arithmetical approach to computability and computation. *Canad. Math. Bull.* , 4:279–294, 1961.
44. M. L. Minsky. Recursive unsolvability of Post's problem of "tag" and other topics in the theory of Turing machines. *Ann. of Math. (2)*, 74:437–455, 1961.
45. M. L. Minsky. *Computation: Finite and Infinite Machines.* Prentice Hall, Englewood Cliffs; New York, 1967.
46. P. S. Novikov. On algorithmical undecidability of the word problem in the theory of groups (in Russian). *Dokl. Akad. Nauk SSSR*, 85(4):709–712, 1952.
47. P. S. Novikov. On algorithmical undecidability of the word problem in the theory of groups (in Russian). *Trudy Mat. Inst. Steklov.* , 44, 1955.
48. J. Pintz, W. L. Steiger, E. Szemeredi. Infinite sets of primes with fast pramality test. *Mathematics of Computations*, 53(187):399–406, 1989.
49. E. L. Post. Recursive unsolvability of a problem of Thue. *J. Symbolic Logic*, 12:1–11, 1947.
50. H. Putnam. An unsolvable problem in number theory. *J. Symbolic Logic*, 25(3):220–232, 1960.
51. J. Robinson. Existential definability in arithmetic. *Trans. Amer. Math. Soc.* , 72(3):437–449, 1952.
52. J. B. Robinson. The undecidability of exponential Diophantine equations. *Notices of the American Mathematical Society*, 7(1):75, 1960.
53. J. Robinson. Unsolvable Diophantine problems. *Proceedings of the American Mathematical Society*, 22(2):534–538.
54. J. C. Shepherdson and H. E. Sturgis. Computability of recursive functions, *J. ACM* 10(2):217–255, 1963.
55. D. V. Shiryaev. Nerazreshimost' nekotorykh algoritmicheskikh problem dlya nevetvyashchikhsya program. *Kibernetika*, no. 1:63–66, 1989.
56. Th. Skolem. Über die Nicht-charakterisierbarkeit der Zahlenreihe mittels endlich oder abzählbar unendlich vieler Aussagen mit ausschliesslich Zahlenvariablen. *Fundamenta Mathematicae*, 23:150–161 (1934).
57. C. Smoryński. *Logical number theory* I*: An Introduction*, Berlin, Springer-Verlag, 1991.
58. A. Thue. Problem über Veränderungen von Zeichenreihen nach gegebenen Regeln. *Vid. Skr. I. Mat.-natur. Kl.* , 10, 1914. Reprinted in: A. Thue. Selected Mathematical Papers. Oslo, 1977, 493–524.
59. M. K. Valiev. On polynomial reducibility of word problem under embedding of recursively presented groups in finitely presented groups. In J. Bečvář, editor, *Mathematical Foundations of Computer Science 1975*, volume 32 of *Lecture Notes in Computer Science*, pages 432–438. Springer-Verlag, September, 1975.
60. A. K. Vinogradov, N. K. Kosovskiĭ. Ierarkhiya diofantovykh predstavleniĭ primitivno rekursivnykh predikatov. *Vychisl. Tekhn. i Voprosy Kibernet.* , Lenigradskiĭ Gosudarstvennyĭ Universitet, Leningrad 12:99–107, 1975.

61. N. N. Vorob'ev. *Fibonacci Numbers*, 3rd ed., Moscow: Nauka, 1969 (in Russian).
62. M. A. Vsemirnov. Infinite sets of primes with Diophantine representations in eight variables. *Zapiski Nauchnykh Seminarov Peterburgskogo Otdeleniya Matematicheskogo Instituta im. V. A. Steklova RAN*, 220:36–48, 1995.
63. S. Yukna. Arifmeticheskie predstavleniya klassov mashinnoĭ slozhnosti. *Matematicheskaya logika i eë primeneniya*, no. 2:92–107. Institut Matematiki i Kibernetiki Akademii Nauk Litovskoĭ SSR, Vil'nyus, 1982.
64. S. Yukna. Ob arifmetizatsii vychisleniĭ. *Matematicheskaya logika i eë primeneniya*, no. 3:117–125. Institut Matematiki i Kibernetiki Akademii Nauk Litovskoĭ SSR, Vil'nyus, 1983.
65. URL: `http://logic.pdmi.ras.ru/Hilbert10`.

Hermann Maurer

Professor Maurer was born in 1941 in Vienna, Austria. He has got a Ph.D. in Mathematics from the University of Vienna 1965. Assistant and Associate Professor for Computer Science at the University of Calgary and Professor for Applied Computer Science at the University of Karlsruhe, West Germany. Since 1978 Full Professor at the Graz University of Technology. Honorary Adjunct Professor at the University of Auckland, New Zealand since October 1993. Honorary Doctorate Polytechnical University of St. Petersburg (1992), Foreign Member of the Finnish Academy of Sciences (1996). Author of thirteen books, over 400 scientific contributions, and dozens of multimedia products. Editor-in-Chief of the journals *J.UCS* and *J.NCA*. Chairperson of steering committee of WebNet and ED-MEDIA Conference series. Project manager of a number of multimillion-dollar undertakings including the development of a colour-graphic microcomputer, a distributed CAI-system, multi-media projects such as "Images of Austria" (Expo'92 and Expo'93), responsible for the development of the first second generation Web system Hyper-G, now HyperWave, and various electronic publishing projects such as the "PC Library", "Geothek" and "J.UCS" and participation in a number of EU projects (e.g., LIBERATION). Professor Maurer research and project areas include: networked multimedia/hypermedia systems (HyperWave); electronic publishing and applications to university life, exhibitions and museums, Web based learning environments; languages and their applications, data structures and their efficient use, telematic services, computer networks, computer assisted instruction, computer supported new media, and social implications of computers.

Not only Theory

When trying to summarize my over 35 years of computer science research I end up with one major very personal lesson that I have learnt, and that might be helpful to some younger researchers: it is such a rewarding and productive experience to work together with other researchers, be it students or top-notch experts that one must not miss it. Of two persons no one is ever "better" than the other. Rather, each one has weaknesses and strengths: the fun and challenge is to determine the right symbiosis. It took me a long time to find this out. Once I had realised the full importance of this for me, my life changed: research turned from hard work to hard fun. There are certainly scientists who have obtained greater insights working on their own than my brain would ever have allowed me to achieve, and I admire them. However, for me and I believe for many researchers the key for success is collaboration. Research can often be compared to solving puzzles. On your own, you may easily get stuck; it is my firm believe that n persons together ($2 \leq n \leq 5$) can solve a problem more than n times faster: research results are "super-linear" in the number of researchers. Thus, this paper focuses more on persons than on results. It is a thank you to all who have worked with me and who have become friends one way or another. I have made an attempt to mention all those I have ever co-authored something with. I apologize to others that I have met, learnt to appreciate and who have helped in different ways: there have been many, and there would not be enough room to do justice to all. Before going on, let me mention one further point: the book containing this paper is dedicated to the theory of computer science. However, I have spent much time also in other areas. For completeness' sake, and since drawing borderlines is difficult I will also report on non-theory stuff, albeit shorter.

The early years (59-71)

When entering university in Vienna in 1959 I was set on studying applied physics. However, the first mathematics class I attended was taught by Professor Edmund Hlawka - a superb teacher and researcher. He turned me around 180 degrees: mathematics it was to be, henceforth. And "clearly" the most beautiful areas of all, the theory of numbers. I took the only two computer science (= programming) courses available in those days in Vienna and progressed rapidly with my mathematics coursework. When I happened to meet Professor John Peck (who later became famous for e.g., his work on Algol 68) from the University of Calgary at the 2nd IFIP congress in Munich in 62 I was ready to accept his offer to go to Canada as graduate assistant for a year or two. While at Calgary, I fell in love with Canada, learnt more about computers and computer science and continued my work in number theory on diophantine equations, i.e. equations where one is interested in integer solutions, only. One of those equations $u^4 + v^4 = x^4 + y^4$ is called "Euler's equation" (like a lot of other equations) and the smallest non-trivial solution known in 62 was $133^4 + 134^4 = 158^4 + 59^4$. It was open whether smaller solutions exist. This seemed like an obvious application for computers. In a four-fold loop one would check for all quadruples of values (u, v, x, y) with $1 \leq u \leq v \leq 158, 1 \leq x \leq y \leq 158$ and $u < x$ whether $u^4 + v^4 = x^4 + y^4$. This brute-force approach (for the four-fold loop) takes over $100^4 = 10^8$ computational steps, too many for the computer in use at Calgary in

1962 (an IBM 1620). It was then that I discovered the power of sorting! Rather than examining 10^8 quadruples I would calculate some $n = 10^4$ values $u^4 + v^4$ for $1 \leq u \leq v \leq 158$ and sort them in ascending order as $z_1, z_2, z_3, \ldots$. This requires an effort of $n \log n$, i.e. roughly 10^5 steps. A smaller solution for $u^4 + v^4 = x^4 + y^4$ would exist clearly if and only if $z_i = z_{i+1}$ for some i, a test that can be carried out in about 10^4 steps. Thus, using sorting, I could cut down the computational effort from some 10^8 to some 10^5 steps, quite feasible on a 1962 computer. The result (sigh) was negative: no non-trivial solution smaller than the one known to Euler exists. Nothing to publish, but a first lesson for me: computers can help in number theory, and sorting is a surprisingly powerful tool in many application areas. This realisation would come in handy years later in the study of data structures and geometric algorithms ...

While continuing research in number theory and obtaining a few new results on the Pellian equation (integer solutions for $x^2 - dy^2 = 1$) I joined the computer centre of the government of Saskatchewan as "system analyst" (May 62 - December 62). Those short eight months of really down-to-earth computing work would prove invaluable later for my understanding of applied computer science. Although not at all related to theory I feel the urge to report two anecdotes. As first job, I was given a huge assembly language program for an IBM 1401/1410 without further explanations and the request "read it so you understand what it does". After two days I was totally frustrated: after reading pages of code I thought I knew what the initial segment of the program would do: nothing but print two columns of asterisks indefinitely, until someone would physically stop the printer. After a restart, 132 dashes would be printed over and over in the same line until again the printer would be physically stopped! When I reported this obviously wrong conclusion (but I had checked it three times!) to my supervisor he was delighted: "Yes, this is what it is supposed to do. This is used to align the forms in the printer appropriately. And the repeated printing of dashes creates a perforation so the forms can be torn off easily." As it turned out, I was thrown into the middle of the first world-wide project to computerize health care. After gruelling months of work, when the team I felt I belonged to by then had finally completed its job one day at 4 a.m., we drove out into the prairies. Someone in our group knew enough about farming, saw a harvesting machine and a ripe wheat field and before we knew it we released our bent-up tension by harvesting that field. A surprised but pleased farmer invited us for a hearty breakfast 3 hours later ...

Back in Vienna, Austria, I continued my Ph.D. thesis work but also needed a job. Werner Kuich, a friend from my freshman years (who later became one of the first computer science professors at the Technical University of Vienna) helped me to get a job with Professor Heinz Zemanek's IBM funded research group whose roots go back to Mailüfterl, the first European transistorised computer built by Zemanek in the late fifties. In this research group I started to learn about compilers (and wrote my first one [2]) and got interested in formal languages and formal description methods. I also made my first scientific contribution by noticing that one (apparently) could improve on the $O(\log n)$ performance of binary searching by making use of the structure of the data by interpolating rather than brute-force halving. After all, if you look in a phone book for, say, "Beran" you don't open it in the middle but closer to its beginning! I was able to prove the superiority of this technique in a paper [1] only for special cases. Ten years later this "interpolation search" was shown to require $O(\log\log n)$ steps even under rather weak assumptions ...

I completed my Ph.D. thesis on "Rational approximations of irrational numbers" [4] under the expert guidance of Hlawka (based on Baker's hypergeometric series that became quite famous later on) in 65, and was starting to work more and more on the formal definition of programming languages in the IBM Lab. Indeed, I am one of the co-authors (together with Kurt Bandat, Peter Lucas and Kurt Walk) of an early version of the formal definition of PL/1 in 65 [3]. However, I found the work at the IBM Lab very frustrating: how can one describe a mess such as PL/1 in a neat, formal fashion? I quit my job and accepted an Assistant Professor position in Calgary. I must say that later I was very impressed when the Formal Definition of PL/1 was indeed successfully completed, and the technique became well-established as Vienna Definition Method: over thirty years later it is still one of the corner stones of formal definition methods. This incredible achievement is due to the genius and quiet persistence of Peter Lucas (now Professor at the Graz University of Technology), and due to the team-guiding skills of Kurt Walk (who, now retired, also teaches at Graz). When leaving IBM I remember one piece of advice from Zemanek that I did not appreciate then, but I do now: Zemanek, who knew that I was still torn between number theory and computer science, told me: "Make up your mind. You can only successfully serve one master, you can only successfully pursue one line of research." Of course, Zemanek was right.

I started to concentrate on formal languages. I compiled a large annotated bibliography and used it to write a German book on formal languages ("Theoretische Grundlagen der Programmiersprachen" [6]) that appeared in 1969. It became a German best-seller. Without even realising this, the book made me well-known in Germany, while I was just being promoted to Associate Professor in Calgary and was working on ambiguity problems in context-free languages. I believe that I gave the first simple proof that $\{a^n b^n c^m \mid n, m \geq 1\} \cup \{a^n b^m c^m \mid n, m \geq 1\}$ is inherently ambiguous [5] and that there are context-free languages that are inherently ambiguous of arbitrary degree. From Werner Kuich I learnt how to see such issues in a more general light using formal power series: we have co-authored a number of papers (e.g. [7]) and I was and I am impressed by Kuich's systematic and careful mathematical approach that often reduced "pages of handwaving" to a few lines of precise proof. Those of you who have ever looked at the lengthy "proofs" of equivalence between PDA's and CF languages in most books (including my own, Hopcroft and Ullman's classical one, or the one by Harrison) and compared it with the one-paragraph proof in the EATCS Monograph in the "Semirings, Automata, Languages" book by Kuich and Salomaa will have felt the same awe that I have felt! I feel almost embarrassed to report that in the late sixties I was convinced that if L_1 and L_2 are CF (like e.g. $L_1 = \{a^n b^n c^m \mid n, m \geq 1\}$

and $L_2 = \{a^n b^m c^m | n, m \geq 1\}$ and $L_1 \cap L_2$ is not (like $\{a^n b^n c^n | n \geq 1\}$ for L_1 and L_2 mentioned before), then $L_1 \cup L_2$ is inherently ambiguous. Indeed I spent weeks trying to prove this conjecture. When I mentioned it to Professor Seymour Ginsburg a few years later he almost had a laughing fit: he immediately wrote down a trivial counterexample! The book "A Collection of Programming Problems and Techniques" [8] that I co-authored with my friend Mike Williams (who later became famous for his work in the history of computing and his supervision of computer exhibits e.g., in the Smithsonian in Washington) appeared in 72, already after I had accepted the position of professor for computer science at the University of Karlsruhe.

Learning to cooperate (71 -76)

My peaceful pace of life changed as soon as I moved to Germany. I was suddenly surrounded by a host of ambitious colleagues and in charge of supervising half a dozen bright Ph.D.'s. Research emphasis was on formal language. My first joint publications in Karlsruhe were with my colleague Otto Mayer [9] (who later became a very successful professor and dean at the University of Kaiserslautern) and his excellent student Armin Cremers (who, after at stint in the USA is now one of the most influential professors at the University of Bonn) and with my Ph.D. student Hans-Peter Kriegel (e.g. [11]) (now professor at Munich). Both Mayer and Cremers have remained life-long friends, and we have co-operated on a number of completely different projects over the years. With Professor Klaus Neumann (who still is professor at Karlsruhe) I wrote my first "non-scientific" paper on computers and the like, and we both enjoyed it. The year 1974 turned into a decisive year for me. I had corresponded with Seymour Ginsburg, one of the "gurus" in formal language theory for a long time, and I had met him a few times at major conferences. When I invited him for two weeks to Karlsruhe for co-operation he accepted to my delight ... and I was in for a new experience. It was Seymour who introduced me to the work style "spend as much time together as you can for 1-3 weeks ... by that time your are bound to have enough results for a nice paper." After working with Seymour for a week I felt like I was married to him: from breakfast to after dinner we would be together and "talk shop". I could not believe how successful this mode of operation can be, nor could I stand it after 6 times 16 hours daily any longer. I decided we (mainly I) needed a break. I took my family and Seymour for a hike to a small lake in the Black Forest. The weather was lousy but anything was OK to get away from formal languages for a few hours. However, no sooner had we got out of the car when Seymour pulled out a piece of paper and we continued working as we walked (me holding the umbrella for both of us) to the lake and back. (I am deeply grateful that my wife Ursula and the kids put up with me in situations like this. If they ever read this: thanks!) Anyway, Seymour and I wrote a number of papers on "grammar forms" (e.g., [12]) on this and later occasions and I am indebted to him for teaching me a kind of co-operation which would be my main mode of operation for many years. Let me tell a little bit more about Seymour

Ginsburg: he does not need an introduction to my generation . He was one of the giants in formal language theory in the US; his books on Automata Theory and the Theory of Context-Free Languages remained classics for a long time, and his numerous papers and his notions of AFL's, AFA's and grammar forms shaped formal language for over a decade. Grammar forms were one of the latest features blooming in formal languages in the US (till about the late seventies): by defining certain morphisms on the production rules of grammars, each grammar (form) G gives rise to a family of related grammars $\mathcal{G}(G)$, and thus to a family of languages $\mathcal{L}(G)$ by defining $\mathcal{L}(G) = \{L(G') | G' \in \mathcal{G}(G)\}$. Thus, grammar forms are a good tool for studying families of related languages. The "fate" of formal language theory in the US is typical for developments there that one can either smile about or disapprove (I do the latter): a group of researchers comes up with some new concepts; they produce good results; they oversell their applicability to get large research funds; many other researchers join the band-wagon; for a while all research outside this area is considered "obsolete" or "irrelevant"; at some stage disillusion sets in; the topic stops to be fashionable; another area starts to be THE in-thing to do; an analogous cycle starts all over again. I do not disapprove of researchers overselling the applicability or importance of their results: this has always been done. Have you e.g., seen the proposal of Leibnitz for a new language as cure-for-all as recounted in Umberto Ecco's book "The universal language"? What I do not like is the strong "fashion trends" that dictate what is "in" and "out" in North America. And if you are not working in an "in" area your chances for research funds, recognition or good job offers are close to nil. I am happy that Europe is less radical, this way.

It was around 1973 when I first met Derick Wood (then at McMasters in Canada, now prestigious professor at the top university in Hong-Kong): we started to co-operate both in formal languages (e.g., [13]) and algorithms for data structures. I will have to say more about Derick later, but let me mention that at this time Thomas Ottmann (now professor at Freiburg) and Hans-Werner Six (still a good friend and now professor at Hagen) joined my group as assistants in Karlsruhe. In a joint paper [15] we introduced a class of search trees that Ottmann later improved to the now classical brother-trees, one of the most elegant search structures ever invented. Indeed, I have always admired, and certainly still do Ottmann's sharp brain, endurance when working on whatever project, and his reliability that is hard to describe if you have not experienced it: if he says "yes, I will do this by date x" you can be sure to have it by date x-2, the latest. During my 5 years as professor in Karlsruhe the other full professor appointed in my institute was Wolffried Stucky: we published material on using syntax-directed techniques for programming purposes [14]. However, what has impressed me most about Stucky was his quiet humour, systematic organisational work, his ability for co-operation and for putting up with me when I had a flare-up of bad temper or some crazy idea. If I have to ever choose someone to run a group together with, Stucky will always be my first choice: the fact that he continued successfully building up the group in Karlsruhe after I left and times where not always that good, and the fact that he later obtained many honours and became president of the GI (the German Computer Society) says also a lot about his friendly yet in the end decisive personality.

During 75 I got to know one other very fascinating person, Harald Hule. He is an Austrian who had discovered his calling for mathematics only at the age of 28, but then studied and completed his Ph.D. in 4 years (!). He went to Mexico and Brazil for some years thereafter and it was in Brazil where I met him: he helped me understand a bit of Brazilian culture and life-style (thanks Harald!). I invited him to Karlsruhe as visiting professor. And although he had never ever worked in formal languages before, within three months he was the main author of a paper on OL forms [22]!

However, the most important thing that happened to me in the period under discussion occurred at Oberwolfach in 1972, this beautiful retreat where mathematicians (and then also theoretical computer scientists) meet in the idyllic setting of a Black Forest village to work on and discuss some special topic for a week. My long-term friend Werner Kuich who I mentioned before, introduced me to Arto Salomaa from Turku, Finland. Salomaa has been shaping theoretical computer science in Europe since around 1960 and, if this is possible, his scientific output and influence continues to grow even more every year. But this is besides the point: the main aspect for me is that we became close personal friends, not just working together but also in many other ways. The last 20 years would not have been the same without my friendship with Salomaa that continues although from a research point of view I have started to follow other paths over the last years. I think this is the right time to say it: I love my profession for the chance it provides to meet, work with and make friends with not just people, but personalities. Some persons you start to respect or to admire. Others you like for their humour or their idiosyncrasies. And with some you develop a bond akin to members of your core family. Arto Salomaa, or "Tarzan" as some of us call him, belongs to the last category: working with him never felt like work but was a pleasure. I love to think back to the beautiful times we had working and relaxing at his kind late sister Sirkka's place Lauttakylä or at Arto's Rauhala (both an hour's drive from Turku), or in the Black Forest, the mountains of Austria, or the old city of Graz in Austria.

One of the highlights has always been visiting Rauhala - Arto Salomaa's country home - with a rustic sauna that may well be the best in the world. I have very very fond memories of my stays, usually also recorded in a "Sauna poem" in Arto's book of visitors (see the contribution "Events and Languages" by A. Salomaa in this book for further examples). My last entry there (April 1997) goes like this:

In Salosauna, once again
I sat; and I saw plain
that life is more than Hyperwave[1]
that everything and all, I have

[1] For the meaning of Hyperwave see the section "Applications ... and some theory (1981-1997)".

accomplished doesn't mean that much
compared to friends, and love, and such.

In Rauhala, the days are sweet
when Tarzan, Jane and peace I meet
so let me sing as praise this song:
I missed this place and you for long
and thanks for sharing once again
some thoughts of joy, of fun and pain.

Work with Arto Salomaa and Derick Wood who was in our group from the beginning was quite productive: we co-operated as a team on over 30 publications between about 1977 and 1983. This period was probably my most interesting time as theoretical computer scientist, so I spend a separate section on it.

Before turning to the next section let me mention one item that I was able to do for the Theoretical Computer Science community that I am proud of and that is almost forgotten: I was EATCS Bulletin Editor from 1977 to 1981, from no. 3 to no. 14. The first two numbers were few-page leaflets, and even no. 3 was still small enough to be stapled. The current "book-like" look for which the Bulletin is now famous for and that helped to grow membership quite a bit started with no. 4 and was perfected later by Rozenberg. Even the circular EATCS symbol dates back to those days when I was Bulletin editor!

The MSW years (1975-1981)

Although I co-authored quite a few results outside formal language theory with then (or later) very prominent researchers, and I also started to dabble in more applied areas, the most decisive influence in those years was the work with Salomaa and Wood (and we were proud to be called the MSW-team after our first successes). Our work started by combining two areas that were "red hot" in those days: grammar forms (as mentioned earlier) and L-systems. L-systems are the invention of the famous late theoretical biologist Aristid Lindenmayer who noted that growth-processes can be described in a natural way by applying production rules much like in ordinary Chomsky-type grammars, but applying the rules simultaneously to all symbols of an intermediate string (representing a linear arrangement of cells of some organism), rather than to just a symbol at-a-time as is done in the usual Chomsky-type derivation process. It is curious that this idea that gave so much impetus to the study of formal languages came from biology rather than computer science. But far-sighted computer scientists such as Grzegorz Rozenberg, Arto Salomaa, Gabor Herman and others soon recognized its importance to language and automata theory; thus "L-theory" started to explode as synthesis between biology and computer science.

The first MSW paper on EOL Forms [17] combined ideas from grammar forms with EOL systems, the "L equivalent of CF grammars", as one might say. The paper laid a solid foundation for what we thought could turn into a major field of study. Two further papers followed. All three received good reviews and were accepted for publication immediately. Then, when working on MSW paper number four [23] we ran into a crisis that I will never forget.

We had, by that time, established our MSW routine: two of us would meet, work on a topic all of us had agreed on, write up a sketch of the paper including all proofs, etc. The third one would critically read and amend the results, and do a first draft of the final version of the paper. This time, it was Arto and myself doing the groundwork on a new paper at Turku. On the second day, our work started to bog down: when Arto came up with a suggestion, I would find a counter-example; when I proposed a possible result he would show that it could not be true. We continued this Ping-Pong "game" for hours, frustration growing. Suddenly we stared open-jawed at a counterexample to the first lemma in our first paper [17], a lemma all our work (and three already accepted papers) had relied on: We checked the proof of that first lemma. It said laconically "Trivial". Intuition had badly tricked us and all referees! I am sure you can imagine how we felt: all previous work going down the drain; three papers based on a wrong result about to appear in print! I was at a complete loss. At this point Arto said something I will never forget: "I think it is time for a long sitting in sauna."

This is what we did. And the Finnish proverb that "Sauna opens your brain" proved correct: after two sittings with excellent löyly it was clear that (a) the critical lemma was indeed wrong and (b) a weaker version of it could be proven and was enough for all our purposes.

We were able to correct the first three papers before they were printed with a sigh of relief. And indeed, the fact that the original lemma 1 was incorrect established that L-forms were not just a variant of grammar forms, not just a cute little new island of knowledge to explore, but a new continent with entirely new phenomena, as striking to us as Africa with its elephants and giraffes must have been to early explorers (if you permit me to stay with my geographical metaphor).

MSW work flourished, and L-forms have become part of today's classical formal language theory.

It is with deep gratitude to both Arto Salomaa and Derick Wood for the wonderful co-operation, never marred by any rivalry, envy or what have you. Wood turned out to be the master of looking at obstacles from so many different directions, chipping away at the problem until it dissolved into nothing or until a real hard core would remain. And then Salomaa would take over, sit and think for a long time and finally say: "Maybe we should try the following: ... ". And then it was already clear that he had an outline for a new route that would eventually succeed.

It was during this time that I also got to know Karel Culik II better, and we started to co-operate. Karel is one of the persons I know with a terrible sharp (but sometimes impatient) brain ("come on, we fill in the details later"), who has an infinite amount of energy, always willing to prove that he is better than you are (in research, tennis, chess, ... you name it) ... and usually he is. Before we did our first joint paper I had learnt to admire him for showing that the DOL sequence equivalence is decidable. If the following report on how this happened is not correct, forgive me, it really does not matter: it is certainly typical for Karel.

Karel had not done much on L-systems, when the wave started. However, when he attended an L-conference he heard a talk on the above mentioned problem: given two words w_1, w_2 and two homomorphisms h_1 and h_2 can you decide if ${h_1}^n(w_1) = {h_2}^n(w_2)$ for all n (where h^n means n-fold iteration of a homomophism h)? This question sounds deceptively easy, yet is quite deep. At the end of the talk, Karel got up and said he could prove decidability. To a stunned audience he gave a sketch of a proof: although intriguing, the proof contained "large holes", as Karel was ready to admit. "But these are just details that I will fill in till tomorrow". Well, the proof was refined next day by him, yet many gaps remained. From there on, Karel kept improving and detailing his "proof" many times, much to the chagrin of some colleagues who got more and more exasperated by having to wade through more and more complex arguments and - at the end - still discovering gaps filled by "handwaving". I remember a letter by Arto Salomaa that he was about to study the last version of (and for the last time) a "proof" of the decidability of DOL sequence equivalence by Karel. However, this time the proof was "water-tight". Karel had indeed solved this very hard problem ...

Thus, it was "natural" that I wanted to start with simple topics such as e.g. [18]. Soon we ended up in deeper things like [20] or [21], the latter also with co-author K. Ruohonen, then one of the many top-notch assistants of Salomaa. Working with Karel was both exhilarating and frustrating: Karel always seemed to see solutions (like in the case of the DOL sequence problem) very fast, but just ignored (or considered trivial) gaps in the proof. In working with Culik I understood the first time very clearly: one can co-operate even if one "functions" very differently. With Culik, my only function was to punch holes in his arguments; he would fix them, I would find problems, he would refine his proofs: he was the brain, I just a humble critic ...

Grzegorz Rozenberg (or Bolgani as his friends call him) whom I had the pleasure to meet first during those years is just about as much the opposite of Karel Culik II as can be. Where Karel can be abrasive, Bolgani is gentle. Where Karel stubbornly pursues one problem at a time, Bolgani sees a vast array of problems and possibilities, too large to explore, so he concentrates on new ideas and fields with a staggering amount of energy and imagination. He is the best "salesman" of scientific ideas I have ever met, presenting difficult material in such a superb way (almost like a magician) that one feels compelled to listen and to appreciate what he has to say. It has been an honour to be accepted by Bolgani as friend: I know that I will not be able to ever repay him for his generosity, and for the open and warm way we have co-operated scientifically and in other ways. It was also through Bolgani that I have co-authored a paper with the famous A. Ehrenfeucht from Boulder [32].

This period of my life has also been rewarding by being able to work with bright young assistants, first at Karlsruhe, Germany, later at Graz, Austria. I have reported about some already above, but I must also particularly mention Jürgen Albert who wrote an excellent Ph.D. thesis, with whom it was a pleasure to co-operate (e.g., [19] and [27] who has been now professor at Würzburg, Germany, for a long time and whose quiet and gentle ways combined with excellent work continues to impress me. One of my last excellent Ph.D's in formal languages (already in Graz) was Werner Ainhirn [26], who later left for work in industry in Germany, but who has returned to Austria in the meantime. As luck has it, we are now cooperating on a substantial applied project.

During my first years in Graz (starting in 1978) I also had the chance to invite and work with many visitors, some of them already famous, some rising stars. My respect, if not awe, for e.g. Maurice Nivat from Paris had always been tremendous: to actually work together with him and publish successfully on rational transductions (e.g., [29]) has been a definite highlight. As my interest slowly extended from formal languages to the theory of algorithms (particularly concerning region location problems) I had the pleasure to get to know John Bentley (e.g.[24] and [28]) later e.g., famous for his "Programming Pearls" who was a hit with the students in Graz when he came to our offices and to stodgy, conservative professor Hermann Maurer on his skateboard; I am particularly proud of the papers [30] and [31] with Thomas Ottmann and Jan van Leeuwen, since I believe they constitute the first systematic approach to provide efficient solutions for the dynamic versions of problems for which up to then good techniques only for static cases had been known. While I also published quite a bit with members of my institute such as R. Frey, V. Haase, J. Stögerer, W. Bucher, G. Greiner, H. Mülner, I. Mischinger, F. Haselbacher, P. Lipp, H. Cheng, J. Theurl (now vice-president of Graz University of Technology), W. Jaburek (who received two Ph.D.'s in Law and in Computer Science and has been an influential force in computer science laws in Austria for now over ten years), and G. Haring (who later became professor at the University of Vienna and Head of the Austrian Computer Society) and short-term visitors such as Detlev Wotschke who later became professor at Frankfurt [33], I.H. Sudborough (professor at the University of Texas at Dallas) [35], D.G. Kirkpatrick [39] from the University of British Columbia at Vancouver, and both famous Franco Preparata and Arni Rosenberg [42], and while I also did some non-scientific publishing on my own and with others including my wife Ursula, I cannot go into detail but will just concentrate on one further aspect: two of my last theory Ph.D.'s in Graz turned out to be particularly talented:

In the process of writing their theses they started to surpass my knowledge and talent in this area and have become leaders in their field: Emo Welzl, now professor at the ETH in Zurich, Switzerland, and Herbert Edelsbrunner, who has been professor at the University of Illinois at Urbana, for some ten years. I am proud to know that I helped them a bit early in their career; it is both exhilarating and humbling to see how fast talented students develop. I got to know Herbert in a second year course on data structures where I followed my books [10], [16], presenting well-known material with well-known proofs. It happened over and over again that Herbert would ask during my classes "could one not also see this in the following way?" ... and such question invariably meant that he had discovered some inaccuracy or mistake in a proof that had been around for years. We started to work together when Herbert was just a bit over twenty years old (e.g. [37], [38], [42]). By the time Herbert was finishing his Ph.D. visitors often came to Graz to see him, rather than me ...

Emo Welzl was initially working more on formal languages and discrete mathematics. Already his early contributions like [34], [35] or [41] showed his talent; his career went rapidly upwards: he became young professor at Berlin, obtained prestigious awards and distinctions and was soon one of the stars in German computer science before moving to the famous ETH in Switzerland.

Looking back at those days when I was mainly working in theory I must say it was a great time; the talent of some students in both Karlsruhe and Graz was indeed impressive and it was a pleasure to see them "grow".

Let me finish this section with one of the more embarrassing stories of my theory days. In [25] we tried to show how to use single public key pairs to allow various overlapping groups secure access to information. The aim (to avoid having a special set of keys for each class of information one is entitled to access) is important, and the paper got quoted a few times. However, two years after it had appeared a gap in our argument was found showing that the proposed method was not safe. Of course, we published a corrigendum (basically much weakening our earlier "results") but I can just hope that nobody has seriously used the method proposed. The fact that in electronic versions one can make an addendum in the spot where the mistake occurred is one of the reasons why I believe today in electronic (Internet) publishing (see [67] or [76]) and http://www.iicm.edu/jucs_annotations, and I am happy that Cris Calude and Arto Salomaa decided that we three together should get started in this direction in 1994. I will have to say more about this in the latter part of the next section.

Let me finish this section with a story of my early theory days that might destroy my credibility as good organiser, but it is time to tell. I was responsible for organising ICALP'79 (the sixth ICALP) in Graz. The reception was to take place in the best setting Graz has to offer: the Renaissance castle Eggenberg, lit by 3.000 candles in crystal chandeliers (electrification would destroy the murals!), local food and wine served by pretty young restaurant-trainees, sit-down dinner with the provincial governor.

As recent arrival I asked a more senior member (let me call him Mr. X) of the organising committee to make sure that we would have a reception on Monday evening. At each meeting I asked Mr. X whether everything was OK with the reception. Every time (I can prove it through the minutes of the meetings!) the answer was "yes", yet I never received a written confirmation from the government.

When the program was already printed I got nervous and phoned the secretary of the governor. "Let me check", he said. "Yes, everything is fine, a dinner reception with the governor is scheduled for your group for Tuesday evening." I was appalled: "You mean Monday evening." "No, sorry, Monday is impossible, there is a concert scheduled in the same rooms."

I was at a loss what to do: we had, of course, another event scheduled for Tuesday! I ran to the office of Mr. X and told him furiously that he bungled the program. He listened patiently, sighed, took the phone, called the ticket reservation agency. "How many tickets have you sold sofar for that concert?" "Two", was the answer. "Great, I am Mr. X., I take the remaining 198 tickets".

Thus, two surprised tourists got a a free dinner after the concert they had booked. And many colleagues complemented me and asked with surprise how I had managed to throw in a high-class concert free of charge ...

Applications ... and some theory (1981-1997)

Around 1981 I started to work more and more in applied areas of computer science. I have never quite given up theory, as recent publications like [68],[70],[71], or [88] show. Yet, my emphasis has shifted. Following the spirit of this book that is dedicated to theory, but also following my intention to mention all the many persons I have had the pleasure to co-author papers with I will give as compromise a fairly tense description of this comparatively long period.

Shortly after I moved to Graz I got involved in a study for the Austrian government whether Videotex (also called Prestel in UK, Minitel in France, Bildschirmtext in Germany, ...) – a TV/telephone/modem-based distributed information system – should be introduced also in Austria. My recommendation was yes, but the network should be seen as a network for special network PC's rather than for "dumb" videotex terminals. In particular, those network PC's should be able to handle vector based colour graphics and the execution of code just downloaded. We called our network PC's then "intelligent videotex terminals" rather than Net Computers (NC's) as they would now be called, and while we called the downloadable executable software "telesoftware" today everyone talks about JAVA (applets), really just a variant thereof. The recommendation to develop such a more modern version of Videotex was accepted by the Austrian Telecommunication authorities. Since no suitable PC's existed (this was before the time of the first IBM PC, remember!) my group got suddenly shouldered with the development of a dedicated terminal. Without any hardware knowledge to speak of I could not have done the job without my brilliant assistant Reinhard Posch who designed the hardware and supervised most of the system software of the device we called MUPID: officially this stood for "MultiPurpose Intelligent Decoder", but insiders know that it stood for "Maurer Und Posch Intelligenter Decoder". One of my best friends said at some stage that it really means "Maurer's Undertaking Puzzles IBM Directors". Anyway, MUPID was quite a hit for a few years (there is no Austrian developed computer that was ever produced in similar quantities, a total of some 35.000, almost 40% in export), but the emergence of the IBM PC's and clones and the departure from Videotex towards more open standards as we see them today in the internet was the end of MUPID as hardware product. However, MUPID team members continued their work by founding a total of 15 new IT-companies in Graz with today some 250 employees. The MUPID years and all the turmoil of going commercial would warrant a separate chapter except that there is not so much theory involved in all this. The most powerful person, and the person who I count as one of my best friends in Graz, is Reinhard Posch. He has been now full professor at Graz University of Technology for over ten years and is internationally recognized for e.g., his work on computer security. Co-operation with him has always been a pleasure, see e.g. [44],[45]. The MUPID and Bildschirmtext efforts brought also interesting publications with other team members, such as Heidrun Bogensberger, Walter Schinnerl, Gerhard Greiner, Walter Jaburek, Helmut Mülner, Günther Soral and particularly Dieter Fellner: Dieter later went to Canada for some time (to St. John's, Newfoundland) and returned as professor to Bonn, Germany. There he has built up a sizeable and recognized group in the areas of computer graphics and electronic publishing. It is a pleasure to say that the contact with the original core "MUPID Team" is still intact, and that co-operations on many levels still continue.

During this time I was also consultant at IIASA (International Institute for Applied Systems Analysis) for two years. This was the Austrian attempt to diffuse the then "cold war" by bringing researchers from East and West together for shorter or longer stints of co-operation in the former summer castle of empress Maria Theresia just outside Vienna. It was there that I met Istvan Sebéstyén from Hungary and Wolf Rauch. The former joined my group in Graz for some time at a later stage, the latter became professor at the University of Graz and in 1997 its president. Sebéstyén was marvelously good in digging up facts and combining them, see e.g., [36] or the paper [40] that we co-authored with J. Charles from the Institute for the Future in California. Wolf Rauch is an absolutely unique combination of researcher, philosopher, organiser and diplomat, and interested in all kinds of questions, see e.g. [43]. It is a real joy to work with him and have him as friend ... and we have done some unusual things together! Like at the Hypermedia 1991 meeting in Graz [57] when we had a public discussion "Pro and Contra Hypermedia": One of us had to take the "pro" point of view, the other the "contra". But the chairperson Jürg Nievergelt from ETH Zurich had the audience vote before we started who would be "pro", who "contra". (I ended up "contra" and it was not easy to stay my ground against an opponent as skilled in arguing as Rauch is!) [59].

My long-term friend Wilfried Brauer (the founder of computer science in Hamburg, and now professor in Munich) sent one of his students, H. Cheng, to Graz resulting in a number of papers, e.g., [46].

My interest in non-mainstream computer science also resulted in a joint paper [47] with Norbert Rozsenich, who as vice-minister for research for over 20 years has been shaping research policies in Austria more than any other individual. I have found Rozsenich's support, imagination and frank criticism always very refreshing ... and I think I learnt also a bit from him how to deal with politics (unfortunately not enough!).

Starting in 1985 my interest in using computers for teaching and learning continued to grow. After early work with H. Huemer, Peter Sammer, and Dana Kaiser, co-operation with the late John Garratt from Control Data, Germany (a fruitful and fun co-operation made easy through John's stamina and humour) started to lead to significant projects culminating in COSTOC (Computer Supported Teaching of Computer science). This was an interesting period consisting of three aspects: (1) Implementational work with e.g., P. Lipp. J. Nagy, John Garratt, and others; (2) co-ordinational work as editor of a series of courseware modules (with prominent authors such as Arto Salomaa, Gerhard Barth, Thomas Ottmann, Peter Widmayer, Herbert Kraus, Henry Shapiro, Egon Börger, Vladimir Stepanek, and Peter Warren ... (just to mention a few), and (3) work on the boundary between applications and theory. For the purpose of this book it is appropriate to dwell a bit more on the last point.

This semi-theoretical work was carried out with e.g., my assistants Fritz Huber (see e.g. [51]), Robert Stubenrauch and Ludwig Reinsperger, but also with my colleagues Thomas Ottmann from Freiburg and Fillia Makedon, then Dallas now Dartmouth, e.g. [48], [50], or [52]. I particularly enjoyed working with brilliant, imaginative and enthusiastic Fillia Makedon: I have learnt a lot from her, particularly looking at things from a point of view as general as possible ... and she also converted me to a fan of her home country Greece (but it takes little to become a fan of Greece: if you aren't one yet, just go there!) A period as adjunct professor at the University of Denver also enabled me to become Ph.D. supervisor of sharp-minded John Buford-Koegel, now at the University of Lowell, Massachusetts [49].

Paul Gillard from St. John's, Newfoundland, and Mike Stone from Calgary, Alberta, came to Graz as visiting professors. Both are long-term friends who have shown me more beauty and serenity in Canada than I can describe in a few lines, but I cannot suppress my urge to mention one or two events. Like when Paul Gillard took me on a multiday fishing trip into real wilderness: as we were camping on an island in the stream, and evening fog was rolling up the river from the sea, the sight of a dinosaur appearing around the corner would have not much surprised me; or like when Mike Stone took me telemark-skiing in 15 foot powder snow in Western Canada and we started to get caught in a blizzard: I reciprocated by taking Mike a year later on skis up a mountain where we had to make our way down through dense, steep forest and on just patchy icy snow .... Both with Paul and Mike we worked together on teaching aspects, e.g. [53] and [56]. Short-term visits of my good friends Gordon Davies and Jenny Preece (then both at the Open University in the UK where Gordon still is) resulted in e.g. [55]. And it was the first longer meeting with Pat Carlson from Rose-Hulman Institute of Technology in the USA that crystallized my "missing organ thesis", later published in e.g. [60]. I believe this thesis is simple yet interesting enough to briefly review it here: our ears are passive instruments (they only receive sounds); we have an active counterpart (our mouth); the eyes are also passive instruments (they can just receive pictures); but we do not have an active counterpart, no "mouth for the eyes", no "picture generating organ" (= the missing organ) that allows to easily convey (mental) pictures from one person to another. Looking at this phenomenon more closely leads to two conclusions: (a) the missing organ is deeply influencing our communicative behaviour and (b) we should try to develop a prosthesis for the missing picture generating organ much as we have done for other missing organs such as wings or gills. Some work on this is still in progress. In connection with this and computers and teaching I also have co-operated with some of my learning-theory and cognitive-psychology colleagues, particularly Ricky Goldman-Segal from Vancouver and Dave Jonassen from Pennsylvania [84].

Despite all efforts that have gone into the development of Videotex and computer assisted instruction all over the world and in Graz neither area managed to achieve a real breakthrough in the eighties. Around 1988 I started to form a group to analyse what future networked multimedia systems would have to look like, networks that would work better than Videotex and would solidly support educational aspects.

Previous work, particularly with Fillia Makedon, Reinhard Posch and the MUPID and COSTOC efforts had yielded some insights. The wave of hypermedia efforts at Brown and MIT yielded further. As luck would have it other factors in the form of three brilliant scientists helped our efforts in Graz, tentatively code-named Hyper-G: Ivan Tomek from Acadia University, a bit later Nick Sherbakov from St. Petersburg, both visiting Graz for an extended period (Nick actually deciding to stay for good) and particularly Frank Kappe, then one of my ambitious and ingenious assistants who soon became project leader. Ivan Tomek's quiet and systematic work brought the theoretical underpinnings of Hyper-G to a good start. Also, it was and is a pleasure to co-publish with Ivan: after a few hours of brainstorming we often end up with enough ideas for more papers than we can possibly handle. Using Ivan's great skills to compose excellent papers once the basic ideas are clear we managed to co-author over a dozen publications within two years (this must be close to a record!), e.g. [54], [58] or the paper co-authored with M. Nassar [63].

Nick Sherbakov brought with him deep knowledge in database theory and data modelling that resulted in a host of valuable ideas and joint papers, some co-authored with P. Srinivasan or Ann Philpott, and others such as e.g., [62], [65], or [72]. The driving force behind the modern JAVA-authoring tool HM-Card [85] is also Nick Sherbakov. It was Frank Kappe's Ph.D. thesis that gave the first fairly rigorous specification of what future networked multimedia systems (like today's Hyperwave, the successor of Hyper-G) must look like. Good introductory papers co-authored by the rapidly growing Hyper-G team around Kappe, including both capable researchers and developers such as Keith Andrews, Klaus Schmaranz, Gerald Pani, Florian Schnabel, Jörg Faschingbauer, Mansuet Gaisbauer, Michael Pichler, and Jürgen Schipflinger are [61], [62] [75] [77], and the book [86].

Parallel with above activities I had the pleasure to help establish the Interactive Information Center (IIC) in Graz in a paper co-authored with famous media "guru" Don Foresta from Paris, the well-known Styrian philosopher Johann Götschl, and Wolfgang Schinagl, the real "motor" behind IIC under whose guidance IIC has developed in four years into a top-notch IT show-case.

Another important stage in my life was my (temporary) move to Auckland, New Zealand, in 1993. With my two first Ph.D. students there, Achim Schneider and Jennifer Lennon, we managed to very successfully pursue a number of topics in networked multimedia. I was particularly impressed by the impeccable work of Jennifer Lennon who has become a very good personal friend, one of the leading personalities in multimedia in New Zealand, and a prolific co-author, see e.g. [64] or [73]. Other publications from my time in Auckland are with Achim Schneider (e.g., [80]), with L. Rajasingham and John Tiffin [66] from Wellington, Julian Harris [74], Barry Fenn [78], Bill Flinn [79], Channa Jayasinha (the IT director of New Zealand's main museum in Wellington) [82], the German student Michael Klemme [89], and particularly Professor Cris Calude who rekindled my interest in theory [68]. Cris and Arto Salomaa were also "responsible" for convincing me to start an electronic journal J.UCS, mentioned earlier. See http://www.iicm.edu/jucs and [67] for more technical information. The further technical development of J.UCS is now much in the hands of one of my top Ph.D. students Klaus Schmaranz, see e.g., [76] but has also stimulated co-operation with my friend Gary Marchionini from the University of Maryland, see e.g., [81].

Arto Salomaa turned 60 in 1994. There were a number of big festivities for this occasion and I was lucky to be involved in two. I hosted an international meeting for Salomaa in Graz where the proceedings where edited with my friends Karhumäki and Rozenberg [71], and I was invited to be co-editor of the Salodays proceedings [69] with Cris Calude and Mike Lennon in New Zealand. I have to say a bit more about Mike: he took me on a number of out of the world tramps (as Kiwis usually call hikes!): two days underground; three days wading in water (in the absence of trails and with dense forests you have to walk in the river-beds); through beautiful NZ South island mountain scenery; up an active volcano with winds raging at over 100 km/h; bivouacking at the snow line just with sleeping bags with no way to get a fire going; and much more. And all this sprinkled with the occasional talk about some mathematical problems. Thank you Mike for being such a terrific guide, friend, ... and cook: even under extreme circumstances Mike manages to whip-up an incredible hot stew in a short time. Mike organised something very unusual for Arto Salomaa: not just a native Maori feast, a hangi, but a very special one where Arto, in a touching ceremony, became member of that particularly Maori tribe: I believe there are very few Europeans who have this honour. This is much deeper and much more serious than it sounds: Arto's tribe now considers Arto a member and will support him, if it came to it, from now on no matter what.

Due to a Fulbright scholarship that I managed to get for Auckland I got to know Professor Suave Lobodzinski from California. His scientific vitae had impressed me. But I had not known that Suave is also a top mountaineer (he has been on Mt. Everest without oxygen), is a dive master, and a dynamic person to a degree that is unbelievable. I am lucky that Suave has let me profit from his friendship and experience in outdoor situations that are borderline for me, yet trivial for him. We are also co-operating on medical applications of Hyperwave, see e.g. [83].

My main interests these days are in using Hyperwave: you see, Hyperwave is the first theoretically sound WWW server: it has a database, search scopes, automatic link maintenance, customisation features and much more. B.T.W. it is free for university institutes (see http://www.hyperwave.com) and ideally suited for educational applications (see http://www.iicm.edu/mankind). This is also the reason why we are co-operating with a number of educational groups, particularly with the one around Professor Manolis Skordalakis from the Greek National Technical University in Athens, see e.g., the paper also co-authored by A. Koutoumanos, N. Papaspyrou, and S. Retalis [87]. Skordalakis is a true Greek friend and gentleman. So it is typical that he proposed the acronym GENTLE (General Networked Teaching and Learning Environment) for a Web based training project and permitted me to use it in the future. Manolis has been leader of a successful European Web-based training project, EONT, for some three years. Working under his guidance has been a pleasure. Thus, we had nostalgic feelings at our last joint meeting in Athens in June 1997; and the description of the last evening is a fitting ending also to this report: here we were, in a roof-top restaurant at the foot of the Acropolis. The red sun setting, a gentle evening breeze stirring, the moon rising over Herodot's ancient theater. And many ideas for future work being discussed with growing excitement.

References

1. H. Maurer: Proposal and Examination of a Table Lookup Technique; Report, IBM Lab., Vienna (1964)
2. H. Maurer: A Stringhandling Compiler Allowing for Basic Stringhandling in Connection with FORTRAN II; Report, IBM Lab. Vienna (1964).
3. K. Bandat, P. Lucas, H. Maurer, K. Walk: Tentative Steps towards a Formal Definition of PL/1; Report, IBM Lab., Vienna (1965).
4. H. Maurer: Rationale Approximationen Irrationaler Zahlen; Ph.D. Thesis, University of Vienna (1965).
5. H. Maurer: A Direct Proof of the Inherent Ambiguity of a Simple Context-Free Language; J.ACM 16, 2 (1969), 256-260. (J)
6. H. Maurer: Theoretische Grundlagen der Programmiersprachen; BI, Mannheim (1969).
7. W. Kuich, H. Maurer: The Structure Generating Function and Entropy of Tuple Languages; Information and Control 19,3 (1971), 195-203.
8. H. Maurer, M.R. Williams: A Collection of Programming Problems and Techniques; Prentice Hall, Englewood-Cliffs (1972).
9. A. Cremers, H. Maurer. O. Mayer: A Note on Leftmost Restricted Random Context Grammar; Information Processing Letters 2 (1973), 31-33.
10. H. Maurer: Datenstrukturen und Programmierverfahren; Teubner, Stuttgart (1974).
11. H.P. Kriegel, H. Maurer: Formal Translations and Szilard Languages; Information and Control 30, 2 (1976), 187-198.
12. S. Ginsburg, H. Maurer: On Strongly Equivalent Context-Free Grammar Forms; Computing 16 (1976), 281-290.
13. H. Maurer, D. Wood: On Grammar Forms with Terminal Context; Acta Informatica 6 (1976), 397-402.

14. H. Maurer, W. Stucky: Ein Vorschlag fr die Verwendung syntaxorientierter Methoden in höheren Programmiersprachen; Angewandte Informatik (1976), 189-196.
15. H. Maurer, Th. Ottmann, W. Six: Manipulation of Number Sets Using Balanced Trees; Applied Computer Science 4, Graphen, Algorithmen, Datenstrukturen, Hanser (1976), 9-37.
16. H. Maurer: Data Structures and Programming Techniques (transl. by C. Price); Prentice-Hall (1977).
17. H. Maurer, A. Salomaa, D. Wood: EOL Forms; Acta Informatica 8 (1977), 75-96.
18. K. Culik II, H. Maurer: Tree Controlled Grammars; Computing 19 (1977), 129-139.
19. J. Albert, H. Maurer: The Class of Context-Free Languages is not an EOL Family; Information Processing Letters 6, 6 (1977), 190-195.
20. K. Culik II, H. Maurer, Th. Ottmann: On two-symbol complete EOL forms; Theoretical Computer Science 7 (1978), 69-83. (J)
21. K. Culik II, H. Maurer, Th. Ottmann, K. Ruohonen, A. Salomaa: Isomorphism, form equivalence and sequence equivalence of PDOL forms; Theoretical Computer Science 6 (1978), 143-173.
22. H. Hule, H. Maurer, Th. Ottmann: Good OL Forms; Acta Informatica 9 (1978), 345-353.
23. H. Maurer, A. Salomaa, D. Wood: ETOL Forms; J. Computer and Systems Science 16, 3 (1978), 345-361.
24. J.L. Bentley, H. Maurer: A note on Euclidean near neighbor searching in the plane; Information Processing Letters 8, 3 (1979), 133-136.
25. K. Culik II, H. Maurer: Secure information storage and retrieval using new results in cryptography; Information Processing Letters 8, 4 (1979), 181-186.
26. W. Ainhirn, H. Maurer: On ϵ-productions for terminals in EOL forms; Discrete Applied Mathematics 1 (1979), 155-166.
27. J. Albert, H. Maurer, G. Rozenberg: Simple EOL forms under uniform interpretation generating CF languages; Fundamenta Informaticae III, 2 (1980), 141-156.
28. J.L. Bentley, H. Maurer: Efficient worst-case data structures for range searching (with J.L. Bentley); Acta Informatica 13 (1980), 155-168.
29. H. Maurer, M. Nivat: Rational bijection of rational sets; Acta Informatica 13 (1980), 365-378.
30. H. Maurer, Th. Ottmann: Dynamic solutions of decomposable searching problems; Discrete Structures and Algorithms, Hanser Mnchen (1980), 17-24.
31. J. van Leeuwen, H. Maurer: Dynamic Systems of Static Data-Structures; Report 42, Institut für Informationsverarbeitung Graz (1980).
32. A. Ehrenfeucht, H. Maurer, G. Rozenberg: Continuous Grammars; Information and Control 46 (1980), 71-91.
33. W. Bucher, K. Culik II, H. Maurer, D. Wotschke: Concise description of finite languages; Theoretical Computer Science 14 (1981), 227-246.
34. H. Maurer, A. Salomaa, E. Welzl, D. Wood: Dense intervals of linguistical families; Computer Science Technical Report 81-CS-08 McMaster University (1981).
35. H. Maurer, J.H. Sudborough, E. Welzl: On the complexity of the general coloring problem; Information and Control 51, 2 (1981), 128-145.
36. "Unorthodox"' Videotex Applications (with I. Sebéstyén); Information Services and Use 2 (1982), 19–34.
37. H. Edelsbrunner, H. Maurer: A space optimal solution of general region location; Theoretical Computer Science 16 (1981), 329-336.
38. H. Edelsbrunner, H. Maurer: On the intersection of orthogonal objects; Information Processing Letters 13, 4/5 (1981), 177-181.

39. H. Edelsbrunner, D.G. Kirkpatrick, H. Maurer: Polygonal intersection searching; Information Processing Letters 14, 2 (1982), 74-79.
40. J. Charles, H. Maurer, I. Sebéstyén: Printing without paper; Electronic Publishing Review 2 (1982), 151-159.
41. H. Maurer, G. Rozenberg, E. Welzl: Picture description languages; Information and Control 54 (1982), 155-185.
42. H. Edelsbrunner, H. Maurer, F.P. Preparata, A.L. Rosenberg, E. Welzl, D. Wood: Stabbing line segments; BIT 22 (1982), 274-281.
43. H. Maurer, W. Rauch, I. Sebéstyén Some remarks on energy and resource consumption of new information - and communication technologies; Information Services and Use 2 (1982), 73-80.
44. H. Bogensberger, H. Maurer, R.Posch, W. Schinnerl: Ein neuartiges - durch spezielle Hardware unterstztes - Terminalkonzept fı Bildschirmtext; Angewandte Informatik 3 (1983), 108-113.
45. W.D. Fellner, H. Maurer, R. Posch: Intelligent videotex terminals for rapid videotex penetration; Videotex Europe, Online Conference, Amsterdam (1983), 155-164.
46. H. Cheng, H. Maurer: Teleprograms - the right approach to videotex - if you do it right; Proc. of the IRE Conference on Telesoftware, London (1984), 75-78.
47. H. Maurer, N. Rozsenich, I. Sebéstyén: Videotex without Big Brother (with N), Electronic Publishing Review 4 (1984), 201-214.
48. F. Makedon, H. Maurer: COSTOC - Computer Supported Teaching of Computer Science; Proc. of IFIP Conference on Teleteaching Budapest 1986, North-Holland Publ.Co. (1987), 107-119.
49. J. Koegel, H. Maurer: A Rule-Based Graphics Editor for Presentation CAI; Proc. of the 2nd Rocky Mountain Conf. on AI, Boulder, Colorado (1987), 133-142.
50. F. Makedon, H. Maurer, Th. Ottmann: Presentation Type CAI in Computer Science Education at University Level; J.MCA 10 (1987), 283-295.
51. F. Huber, H. Maurer: Extended Ideas on Editors for Presentation Type CAI; IIG Report 240 (1987).
52. F. Huber, F. Makedon, H. Maurer: Hyper-COSTOC: A Comprehensive Computer-Based Teaching Support System; J.MCA 12 (1989), 293-317.
53. P. Gillard, H. Maurer: Tiny CAI Tools -Giving Students "the Works"; J.MCA 13 (1990), 337-345.
54. H. Maurer, I. Tomek: From Hypertexts to Hyperenvironments; e & i, Special Zemanek-Issue (1990), 614-616.
55. G. Davies, H. Maurer, J. Preece: Presentation metaphors for very large hypermedia systems; J.MCA 14 (1991), 105-116.
56. P. Gillard, H. Maurer, M.G. Stone, R. Stubenrauch: Question-Answer Specification in CAL Tutorials: Automatic Problem Generation does not work; Proc. 6th Symposium Didaktik der Mathematik, Klagenfurt/Austria, Hölder-Pichler-Tempsky (1990), 191-197.
57. H. Maurer (Ed.): Hypertext/Hypermedia'91, Proc. of Symposium, IFB 276, Springer Pub.Co. (1991).
58. H. Maurer, I. Tomek: Hypermedia - from the Past to the Future; LNCS 555, Springer Pub.Co. (1991), 320–336. (P)
59. H. Maurer, W. Rauch: Pro und Contra Hypermedia; Computerwoche 43 (1991), 73-76.
60. P. Carlson, H. Maurer: Computer Visualization, a Missing Organ and a Cyber-Equivalency; Collegiate Microcomputer X, 2 (1992), 110-116.

61. F. Kappe, H. Maurer, G. Pani, F. Schnabel: Hyper-G: A Modern Hypermedia System; Proc. Network Services Conference (NSC)'92, Pisa, Italy (Nov. 1992), 35-36.
62. F. Kappe, H. Maurer, N. Sherbakov: Hyper-G - A Universal Hypermedia System; J.EMH (Journal of Educational Multimedia and Hypermedia) 2,1 (1993), 39-66.
63. H. Maurer, M. Nassar, I. Tomek: Optimal Presentation of Links in Large Hypermedia Systems; Proc. ED-MEDIA'93, AACE, Virginia (1993), 511-518.
64. J. Lennon, H. Maurer: Lecturing Technology: A Future with Hypermedia; Educational Technology 34, 4 (1994), 5-14.
65. H. Maurer, A. Philpott, N. Sherbakov: Hypermedia Systems Without Links; J.MCA. 17,4 (1994), 321-332.
66. H. Maurer, L. Rajasingham, J. Tiffin: They Just Sold New Zealand; NZ SCIENCE Monthly (March 1994), 6-7.
67. C. Calude, H. Maurer, A. Salomaa: JUCS: The Journal for Universal Computer Science and its Applications to Science and Engineering Teaching; Report No. 91, University of Auckland (March 1994).
68. C. Calude, H. Maurer: Pocket Mathematics; Proc. Salodays in Auckland, The University of Auckland (1994), 25-29; Mathematical Aspects of Natural and Formal Languages (Ed. G.Paun), World Scientific Series in Computer Science, vol. 43, World Scientific- Singapore (1994), 13-41.
69. C. Calude, M. Lennon, H. Maurer: Salodays in Auckland (Eds.); Auckland University (1994)
70. F. Kappe, H. Maurer: Theory as Basis for Advances in Hypermedia; RAIRO - Theoretical Informatics and Applications 28, 3-4 (1994), 201-211.
71. J. Karhumki, H. Maurer, G. Rozenberg: Results and Trends in Theoretical Computer Science (Eds.); LNCS 812 Springer Pub.Co. Heidelberg/New York (1994).
72. F. Kappe, H. Maurer, N. Scherbakov, P. Srinivasan: Conceptual Modeling in Hypermedia: Authoring of Large Hypermedia Databases; Proc. Hypermedia'94, Vaasa, Vaasa Institute of Technology (1994), 294-304.
73. J. Lennon, H. Maurer: MUSLI – A MUlti-Sensory Language Interface; Proc. ED-MEDIA' 94 (best paper award), AACE, Virginia (1994), 341-348.
74. J. Harris, H. Maurer: HyperCard Monitor System; Proc. ED-MEDIA' 94, Vancouver, AACE, Virgina (1994), 246-250.
75. F. Kappe, K. Andrews, J. Faschingbauer, M. Gaisbauer, H. Maurer, M. Pichler, J. Schipflinger: Hyper-G: A New Tool for Distributed Hypermedia (with); Proc.Distributed Multimedia Systems and Applications, Honolulu (1994), 209-214.
76. H. Maurer, K. Schmaranz: J.UCS- The Next Generation in Electronic Journal Publishing; J.UCS 0, 0 (1994), 118-126; Computer Networks and ISDN Systems 26 (1994), 563-569.
77. K. Andrews, F. Kappe, H. Maurer, K. Schmaranz: On Second Generation Hypermedia Systems (with); J.UCS 0,0 (1995), 127-136.
78. B. Fenn, H. Maurer: Harmony on an Expanding Net; Interactions (October 1994), 26-38.
79. B. Flinn, H. Maurer: Levels of Anonymity; IIG Report No. 387, Graz/Austria (1994); JUCS 1, 1 (1995), 35-47. (J)
80. H. Maurer, N. Scherbakov, A. Schneider: HM-Card: A New Hypermedia Authoring System; Multimedia Tools and Applications 1, Kluwer Academic Publishers, Boston (1995), 305-326.
81. G. Marchionini, H. Maurer: The roles of digital libraries in teaching and learning; C.ACM 38, 4 (April 1995), 67-75.

82. C. Jayasinha, J. Lennon, H. Maurer: Interactive and Annotated Movies; Proc. ED-MEDIA'95, Graz (1995), 366-371.
83. S. Lobodzinski, H. Maurer: Hypermedia Network Architecture for Digital Echocardiography; Medical Imaging 1996: PACS Design and Evaluation: Engineering and Clinical Issues (R. G. Jost, S.J. Dwyer; Eds), Proc. SPIE Vol. 2711, 214-221.
84. D.H. Jonassen, R. Goldman-Segal, H. Maurer: DynamIcons as Dynamic Graphic Interfaces: Interpreting the Meaning of a Visual Representation); Intelligent Tutoring Media , vol. 6 (3/4) (1996), 149-158.
85. H. Maurer, N. Scherbakov (Eds.): Multimedia Authoring for Presentation and Education - The Official Guide to HM Card; Addison-Wesley, Bonn, 1996.
86. H. Maurer (Ed.): HyperWave: The Next Generation Web Solution; Addison-Wesley Longman, London (1996).
87. A. Koutoumanos, H. Maurer, N.Papaspyrou, S. Retalis, E. Skordalakis: Towards a Novel Networked Learning Environment; Proc. WebNet'96, San Francisco, AACE (1996), 267-272.
88. K. Andrews, H. Maurer, N. Scherbakov: Browsing Hypermedia Composites: An Algebraic Approach; Proc. WebNet'96, San Francisco, AACE (1996), 348-353.
89. M. Klemme, H. Maurer, A. Schneider: Glimpses at the Future of Networked Hypermedia Systems; Journal Educational Multimedia and Hypermedia, AACE, 5, 3/4 (1996), 225-238.

Note. This is a very partial list of papers that I have co-authored with some of the persons mentioned in the body of the paper. For a full list see http://www.iicm.edu/maurer.

Grzegorz Rozenberg

Professor G. Rozenberg received his Master and Engineer degree in computer science in 1965 from the Technical University of Warsaw, Poland. In 1968 he obtained Ph.D. in mathematics at the Polish Academy of Sciences, Warsaw. Since then he has held full time positions at the Polish Academy of Sciences, Utrecht University, The Netherlands, State University of New York at Buffalo, U.S.A., and University of Antwerp, Belgium. Since 1979 he has been a professor at the Department of Computer Science of Leiden University, The Netherlands, and an adjoint professor at the University of Colorado at Boulder, U.S.A.
Professor Rozenberg was the President of the European Association for Theoretical Computer Science (EATCS), 1985-1994, and he is currently the chairman of the Steering Committee for International Conferences on Theory and Applications of Petri Nets. He has published about 300 papers, edited about 30 books, and written 3 books. He is the editor of the Bulletin of the EATCS, the editor of the series Advances in Petri Nets (Springer-Verlag), and a co-editor of the Monographs in Theoretical Computer Science (Springer-Verlag). He has been a member of the program committees for practically all major conferences on theoretical computer science in Europe. He is also involved in a number of externally funded research projects on both national and international levels.
He is a foreign member of the Finnish Academy of Sciences and Letters, a member of Academia Europaea, and he holds a honorary doctorate of the University of Turku, Finland.

The Magic of Theory and The Theory of Magic

I will recall here, and reflect upon, the places I have been, the events that I have witnessed or participated in, and the people that I have met. I will not discuss (in any depth) my ideas about science, or technical aspects of my scientific work, since they are explicitly or implicitly visible in my papers and books.

Warsaw, Poland

I have received my basic education in Warsaw, Poland. The high school I attended had (as usual) bad teachers and good teachers; as a matter of fact the bad teachers were in the majority. I have always had a bad memory and this has determined many choices in my life, like, e.g., the choice of topics I did like in school and those that I didn't. For example, we had a bad teacher of chemistry, and moreover it seemed to me then that the main aim of the chemistry classes was to memorize difficult names of molecules and compounds. Thus, I did not like chemistry at all. On the other hand, we had a fantastic teacher of mathematics, and at a very early stage I realized that you don't have to remember things in mathematics because you can deduce "everything". Thus mathematics has become one of my most favorite topics. I also did like very much literature and poetry. I wrote poetry during the high school and during my studies – most of it is lost by now.

Some of the books that I had been reading when I was in school, were books that were not available in Poland after the war, for example westerns. The way to get these books and to read them was very memorable. You could borrow such a book (or a hand written copy of it) for a specific period of time, and then you had to rewrite it by hand. You had to return the book to the owner, but in this way you were in a possession of the manuscript that you could read several times, and moreover you could exchange it with other boys for their manuscripts. I surely rewrote many books during my school times.

I had two passions during the school. One of them was playing cards, and the other one was fencing.

Playing poker was very popular after the war in Warsaw. You had to pay with cash, or with some of your valuable possessions (like pocket knife, or a key ring). I remember that I could not concentrate on a poker game, because when I would get five cards into my hands, I would start thinking about various arrangements of the cards so that something "unusual" would happen. This must have been very annoying to my co-players, and so I was excluded from playing poker. However, I have been inventing some nice tricks with cards, and at some point I started to show them to my friends. Then I regained their respect because I could do something that none of them could. I remember being very proud of this.

Already at the beginning of the high school I was accepted into a fencing club. Fencing became my most favourite sport. One should know that in the '50s and in the '60s Poland was one of the very top countries in the world as far as fencing was concerned. I had been lucky to get into the hands of a trainer, a Hungarian by origin, who was also the trainer of the national team. He was very encouraging towards me, and so I must have been really good. I remember that one of the unpleasant consequences of becoming a student, after the high school, was the choice between studying and fencing, which I had to make if I wanted to be good in either of the two. Fencing was very useful for me because I have often walked in Warsaw in the evening with a stick in my hand, which I could use to defend myself if attacked by some thugs. As a matter of fact, this has happened several times, and I could defend myself quite well.

Yet another passion of mine was dancing, which was quite natural because it has also involved the footwork which is so important in fencing. Later, during my studies, I was working in a jazz club and in a student club, where dancing was the main activity; dancing jazz or rock and roll was considered very cool then.

I decided to study electronics because it looked to me like a very nice way of combining physics and mathematics on the one hand and the technology on the other. At that time, the word "electronics" was synonymous with the most modern technology. The study of electronics was very popular and so the entry exams were very competitive – only one in ten candidates has been admitted. The entry exam was really hard.

A big part of the study of electronics was very interesting, but there were a lot of classes, especially in the first half of the studies, that were totally useless in any sense, and taught by totally unintelligent teachers. I did however enjoy many classes – the mathematics and physics programmes were very good, and I liked very much information theory, design of electronic lamps and transistors, electronic circuit design, analogue computing, and all kinds of laboratories.

The Soviet Union was the biggest pirate in the market of translation of books – they were translating many books published in the West without any permission whatsoever. That must have been awful from all legal points of view, however it was wonderful for Polish students and scientists. We could buy translations of all kinds of scientific books published in the West and this was the only access to these books we had. I still remember the joy of buying such books (and starting to read them in the bookstore already). Obviously we never knew that producing these books by Soviet Union was illegal – it was axiomatic that everything that Soviet Union did was legal and admirable.

I have been a big lover of theater and during the studies I even organised a Society of Student Fans of Theater – I remember that quite many times we, the students from the society, were in the big majority in theater. There were many terrific actors in Warsaw, and very often a performance, of even a classic play, was a symbolic protest against the system – the audience always knew that!

What was really great during these years in Poland were jokes. Jokes formed a natural survival tool in the system of total arrogance and stupidity. It was quite dangerous to tell good jokes – but one always knew with whom one could exchange them. As a matter of fact, the best jokes were the political ones – here is one. "The biggest factory in the world was built in one of the communist countries. Many visitors are coming to see it. A very proud mayor of the city where the factory was located is driving one of the important visitors to the factory. When they arrive at the entrance gate, the mayor says: "This is the biggest factory in the world, it employs about 15000 people". Because of the noise of a passing truck, the visitor has not heard the end of the sentence, and so he asks "How many people work there?" The mayor says "Oh! you mean work", and then he adds "May be two or three". This is an example of a beautiful abstraction: the communist system has destroyed the work ethics – factors other than the quality of your work were determining your position, your salary was the same whether you really worked or you were pretending that you work.

Students could do the army service in such a way that (for a number of years) one day a week was spent (in uniforms) at army barracks in Warsaw. Also one or two summers were spent in some army unit in a country. There were exams taken each semester, and failing them was really dangerous because this could mean the regular army service after the studies. My relationship with the army was simple and symmetric: we hated each other! I often had nightmares that I will have to do the service after my studies. Fortunately, this has not happened, which (at least partially) was due to the fact that I had to give private lessons (mostly mathematics) to the military that were following (were forced to follow) some evening classes in order to get some kind of certificates. It was a clear blackmail, but I accepted this as the best way out for me from the awkward situation that I was in. As a matter of fact, being a good student in science has helped me in a number of ways. I was giving private lessons in mathematics and physics already in high school – this provided me with a nice way of earning money; this continued also during my later studies. It also saved me from the army service. Moreover, during studies I could do assignments (also optional) quickly and well, and so I was allowed to skip some of the exercise classes which were scheduled for each week. In this way I could collect free days which I was then using for visiting my girl friend Maja (now my wife) who lived in a different city.

I chose to specialize in "mathematical machines" – this is how computers were called then in Poland, and it was a very good choice indeed. I was in love with all the stuff that we were learning, especially the engineering aspects of computer architecture. I did not like programming so much because totally stupid paper-punching mistakes made programmes totally unusable. I still remember spending a lot of time with paper punchers (using Hollerith code) for correcting mistakes.

I wrote my master thesis on the theory of algorithms (for the engineers!) – it looked very challenging for me then. My advisor, the director of the Institute for Mathematical Machines, was very fond of my thesis and was showing it to many people (even when it was not completed yet). Somehow in this way I got into contact with Andrzej Ehrenfeucht who worked then at the Mathematical Institute of the Polish Academy of Sciences. This has changed my life in many ways.

In 1964 I was offered a position at the Institute of Mathematics of the Polish Academy of Sciences, IMPAS, in the group of Mathematical Logic headed by A. Mostowski. I had been also offered a position in the Electronics Department of the Technical University – my advisor wanted to keep me there. I chose the offer from the Mathematical Institute, because, especially through my contacts with Andrzej, I had discovered the **MAGIC OF THEORY**.

Mostowski was a very kind man of very good manners. He always had time whenever I wanted to talk to him. I still remember the interview when he offered me a job. Looking through the papers he said at some point: "I see that you will be the youngest member of the Institute", he paused and then he continued "but this problem will resolve itself with time". Other members of the group were Andrzej Ehrenfeucht, Grzegorczyk and Zdzislaw Pawlak. I had a lot of contacts with Andrzej and Zdzislaw, and much less with Grzegorczyk. The years in the Institute were very exciting for me. It was great to see famous people like Kuratowski, Mostowski, Łoś, and Sierpinski around. But the main reason for my excitement was the fact that I was coached by Andrzej and Zdzislaw who were always there to talk to me. Andrzej a real virtuoso of theory, and Zdzislaw, a real master of capturing the essence of various applications and translating them into elegant models.

Several books and papers influenced me very much at the beginning of my scientific development. The books were: "Abelian categories" by P. Freyd and "Computability and Unsolvability" by M. Davis. The papers were "Finite automata and their decision problems" by M. Rabin and D. Scott, "Linear automaton transformations" by A. Nerode, "Formal properties of grammars" by N. Chomsky, "The algebraic theory of context-free languages" by N. Chomsky and M.P. Schützenberger, and "On formal properties of simple phrase structure grammars" by Y. Bar-Hillel, M. Perlis, and E. Shamir. Especially the last three papers had a lasting effect on me: I became a "grammar man" and remain so until today. The Mathematical Institute had a wonderful library (where the wife of Mostowski was the main librarian) – I found the above books and papers in our library. Later I got my own copy of M. Davis' book – I do not remember how I got it, but it was then one of my cherished possessions.

I had already discovered automata theory during the last year of my study and together with a friend of mine, Pawel Kerntopf (who was an assistant professor), we organized a seminar on automata theory at the electronics department. Some of the very good students joined it, together with some young members of the scientific staff. I had been working in parallel on category theory and automata theory. The paper by Rabin and Scott had led me to consider multitape automata. Through my work on axioms for the category of relations I got into contact with Eilenberg. I was very impressed when I met him, and I was very proud to learn that he was also working on the axioms for the category of relations with a considerable overlap with my work. I remember that he was very encouraging towards me then.

Working at IMPAS gave me an opportunity to meet some well-known scientists. Thus, e.g., I met Solomon Marcus, who was visiting Pawlak in Warsaw. We talked about contextual grammars, which he started to develop then – by now the theory of contextual grammars is well recognised in linguistics and formal language theory. As a matter of fact, many years later, in the mid-90's, I have worked quite intensively with Gheorge Păun, Arto Salomaa, and Andrzej Ehrenfeucht on the theory of contextual grammars. When Marcus was in Poland I took him to the city of Lodz, where Maja lived. I offered Marcus a number of possibilities to spend an evening together with Maja and her parents. He explicitly asked me to take him to see an operetta, which I did. During the play I had been quite worried that Marcus did not understand the songs - everything was in Polish. When I told him (after the show) about my worries, he told me that this is exactly what he wanted. He was writing a book on the mathematical theory of theater, and this particular evening he was just conducting an experiment: how much can one get from a play without understanding the language!

Nowadays I am working on DNA computing which also involves the planning of experiments. It looks like I may have some talent for doing this. If so, then the roots of this talent may lie in my Warsaw times. At some point I was convinced that my phone was bugged by the police. In order to check whether it was so, I had decided not to pay my telephone bills for some time. Usually, if one would not pay the bills even for a short period of time, the telephone would be cut off. Since the police were apparently interested in listening to my conversations, my phone was not disconnected – this experiment had confirmed my conjecture that the phone was bugged!

Marcus invited me to Bucharest to present my results in formal language theory. I was very happy to go there to continue our discussions on contextual grammars, and other topics. I went by train. The first border to be crossed was the border between Poland and Soviet Union – it certainly did not look like the border between two countries belonging to the same political block. It looked much more like an entrance to a concentration camp with all kinds of grotesque fortifications. I had been working on my lecture for Bucharest, when the Russian border police entered my compartment. They asked for the passport and other travel documents that I had for this trip. They certainly did not like my face. I had a book by S. Ginsburg on context-free languages on the seat next to me. The Russian officer (apparently the head of the trio that have entered my compartment) took the book and noticed the word "New York" on the first page. He asked then why do I have "This American Book". I have explained to him that I am going to Bucharest to give a lecture and that this is a mathematical book that I was using for preparing my lecture. "Aha!" he said, and asked immediately "do you have more of this type of American propaganda with you?" Even though I have answered in the negative, he went through all my luggage with amazing precision. He took away every single piece of paper I had (including all material for my lecture), each time asking me whether I need this specific piece. Each time I had answered "yes", and each time he would say "Aha!" and confiscate this piece of evidence! It is a pity that this scene was not filmed, it certainly looked like a grotesque ritual. After he took everything I had, the trio has disappeared – I was very relieved because they gave me back my passport. My first thought was to go back right away to Warsaw, however, I was afraid to leave the train: I could be arrested. I had been quite tense during this incident. I forgot to say that there was one more passenger in my compartment. He was working at a technical university somewhere in Soviet Union. We had talked quite a lot before the train came to the Russian border and so there was some "bond" between us. When I looked at him after the Russians left he was totally pale. Then, after the train left the station, he has asked me how could I be so careless and not to hide the book by Ginsburg. He then showed me that he had on him an American book on calculus strapped by his trousers' belt to his back under the shirt. When I saw this, I had an attack of hysterical laughing, and very soon we were both laughing like crazy for some time. I was again in a good shape. After I had returned to Warsaw, I reported the incident to J. Łoś who was the director of IMPAS. He wanted to make all kinds of official protests and actions, but I have asked him not to do anything because, after all, it was such a minor incident in the ocean of idiocies happening all the time.

Utrecht, The Netherlands

We (i.e., my wife, my 7 months old son, and myself) left Poland in 1969 – this was the first real opportunity for us to leave. I had an offer from the Mathematical Institute of the University of Utrecht for a visiting position – this was very important for me then. One other reason to go to Holland was that I met a wonderful Dutch couple – André and Rita, and so I felt that I have already friends there (which was psychologically very important to us). Henk Barendregt (today the "world champion" in lambda calculus and functional programming) was very instrumental in getting the offer from Utrecht. I was "appended" to the group of Dirk van Dalen, but Dirk made it very clear from the beginning that I am an independent "singleton" group. The support by Dirk and by A. van der Sluis was very important for me.

Shortly after my arrival, Dirk gave me his paper on context-sensitive Lindenmayer systems (L systems). I read the paper with great interest, and right away got interested in L systems. Dirk introduced me to Aristid Lindenmayer, and from the moment we met we began our scientific cooperation. We became very good friends and this friendship has extended to our families. Aristid's wife Jane, an artist, is a wonderful person. Their home was in a village very close (biking distance) to ours and we have often visited each other. We have had endless scientific discussions with Aristid – only a small part of these discussions has been reflected in our papers. Through my work with Aristid I have discovered the excitement of interdisciplinary work. Working with Aristid was a real pleasure – I have learned from him that elegant abstractions from the things we observe in nature may be as beautiful as nature itself. Although he was a biologist he was in constant search for formal theories. This reminds me of the following story. Together with Aristid we had organized a conference on "Formal Languages, Automata and Development". Stan Ulam, a very famous mathematician, was one of our invited guests. He told the following joke. Two scientists arrive in the evening to a city for a conference. They learn that in the morning there is quite an official opening, and they are supposed to wear shirt and tie. They try to find a laundry to wash their shirts, but it is quite late and it looks like all laundries are closed. They are already quite desperate, when they notice a big lighted window with signs "Laundry". They enter the place quite happy, put the shirts on the counter and ask the man behind the counter to wash them. But the man says: this is not a laundry, we just sell signs. This was quite a subtle comment by Stan on the use of theory.

In the first two years of my stay in Utrecht, I had given/organized many lectures on theoretical computer science and organized many seminars covering quite a broad scope of theoretical computer science. The seminars and lectures were national events – Holland is a small country and traveling to lectures/seminars at other universities (cities) is quite usual. The support by the department allowed me to invite many outstanding speakers from various European countries and from the US. In this way I met e.g., Maurice Nivat, Mike Harrison and, most importantly, Arto Salomaa.

With Arto we had a special bond from the moment that we have met. Since then we have written very many papers, written and edited many books, organized many conferences, and talked to each other an uncountable number of hours. Our scientific cooperation is natural in the sense that just being together leads to doing research and writing papers. It is amazing how, when being together, we always run out of time before even reaching many of the topics we have planned to cover. As a matter of fact, when I met Arto for the first time at the Schiphol airport to bring him to Utrecht, we were so much engaged in a conversation that I drove to Amsterdam rather than to my home in Utrecht !! Arto has a photographic memory. If I want to recall an event from the past, I mention it to Arto, he switches on his internal video (as we say), and then he can tell all the details of the event. Since I have very bad memory, we have already agreed that Arto will write my memoires! Our families are very close. The hospitality of his wife Kaarina made my stays at their country farm, Rauhala, unforgettable. Kaarina is a real cat lover, and I became good friend also with her cats. Through my stays at Rauhala with the "best sauna in the world", I have learned the pleasures of the Finnish sauna, and I am a (very proud) holder of sauna records for foreigners. I am also a good friend with Arto's children, daughter Kirsti and son Kai. Kai is by today a very good scientist; as a matter of fact I have been a co–author of Kai's first published paper. His wife Suning is a chemist and I hope to cooperate with her on molecular computing.

Also Mostowski, Pawlak and Eilenberg have visited me in Utrecht. Eilenberg was a collector of some kind of figurines – particularly valuable were those from Indonesia. Because of the historic ties of Holland with Indonesia, there were many sources of these figurines in Holland, especially in Utrecht. So we have visited quite a number of rather unusual places and people in Utrecht during his stay.

I have also met Joost Engelfriet, he was attending some of my seminars in Utrecht. We then started our cooperation, and have written many joint papers since then. When I settled in Leiden, Joost joined my group. He is technically very strong – I remember that many years ago I told Joost that I like to work with him because "usually I give him a lemma and then he returns a theorem".

When I arrived in Utrecht, there was already a group of researchers working on L systems. Besides Dirk and Aristid it included also Jan van Leeuwen and Paul Doucet, both from the Mathematical Institute. Paul wrote his Ph.D. thesis on the theory of L systems, but left science afterwards. Jan I believe was still a student when I arrived. He is an example of a prominent scientist who was a major contributor to the theory of L systems, and later became a leading figure of theoretical computer science. Another such example is Mogens Nielsen from Aarhus, Denmark. I remember to be very much impressed by his English, especially his pronunciation. I met him many times during my visits to Aarhus (when Arto stayed there for two years) and we are good friends now. I met him, as well as Sven Skyum from Aarhus and Arto, also at one of the famous Oberwolfach meetings (on formal language theory). We all then wrote a very widely referenced paper on L systems, also very unusual because its second part was published before the first! Mogens made major contributions to formal language theory and by today he is one of the top researchers in concurrency.

Also in Oberwolfach I met many researchers who at that time were leaders in theoretical computer science in Germany, for instance, Günter Hotz and Wilfried Brauer. I became good friends with Wilfried and his wife Ute.

Shortly after arriving in Holland, I have visited Technion in Haifa, Israel, where I met Shimon Even and Azaria Paz. We became good friends and the friendship extended to our families. I was amazed to learn during a visit by Shimon and his wife Tamar to Holland, that Shimon is a very good car mechanic. With Azaria and his wife Erela we still maintain very close contacts; as a matter of fact their daughter Sharon was adopted by me. A couple of times Maja has accompanied me to a conference just because Erela was also coming there.

Buffalo, U.S.A.

In January 1971 I went to Buffalo, New York, U.S.A. to work in the Department of Computer Science – I have been invited by Gabor Herman. The encounter with the legendary bad winter weather was quite remarkable. I bought a used Ford Mustang very soon after our arrival, and on the first ride home the snow "got loose". First I heard on the car radio that "the visibility is approaching zero; please find a safe parking place", and then indeed within minutes I could not see "anything" – fortunately I was then close to our apartment building and made it just on time. Our apartment building was quite close to my office which was really important in the winter. In the coffee room of our department we had a kettle with very hot water available all the time. Whenever one could not open the frozen car doors one would use the kettle to pour the hot water over the slot of the frozen drivers door, and get inside the car in this way. Then, however, quite often the door would not close properly, and so one had to keep the door locked using a rope fixed to the drivers seat! – driving in the winter in Buffalo required some ingenuity.

Gabor was a very intense researcher, who was then totally devoted to L systems. Together with three Ph.D. students, we had in Buffalo a very active group developing the theory of L systems. At some point we have decided to write a book on L systems. This was certainly a good decision because the book brought many gifted researchers to L systems. We had regular weekly "book meetings" with our three Ph.D. students going through (the many versions of) the manuscript. Two of our Ph.D. students were Chinese and one was British. Gabor came to Great Britain from Hungary when he was a young boy. His English was very good and he was a very conscientious writer. Very often during our book meetings, Gabor would have long arguments with Adrian Walker (our British student) about the placing of commas in the text. Gabor would cite all kinds of rules about the commas, while Adrian would simply say that a comma does or does not fit in a given place. I usually supported Adrian because British was his mother tongue and then he "simply knew" where to put commas. One time the discussion became very animated and Gabor insisted on his "commas decisions". I proposed then that we place at the end of the book an "Epilogue on Commas" – it would consist of lines of commas and the following comment: "The above commas were removed/added by Adrian Walker from/to the original manuscript of the book". Everybody laughed, and the COMMA PROBLEM has been resolved forever.

The stay in Buffalo was my first American adventure. I quickly learned that my American colleagues work much harder than my Dutch colleagues. It was quite usual to stay at your office late at night; often a group of "night workers" would drive together around the midnight to have coffee and doughnuts at one of the "breakfast places" opened 24 hours a day. I also learned that the best graduate students were Chinese – as a matter of fact my first Ph.D. student in Buffalo was K.P. Lee, a very clever and nice person.

Buffalo was not a nice city. As a matter of fact the biggest asset of Buffalo was its proximity to Toronto. We drove there quite often – I have family there, and the city is very nice – it was then relatively safe. Buffalo was quite a crime city. Maja worked at a research institute in downtown Buffalo. At some point the institute built a new garage, which was two blocks away. We were quite shocked when Maja got a letter from the administration of the institute asking employees not to walk to the institute from the garage, but rather to use shuttle busses – the crime downtown was the reason.

I have driven an uncountable number of times to Niagara Falls – almost every visitor we had wanted to see the falls. At some point I was certainly qualified to be an official guide for the Niagara Falls.

Rohit Parikh was in the Mathematics Department which was on the same campus. He is a very fast thinker, but also a very fast talker and extremely fast writer (on the blackboard). He was also a supplier for us of a very good yoghurt (cultured on special bacteria from India). I have enjoyed very much his company. Tony Ralston was the chair of our department – he was very active and visible in many professional organizations such as AFIP and IEEE. We often had discussions about the nature and the future of computer science. It is amazing how computer science of today differs from the predictions of then. In particular, Tony was convinced that discrete mathematics is "the only" kind of mathematics needed for computer science – this opinion was shared by most of computer scientists then. It suffices to have a look today at neural networks, or computer graphics to see that this prediction has not been confirmed.

Also from the scientific point of view the proximity to Canada was a big asset of Buffalo. In this way I have met Derrick Wood who was at McMaster University in Hamilton, and Andy Szilard from the University of Western Ontario in London, Ontario (the REAL London, as they say there).I have written a number of papers with Derrick. By now he has moved to Hong Kong.

During my stay in Buffalo I was invited by Seymour Ginsburg to Los Angeles – Seymour was at the University of Southern California. There was some kind of a "rule" in the U.S. at that time, that if you wanted to join the top of formal language theorists in the U.S., then you had to be "initiated" by Seymour. I was thus very pleased by the invitation and moved for six weeks (or two months) to Los Angeles. Working with Seymour was a very interesting experience for me, to see the way he approached research and the way he wrote papers was very instructive. The following story says a lot about addictions. I was a passionate pipe smoker – I was preparing tobacco mixtures myself, and was "insmoking" new pipes also for my friends. As a matter of fact I have even written a book on pipe smoking ("The art and science of pipe smoking"). This is one of the two books that I have written and never published (I quit smoking in 1979, and there was a real danger that finishing the manuscript then could lead me back to pipe smoking). Anyhow, even when I was smoking I had this theory that if I have a pause in smoking (say one month) on regular bases, then my organism will clean up all the negative effects of smoking, whatever they were. The stay in LA coincided with one of my non-smoking periods. We were working with Seymour on control languages for TOL systems. The work was going well but we could not get a real theorem. I decided then to shorten the nonsmoking period and drove to a very good shop with pipes and tobacco in Westwood where I bought a pouch of really good "Turkish Melange". We had "the theorem" next day.

Antwerp, Belgium

Shortly after coming back from Buffalo to Utrecht, I received an offer from the mathematics department of the Antwerp University (UIA) in Antwerp, Belgium to build up computer science there. The offer was attractive, but I did not want to cut the academic ties with Utrecht. Thus, for the first two years I was one day a week in Antwerp (and the rest in Utrecht), and later I was one day a week in Utrecht and the rest in Antwerp.

Antwerp is a beautiful city and I enjoyed being there very much. My hotel was next to the famous cathedral in the centre of the city, with an abundance of excellent restaurants and coffee shops. The food in Antwerp was superior – Belgians really know how to enjoy life. I was quite happy there. I had 3 Ph.D. students there, all very good – two of them, Dirk Vermeir and Dirk Janssens, are university professors now. With Dirk Vermeir we worked on the L systems of finite index – his Ph.D. thesis is a definitive treatment of this topic. The following happening was the beginning of my friendship with Dirk Vermeir. Shortly after he began work on his Ph.D. under my supervision, we were having lunch together with a number of faculty members in the university restaurant. At some point

Dirk asked me about my age. Turning my answer into a joke I said "sixty" (I was then about 34 years old). Then Dirk said very politely and in a good faith "You certainly do not look older than fifty years old". This was the beginning of our friendship, and I have not been making this kind of joke since then! With Dirk Janssens we worked out the theory of node–label controlled (NLC) graph grammars. I consider the theory of NLC grammars, generalized later by Joost Engelfriet, George Leih and Emo Welzl, to NCE graph grammars, to be an important part of the theory of graph grammars.

During the Antwerp days I have also worked on two–level grammars introduced by van Wijngaarden. I remember that van Wijngaarden visited (unannounced) my seminar on 2-level grammars at UIA. I had been mainly interested in the theoretical aspects, and in particular in the descriptional mechanism behind them. This led to the formulation of cooperating grammar systems. Many years later I learned that they have been reinvented with a different underlying motivation and that there was a great interest in them. As a matter of fact I resumed the work on the theory of cooperating grammar systems in 1995, mainly in cooperation with Gheorge Păun and Arto.

Even when my main job was in Antwerp I lived in Bilthoven – a small village close to Utrecht (the reason was that it is very close to the academic hospital where Maja works). This meant a lot of driving, about 2 hours one way on the main Benelux highway with a lot of wet (rain or snow) days per year – this was the negative aspect of working in Antwerp.

As a matter of fact I've been car driving a lot in my life. I've seen enormous number of driving accidents and that makes me an alert driver. Since I am "mentally absent" when doing science or magic, I have a rule that I never think of (or discuss) science or magic when driving. A colleague of mine said that, since I spend so many hours driving a car, I may miss in this way an opportunity of proving nice results. "It may be true" I replied "that there is a chance that I miss in this way some very good results. But the chance, that I would not be able to write them up is much bigger." The usefulness of this rule was confirmed when Mathias Jantzen from Hamburg was visiting me a long time ago. He is interested in magic, and when we were driving to my home in Bilthoven, he has asked me a question about (some technical aspect of) magic. Somehow I forgot my rule, and got into discussion with him. Within few minutes we had a small accident, bumping (fortunately "softly") into a car in front of us that slowed down very suddenly. This was a pointed reminder about my rule, and I have obeyed it very strictly ever since.

Leiden, The Netherlands

In 1979 I accepted a professorship in theoretical computer science at Leiden University in Holland. This became the last station in my search for a place to settle.

Leiden University is very old, with very rich traditions and well known all over the world. Unfortunately computer science was somewhat lost within the mathematics department. The situation is quite different now. In recent years we have recruited some (very) young professors, and by today we have a really dynamic department of quite a broad scope and independent from the mathematics department, although we cooperate in various matters (both mathematics and computer science belong to the quite large Faculty of Mathematics and Natural Sciences).

I have a very nice group in Leiden: Joost Engelfriet, Jetty Kleijn and Hendrik Jan Hoogeboom. This is the tenured staff; then we also have a number of Ph.D. students – five of them at present. I have written about Joost already; I may add here that he is an excellent teacher. Jetty and Hendrik Jan are both worth their weights in gold. They are very much responsible for the nice functioning of our group. Jetty is an expert in formal language theory and in concurrency. She works more now on concurrency, especially Petri Nets and trace theory. Also, she is very skillful in combining her scientific career, teaching and administrative duties, and family life. Hendrik Jan likes to work on problems of combinatorial nature. He is an expert in formal language theory and trace theory, and in the last two years he has joined the DNA Computing crowd. All three, Jetty, Joost and Hendrik Jan were also my Ph.D. students – this is quite unusual, but I believe that this creates a special bond between us. Together with our efficient secretary Marloes, we form a nice and efficient group. Since I call my Ph.D.'s my children, we really form a family.

A propos Ph.D. students: I still do not have a universal algorithm for guiding them. Some of my Ph.D.'s wrote their theses with very little guidance (interference ?) from me (as a matter of fact I wrote my thesis totally on my own). With other students I've worked very intensely, writing many papers together. And then there were Ph.D. students "in-between". Typically I like to give to the beginning Ph.D. student a hand-written manuscript about a specific technical topic. Such a manuscript is only a rough outline of some ("initial") results and then I suggest that a student starts from this manuscript and works towards a paper. A big majority of beginning Ph.D. students know very little about writing papers, and so my initial coaching consists of enforcing many iterations of the first paper they write. This takes much of my time, but I always hope that this is a good investment – and usually the papers they write later require much less corrections.

I have been very healthy until about 1987 when I started to have serious back problems. They were first sporadic and located only at the lower back, but then with time the pain periods became longer and I got also neck problems, and carpal tunnel problem in both wrists. All of these have greatly influenced my functioning in recent years. Particularly unpleasant are periods when I cannot write. However I think that I manage this problem quite well: I read medical literature and even develop my own exercises. I am quite proud of it as this also supports my contention that "all that I have, I got on my own". In connection with my pain problems I have invented the following definition of optimists and pessimists among people with chronic pain (I am an optimist). When there is a bad period, an optimist is taking it well since he/she looks already forward to a good period that will follow. When there is a good period, a pessimist does not enjoy it because he/she knows already that a bad period will be coming. My doctors like this description very much. During bad periods, my group in Leiden is tremendously helpful and so I function well at the university – our secretary Marloes became literally my right hand (which is interesting because she is left handed). Maja (who is herself a medical doctor) takes care of me very well. I do accept invitations also for longer trips because Maja goes with me and she enjoys traveling very much. Traveling is very important because one learns so much about "the rest of the world" and gets a better perspective on one's own environment. For example I remember the course on Petri Nets that took place in Campina Grande in Brasil in 1996. There were students that traveled about 5000 km one way – 4 days by busses. And they were very enthusiastic, "eating up" every word uttered by the lecturers. It is then difficult to understand when a student from Leiden does not go to an interesting lecture in Amsterdam because it is too far (30 km).

The technological advances can quite soon solve my problems with writing. Already now there exists software that transcribes spoken dictation into a text on the computer screen. I have purchased one produced by Dragon Systems called "Naturally Speaking" (NS for short), however by mistake they have delivered the British English rather than American English version. The distributor in Holland has acknowledged the mistake, and because it takes a long time before they can deliver another version, I was given the British English version in the meantime, just to get acquainted with the software. My dutch-polish accent trained in the U.S. for so many years has little in common with the British pronunciation and this showed quite strongly in my use of the software. First I thought that I would be able to use NS to write this contribution. However I gave up after some most unexpected mistranslations. Here is an example of what has happened several times. I have dictated: "I have entered the Technical University of Warsaw in 1959" and NS wrote: "I have entered the technical uterus to war so 9059". Another time, when I was training NS to better recognize my voice, Maja has entered the room. I said then "I am dictating this text for my lovely wife", and NS wrote "I am dictating this text for my laughing wife" which was pretty good but a bit funny. I then trained NS to distinguish my "lovely" from "laughing", and after a short while NS translated my sentence into "I am dictating this text for my salary wife". Maja said then "This is certainly correct"! Clearly the root of these mistranslations is that NS ALWAYS "says something", even if it does not "understand" what has been said. I see this sort of situation quite often with people, and this often leads to problems.

Boulder, U.S.A.

My first trip to Boulder was in 1971 (from Buffalo) – to visit Andrzej Ehrenfeucht. Since then I have traveled to Boulder two or three times each year. Hence, the Boulder track was (and still is) interleaving with the rest of my life. As I have already written Andrzej has changed my life in many ways. He was the first one to show me the magic of theory so that I have switched from the engineering part of computer science to theory. Then through our collaboration for about 30 years, I have learned a lot from him both about science and about life in general. He knows so much about so many interesting matters, like, e.g., dinosaurs, geology, history of science, butterflies, spiders, ... We have written together more than a hundred papers, and still each new one is an exciting adventure. Just being together with Andrzej and his life partner Pat Baggett is wonderful.

Boulder is a beautiful place and it became my second home. A wonderful aspect of Boulder is its big distance from Leiden. It was definitely much better some time ago when fax and e-mail had not been in (such a) use yet. It was easier then to protect myself there from all "unscientific contacts". But even today being in Boulder allows me to distance myself from the daily stream of events in Leiden, and to reflect (in peace) about various matters of life, science and magic. I must have been a monk in my previous life because I really like long retreats (although I miss my family during longer stays in Boulder). Also, I have a very good medical care there – with quite many health problems haunting me in the recent years, this is a very important aspect of my life in Boulder.

After so many years of commuting to Boulder, I must be by now a part time American. Boulder is my second home – I always stay in the same apartment in the University Club. One of the benefits of staying at the University Club is a possibility of meeting interesting scientists that stay there while visiting one of the departments. In this way I met, e.g., Paul Erdös and had interesting discussions with him.

I have many friends in Boulder – some from the university and some from outside the university. I cherish very much my friendship with Mycielski and Malitz families – Jan Mycielski and Jerry Malitz professors at the department of mathematics. Also Mike Lightner and Linda Lundbeck are very close friends. Mike is a professor of electrical engineering and computer science and also a very serious musician. Linda is a professional musician. They both play recorders and specialize in old music. From time to time we have an "M and M" evening (of Music and Magic) – they play music and I perform magic.

Lloyd Fosdick who essentially created the computer science department in Boulder is a very good friend. He is retired now and lives in Turkey. It is a pity, because I miss very enjoyable dinners we used to have with him and his wife Riki. I also enjoy my "regular" dinners with Hal Gabow. With Mike Main we have worked together on graph grammars, and I have worked now for several years already with Skip Ellis on Computer Supported Collaborative Work (CSCW). I enjoy working with Skip also because through him I keep in touch with important developments in the applied aspects of computer science.

I have also very good friends outside the university. Many years ago I was "adopted" by the Verdoner family – Otto has Dutch roots and Daisy has Greek roots. They are wonderful people and I enjoy very much to be an uncle magician for their two sons.

EATCS

I have devoted a big part of my life to service for the computer science community. My involvement with the European Association for Theoretical Computer Science (EATCS) falls under the cliché of the exemplary military career: "from soldier to general". I became an EATCS member quite early, then I was a Treasurer, the editor of the "EATCS Bulletin", one of the editors of the "EATCS Monograph Series", a Vice-President, and the President. One certainly has to be motivated to devote as much time as I did (and am still doing) serving the community. Without this work my list of publications would certainly be much longer. My main motivation is rooted in my Warsaw time, where the feeling of isolation (implying "not belonging to the community") was very frustrating. Thus, bringing together scientists from various countries and various disciplines of science is for me a noble cause, certainly worth devoting a considerable part of my life.

The responsibilities of the EATCS President are perceived differently by different people. The following story is really true. During the ICALP (the annual conference of EATCS) in Madrid in 1991, I was chatting with a group of friends in a break between lectures. A young man approached our group and asked "are you professor Rozenberg, the EATCS president?". I answered "yes" and then he said: "It is already the second day when there is no toilet paper in the men's toilet on this floor. Could you do something about it", and this was not a joke. I certainly arranged for more toilet paper!

I have benefited from my EATCS work. First of all, there is an enormous satisfaction of doing something useful for so many people. Then, I've met in this way many interesting people. Also, I have gained a very good understanding of the nature of the community of scientists working in theoretical computer science (TCS). I certainly understand quite well the differences between the "cultures" of scientists from different countries. This helps me a lot in understanding the problems of creating the European Community.

I have also cooperated a lot with my American colleagues working for SIGACT. I think that the main difference between the American and European TCS scene lies in the "heterogenity" of Europeans. While in the US it is quite typical to have just a few "fashionable" trends (with a small number of "gurus") followed by "everybody", it is difficult to see how, e.g., British could influence French not to do formal language theory (which is beautifully developed in France) because it is not anymore fashionable in Great Britain! In my opinion, this diversity is the strongest guarantee of health of TCS in Europe.

I think that, in general, the advice of doing something (e.g. within EATCS) because "Americans do it" is not very valuable (in the same way as SIGACT doing something just because Europeans do it would not be valuable either). I remember well that when I was the EATCS Treasurer, my American friends were giving me a lot of advice on how to run EATCS finances; in the meantime SIGACT went broke a couple of times, while EATCS finances were very healthy.

As a matter of fact, my general advice is: do not follow too much advice from someone else!! It leads very easily to the loss of originality – it is often better to make a mistake and to learn from it. This reminds me the following real story. The first car that I bought in Holland in 1970 was a quite old Volkswagen Beetle. It still had a choke and I often had difficulties in starting it. A good friend from the US, a mathematician, was staying in Holland for a couple of years then, and he was a very skillful car mechanic. He was often giving me a demonstration of the use of choke. He was trying to teach me how the sound of the motor (reacting to the pulling or the pushing of the choke bar) was guiding the use of the choke. We concluded that my hearing is somewhat impaired, and so I do not hear "the subtleties". At some point I had to bring my car to a garage for a total overhaul. When I was in the garage to pick up my car, a conversation with the car mechanic (who took care of my car) somehow led to the choke in my car. He told me then (and showed me) that the choke in my car was disconnected by the previous owner, the choke bar was just "hanging there" not connected to anything!

EATCS Monograph Series (recently splitted into Monographs and Texts) is an important and successful activity of EATCS. It has perhaps the most friendly editors–publisher team. Wilfried Brauer, Arto Salomaa and myself are the editors, and Dr. Hans Wössner and Mrs. Ingeborg Mayer are responsible for this series on behalf of Springer Verlag. Once a year, for at least twelve years already, we meet in my home for a day discussing all the matters pertinent for the series. Through these years I have developed a recipe for the *Monograph Soup* which is our traditional lunch (as a matter of fact this recipe is coauthored by Maja). For the evening dinner we are often joined by Wilfried's wife Ute and Maja.

DADARA

I have a wonderful family. I owe a lot to my parents who I really admire – my father died in 1996, my mother is still alive. One reads and hears a lot about heros, one sees them in movies, however I've never met one except for my mother. She is a real hero – one could certainly write a book about her life. I have known my wife Maja since 1962; we were married in 1967. She is a wonderful life partner and my life without her would certainly be much less satisfying (difficult for me to imagine). She has created for us a wonderful nest, no wonder that Maja in Finnish means "a cozy home". Then I have two brothers of choice, Andrzej and Arto. It is difficult to imagine better brothers (an interesting aspect of this brotherhood is that Andrzej and Arto have never met).

My son Daniel is very special to me both emotionally and intellectually. He is an artist, very well known internationally. He was born with a pencil in his hand (Maja always says that this was painful!). We never had a problem of finding something interesting for Daniel to do "because he is bored" – he was always either drawing or reading a book. I remember that when we were in Buffalo and Daniel was in a kindergarten, he was often bringing home all kind of words and expressions that we did not know. At some point we realized that these were Japanese words that he was picking up during lunches with a Japanese girl from the kindergarten and her mother! In this way his talent for foreign languages manifested itself for the first time. By today he knows six or seven languages.

Daniel was not very much impressed by the fact that I was a university professor and perhaps a known scientist. But I had his respect as demonstrated by the following story. Once I came home from my office early in the afternoon and found Daniel and his school friend Ferdie in the hall of our house (Daniel was 14 years old then). Daniel introduced his friend to me, and I went directly to the kitchen to have a glass of water. Since the door was open, I could hear the following conversation. Ferdie: "What is your father's occupation?", Daniel: "He is a university professor", then he paused for about 5 seconds and added "but he is not stupid, he is a very good magician". This story has become well known among my friends – professors and magicians.

Daniel began his professional career, under the name DADARA, as a cartoonist. His cartoons, almost always without titles, were amazing. He has a wonderful sense of humor and a very good (drawing) hand: in few strokes of a pencil he can make a deep comment about the life around us. He is a very prolific artist – he drew thousands of cartoons which made him well known. For example, he was a cartoonist for an established Italian monthly "Linus" (it was nice for me to see his name listed in Linus with cartoonists such as Gary Larsen and Altan), and he became well known in Italy. Once, after giving a lecture in Italy, I went for a dinner with a group of professors. At some point during the dinner I have heard the name "DADARA" mentioned by two Italian colleagues (in their conversation in italian) sitting to the right of me – I did not know them. I then interrupted their conversation and said "I know DADARA". They continued their conversation in Italian, so I have asked my Italian friend sitting to my left what they said. He told me that their comment was that I was bragging. Hence, a couple of minutes later, I again interrupted their conversation and said: "DADARA is my son". They were laughing and continued their conversation. When I asked my friend what was their comment now, he told me that one of them said that I am drunk. I found this very nice (I do not drink alcohol at all), and did not interfere with their conversation anymore.

I am fascinated by DADARA's creativity – it is amazing to see what kind of visions he has in his head! I have always liked a company of creative people, and the fact that he is my son adds an extra dimension to my admiration of his art. He is an example of a wonderful genetic mutation: neither Maja nor I have any talent for drawing or painting, this is also true for our families, and then we get such a talented son.

DADARA does not like labels – since everybody called him a cartoonist, he has switched to painting, and very soon everybody admired his paintings. Several years ago DADARA was designing flyers for all kinds of big events in Amsterdam. His flyers became very popular (collector items) and the press pronounced him "the king of flyers". Since he does not like labels – he quit designing flyers for quite a long time. This drive for change, the evolution of style, is very characteristic for DADARA. While at the beginning I was worried (although I never said it) that he abandons a style that is so nice, now I am looking forward to the changes that surely lead to pieces of art that are even more beautiful and more intriguing. Quite often I found out that a painting by DADARA expresses a feeling (an emotion) that I had "inside me" for a long time but could not express.

DADARA lives in DADALAND which is in his head and which also contains the art that he has created. Also his beautiful atelier in Amsterdam, where each small detail was thought of and created by him, is a material extension of DADALAND.

I think that analyzing why an artist (or a scientist) is so creative does not make much sense – it is simply in his/her brain. Observing DADARA supports this point of view. In the time (not much) that is not used for science or magic (or mundane matters of functioning), I study the paintings by Hieronimus Bosch – the Dutch painter living at the end of the 15th and the beginning of the 16th century. I like Bosch paintings very much, and I have been planning already that after I retire I will travel visiting "all" museums where they have his original paintings. For me, Bosch is the pure genius much ahead of his time. He is the real father of surrealism. I remember visiting once the museum of Salvatore Dali in Figuera, Spain. There I have seen sketches by Dali from his young years, when he was "training his hand for the paintings to come". Many of these sketches were simply figures from paintings by Bosch. Although we know very little about Bosch (we even don't know when he died!), there is an abundance of theories explaining why Bosch was so original and creative. For me they do not make much sense. There is a simple explanation: he had all of these fascinating and intricate scenes in his head – the Boschland.

The Golden Triangle Plus

I have done research in various areas of TCS. At some point my research has converged to three areas – I like this combination very much and so I call it a golden triangle. The areas are: 1) formal language and automata theory, 2) the theory of graph transformations, 3) concurrency. Let me comment on each of them.

1) Formal language and automata theory (FLAT).

This area forms the backbone of TCS in the sense that many other areas have their roots in it, and it is perhaps the best developed and the most mature area of TCS. The Handbook of Formal Languages (in 3 volumes) published by Springer Verlag in 1997, which we have edited with Arto, demonstrates an amazing richness and scope of FLAT.

Although I have worked on various problems in FLAT already in Poland, my first big FLAT adventure was the theory of Lindenmayer systems (L systems). The theory of L systems is a beautiful research area – it is full of Perfect Research Problems, i.e., problems that are elegant, very easy to state and very challenging to solve. Moreover many problems in L systems are very well motivated. The theory of L systems is quite special to me, since through scientific collaboration

I've met interesting people, many of whom are (were) my very good friends. First of all, Aristid Lindenmayer himself – as I have mentioned already, through my work with Aristid I have discovered the pleasures of interdisciplinary research. Then, Arto Salomaa – I have written about him already. Arto must be the best "producer" of brilliant Ph.D.'s within FLAT – the group of his Ph.D.'s includes T. Harju, J. Honkala, J. Karhumäki, J. Kari, L. Kari, M. Linna, V. Niemi, M. Penttonen, and K. Ruohonen. I have profited from the quality of this group, as I have collaborated with some of them. I was an "official opponent" for Juhani's Ph.D. thesis defense – I am quite proud of this because he is by today one of the leading researchers in FLAT, and he could certainly be my official opponent today! Juhani is also an ornithologist – he disappears for about a month (I think in June) in Finnish forests where he rings owls. A long time ago during one of his visits in Holland he gave me some pictures (taken by him) of young owls in their nests – the pictures were beautiful and I've fallen in love with owls. I started to collect "everything" about owls: books, figurines, the real stuffed owls, ... By today I am an owl lover and have a collection of more than 1000 owls. One of the attractions of owls for me is that they look to me like "the magicians of nature".

Arto has introduced me to Hermann Maurer in 1974 during the ICALP (International Colloquium on Automata, Languages and Programming) in Saarbrücken. He is one of the most talented people that I have met: he has been active and very successful in many disciplines of computer science ranging from very theoretic to very applied. His Hyper–G (Hyperwave) is acknowledged to be one of the most important WWW systems. Hermann is producing at the end of each year a report on his activities during the given year and sends it to his friends. Together with Maja we always look forward to these reports and enjoy reading them enormously. My visits to Hermann, first in Karlsruhe and then in Graz, were always very memorable; we had excellent time with his wife Ushi and children Stephan and Claudia. We have written together a number of papers in FLAT.

In recent years I have cooperated a lot with Gheorge Păun mostly on topics from FLAT and DNA Computing. Gheorge is very gifted and extremely efficient researcher – he is certainly one of the stars of FLAT and DNA Computing. Somehow he reminds me myself when I was still in Poland: total involvement in science gives you some encapsulation against all that happens around you. Gheorge is one of former students of Solomon Marcus. The list of former students of Marcus working now in TCS is very impressive – it includes Cris Calude, Lila Kari, Alexandru Mateescu, and Gheorge, all very well known now. I am a friend with all of them, and have written papers with Lila, Alexandru and Gheorge. I was an "official opponent" of Lila for her Ph.D. thesis defense. She is a very talented researcher – already one of the most visible scientists in DNA computing. I certainly have some affinity with Romanians. As a matter of fact I remember that Sheila Greibach told me, when I met her for the first time, that she thought that I am Romanian.

2) The theory of graph transformations.

Graphs are "everywhere" in computer science, and graph transformation is one of the important paradigms of computer science spread throughout many of its disciplines. Graph grammars (where one rewrites graphs rather than strings as is the case in classical FLAT) originated already in the 60's, motivated by considerations about pattern recognition, compiler construction and data type specification. Since then the list of areas which have interacted with the development of graph grammars has grown considerably, and methods for transforming (sets of) graphs other than grammatical are investigated now.

My interest in graph grammars begun around 1973 when I met Hartmut Ehrig from Berlin. Hartmut was one of the pioneers of graph grammars and his enthusiasm about graph grammars was quite contagious. We became good friends and together we have organized quite a lot of activities in the area of graph grammars. Hartmut is a very serious scientist and very reliable in both scientific and organizational matters. Through Hartmut I met Hans Jörg Kreowski who was then his Ph.D. student. Since then we have collaborated and become good friends. Together with Manfred Nagl and Hartmut we organized the first workshop on graph grammars, it initiated a series of workshops which form the focus of research in the area of graph transformations. Manfred was responsible for important applications of graph grammars, and the three of them (Hans Jörg, Hartmut and Manfred) have greatly contributed to the development of graph grammars in Europe. My first big adventure in graph grammars was the theory of node–label controlled (NLC) graph grammars which we have developed in Antwerp with Dirk Janssens. The theory of NLC grammars generalized later by Joost Engelfriet, George Leih and Emo Welzl to edNCE graph grammars (graph grammars with neighbourhood controlled embedding and dynamic edge relabeling) became an important part of the theory of graph grammars. With Dirk we have also developed a theory of graph grammars modeling massive parallelism as represented, e.g., by actor systems. Dirk has later generalized actor grammars to the so called EMS systems which became one of the main models of concurrency within the framework of graph transformations. During one of my trips to Graz to visit Hermann Maurer I met Emo Welzl. We started our cooperation even before we met, through Hermann: we have written a paper on picture languages which became quite popular. With Emo we worked on NLC grammars. He came in 1983 for a year to Leiden where he was a post doc in my group. We became very good friends (he is also a good friend of my son Daniel), I have "adopted him", and by now he is one of my kids. We have worked out together the theory of boundary NLC grammars. Emo is now a professor at the well known ETH in Zurich – he is one of the young stars of TCS in Europe.

For many years now we have been developing with Andrzej the theory of 2–structures which forms a useful framework for considering decompositions and transformations of graph–like structures. Tero Harju has joined us quite early in this research and so most of the theory has been developed by the three of us (although Tero never met Andrzej). Tero is one of these brilliant Ph.D.'s of Arto. He is a wonderful research partner not only because he is a good mathematician but also because he has a very good sense of humor. We certainly laugh a lot when we work together. Unfortunately he is quite addicted to cigarette smoking. However, he never smokes inside his own or someone else's house. He goes outside and paces around smoking. My neighbours call him "a walking chimney". One says that Finns do not talk much (Arto is an exception) – but we have spent countless hours with Tero talking about so many things.

3) Concurrency.

In 1976 I have attended the "Advanced Course on Petri Nets" which took place in Hamburg, Germany (West Germany then). The course has been an important event in my life as there I've fallen in love with Petri Nets and in this way I discovered the larger area of concurrency. During this course I met Carl Adam Petri and have been very impressed with his way of thinking about concurrency. I also met there Antoni Mazurkiewicz from Warsaw and P.S. Thiagarajan, or simply Thiagu for his friends, from the group of Petri at GMD in Schloss Birlinghoven near Bonn. Although I have remembered Antoni Mazurkiewicz from my Warsaw times, we never had any contact there. We have spent many hours together in Hamburg and became very good friends since then. Later I also met his wife, Violetta who is a professor of linguistics at Warsaw University. She is also a world class astrologer – it must be that astrologers and magicians understand each other very well, because we became very good friends. The friendship extended to our families during the one year stay of Antoni in Leiden (for the academic year 1985–1986). Antoni introduced me to his theory of traces, known worldwide as the theory of "Mazurkiewicz traces". It is a very elegant and fruitful way of thinking about concurrency. My group in Leiden has a lot of expertise in traces – Hendrik Jan Hoogeboom and Jetty Kleijn are the real experts in both syntactic and semantic aspects of trace theory.

Thiagu was the real expert on Petri Nets at GMD. He "had nets in his blood" and in particular was a real master of counterexamples. You gave him a conjecture about nets, and if it did not hold then there was a big chance that Thiagu would find a counterexample. Thiagu has also a wonderful sense of humor and always knows a lot of gossip – we have exchanged many jokes and gossip. He is certainly a very stimulating research partner and a very lively friend to have around.

It was really for Petri, Antoni and Thiagu that I have fallen in love with Petri Nets and trace theory. Through the years together with Thiagu, Wolfgang Reisig, Hartman Genrich, Kurt Jensen and many others we have built up a real community of researchers in Petri Nets centered around the European Workshop on Application and Theory of Petri Nets promoted later to the International Conference on Application and Theory of Petri Nets (ICATPN). This became a very well functioning community composed of users, practical designers and theoreticians of concurrent systems. As with EATCS, I have invested a lot of time and energy in the Petri Nets community (I am the chairman of the Steering Committee for ICATPN). It was certainly worth it.

From 1983 until 1993 I had been involved in the Dutch project on concurrency, initially known as LPC (Landelijk Project Concurrency) and then as

REX (Research and Education in Concurrent Systems). This project was based on the cooperation of the group of Jaco de Bakker from Amsterdam, the group of Willem Paul de Roever from Eindhoven, and my group from Leiden. The cooperation was very successful, and the project has certainly contributed to the progress of research on concurrent systems in Europe. With Jaco I have cooperated also in a number of other activities – I find him to be a very professional colleague, and look forward to some future activities where we will cooperate.

4) Plus.

I have also worked (and published papers) on interesting topics from outside the golden triangle. In particular for several years already I work in two new areas that I find interesting

– CSCW (Computer Supported Cooperative Work).

This is a very applied area that can use some good theory, which is not there yet. I have worked here with Skip Ellis, one of the leaders of CSCW and with Andrzej. One can apply here some known models like Petri Nets (modelling Workflow Systems), but clearly new models describing protocols of structured cooperation are needed.

– DNA Computing.

This is a beautiful area. It is often traced back to some remarks of R. Feynman on computing with molecules, but there are really two founders of the area. One is Tom Head who already in his paper from 1987 presented a theory of computing based on the use of restriction enzymes. This work was theoretical and not supported by experiments then. The other one is Len Adleman who published in 1994 a paper in "Science" where he outlines a method for solving the Hamiltonian Path Problem based on the Watson Crick complementarity principle for DNA molecules. He also gave "a proof of the principle" by implementing his method in the laboratory (for a small graph). This is a fascinating area: the fact that you can actually compute with molecules is very exciting, independently of whether or not one can beat NP-complete problems. Although some techniques of formal language theory can be applicable in this area (after all DNA molecules are "double strings") it is becoming clear that new models are needed. A characteristic feature of this area is its interdisciplinary character – something I love from the time of my work on L-systems. At the workshops on DNA Computing one meets biologists, chemists, physicists, engineers, mathematicians, computer scientists, ... I am now quite involved in DNA Computing and this is already rewarding: I have learned a lot in recent years about genetics. It is a pity that Aristid Lindenmayer is not anymore with us – he cannot witness the formation of yet another area of computing motivated by nature.

My optimism about the future of DNA Computing is also supported by the fact that my new collaborators, Hans Kusters and Paul Savelkoul from Amsterdam, and Paul Hooykaas, Kees Libbenga and Herman Spaink from Leiden, are very enthusiastic about this area of research. They are all biologists, except for Paul Savelkoul who is a medical microbiologist. We have now formed in Leiden the Leiden Centre for Natural Computing where we have three research programs: molecular computing, evolutionary algorithms and neural computing. I am the director of the molecular computing program, Thomas Bäck and Joost Kok are the directors of the evolutionary algorithms program, and Joost Kok is the director of the neural Computing program. Thomas and Joost are my colleagues from the Leiden Centre for Advanced Computer Science – they are prominent researchers in their research areas. The biology department participates in the molecular computing program. The future of computing in Leiden looks very natural indeed.

Magic

Magic is a way of living in the same way that science or music is – you go to bed with an idea of an illusion, and you may wake up with a solution. It is certainly very creative and, as in science, you invent your own material where you sometimes build on the work of others. As a matter of fact there is a great similarity between magic and science. Magic teaches you that things do not have to be the way they seem to be at the first glance. The blue deck of cards turns out to be red, or a card that you have seen in the deck is actually not there. DO NOT ACCEPT THINGS ON THEIR FACE VALUE: QUESTION EVERYTHING (your senses may be deceptive, your thinking may be deceptive). But this is really one of the main rules of researchers in science! Another common feature of scientists and magicians is that you make friends very easily. You go somewhere, give a lecture to a group of scientists (or have an "individual" session with a scientist), a discussion follows, and very soon you have new friends – you go to a restaurant together, and between the dishes you use the napkins to scribble some scientific ideas. In the same way, you go somewhere, you give a show (or you have an individual presentation for a magician), a discussion follows, and very soon you have new friends – you go to a restaurant together, and between the dishes you use free spots on the table cloth to demonstrate some ideas on cards and coins. One of the advantages of my involvement in magic is that in this way I have an access to a totally different community of people, which is very interesting. Many of my good friends are magicians.

Magic is a performing art, and modern magic is not so old, it dates back to the last century. Many performing artists consider magic to be the Queen of performing arts, because so many branches of performing arts do contain an element of illusion. If you go to a theater and the lights go down, the curtain goes up, and you see a decoration which represents, e.g., a house in Paris, then if the play is good, you forget that this is a decoration – you simply are in Paris! (but this is an illusion). Illusion is the essence of magic: when magician comes, illusion is there, and when he leaves, illusion disappears.

Magic is divided in categories, and close up magic is what the name says: there is no distance between the magician and a spectator, you see everything done just under your nose, there is nothing to hide. Thus, many magicians consider close up magic to be the Queen of magic. A close up magician may use different media to invoke illusions: playing cards, coins, rope, timbles, ... (I use playing cards because I love them). There is a great similarity between close up magic and mathematics: good illusions are "compact" and elegant. If your building up for the climax is too diffused and "illogical", then you will not catch (control) the attention span of a spectator. If your magic is elegant, then this elevates the senses of a spectator to a different level of perception – it is not anymore the level of "solving a puzzle". And then, the most beautiful effects are unexpected, just like in mathematics. Paul Harris, one of the giants of modern close up magic, calls magic "the art of astonishment" and that's what it is.

Up to about ten years ago I was keeping my magic and my science lives separated. Then I have realized that performing at scientific workshops and conferences which I attend gives me more opportunities to perform for bigger audiences (because of the chronic lack of time I would not have such opportunities otherwise). And so I do quite a lot of "table hopping" (moving from table to table to perform) at conferences now. As a matter of fact this mixing of disciplines goes also the other way around. I have been a couple of times in New Zealand which is a beautiful country. It has some very good magicians, and already during my first stay there I became a sort of "honorary member" of one of the magic societies there. I am a very good friend with some of the very good magicians there like Alan Watson and his magic family and Ken Ring. Anyhow, when I was there in 1997, they learned that I was lecturing in New Zealand on DNA Computing and so before my show ("Story telling with playing cards") they asked me to explain what the DNA Computing was. Hence, before the start of my show I gave about 20 minutes lecture on DNA Computing for magicians, and (judging from the response) it was successful!

One of the problems that a magician may encounter is the attitude of some spectators. Some of them will sit there very, very tense, watching every move of your hands, ready for "catching you". I like to explain (at the beginning of a performance) that such an attitude makes no sense. One should relax and enjoy, because there is nothing there to catch you on – illusions have no explanation. If, by any chance, one will catch a glimpse of "something" (technical), then this particular effect was not an illusion! This logical argument relaxes my spectators, also scientists. I remember that when I gave this explanation to Mike Rabin, Mike called this the "Rozenberg Principle": an event observed disappears!

As I wrote already, my stays in Boulder are wonderful because I am there much more the master of my time, and in particular I have much more time for magic (late nights are often taken by magic). I have been very lucky by finding there a group of friends which are very good magicians: Gene Gordon, Bob Larue, David Neighbours and Lamont Ream (David and Lamont are among world top magicians). Together with Gene and David, often joined by Bob, we have regular card sessions in my apartment (we often joke that the University Club will become really famous in the history of magic!), and with Lamont we have magic sessions in his apartment in Longmont. With Bob and his wife Mel we are now also family friends – we hope to have them visit in Holland soon.

The symbiosis of science and magic in my life has also some frustrating aspects. If I get very involved in science for a longer stretch of time (e.g., writing a paper), then quite often at the end of such a period I get worried that I have been "out of magic" for "so long", and there I have so much to do. The same holds the other way around. So these frustrations control for me the interleaving of science and magic. As a matter of fact, one of the biggest frustrations I had, happened many years ago. I was admitted then to a magic class coached by Tomy Wonder, that was meeting about once a month. Unfortunately, because of my traveling in service of science (conferences, lectures, etc.) I missed most of the classes. I lost a unique chance because he was a wonderful coach, and by today he is one of the most famous magicians in the world.

As a magician, one can ponder over questions that are not totally rational. For example, was magic predestined for me? Nobody knows, but I wonder: the maiden name of my mother is Zauberman (meaning magician) and the name of my wife is Maja, which means "illusion" in Hindu (I certainly did not know it when marrying Maja). Anyhow all my magician friends are very jealous that I am married to illusion!

There is quite a lot of literature on the **THEORY OF MAGIC**. The aim of it is mostly to provide some guidelines, some principles that enhance the performance of a magician. It is mostly concerned with things like perception of events by a spectator, or the psychology of a spectator, or the aims and the strategy of a performance. The difference with TCS is that the theory of magic stays rather close to its applications (in a performance). In TCS we abstract from a phenomenon in information processing and formulate some notions, which then lead to other notions, and so we have "an iterative build up" of a theory that can end up quite far from the initial application (motivation) area. In the theory of magic you stay really close to the application area. This is the way I see the difference. Anyhow, I am really NOT interested in the theory of magic – in magic my interest lies in the heart of applications: the performance.

I have a wonderful family, I have written many papers, I have shuffled many decks of cards. Life has been good to me.

Arto Salomaa

Professor Arto Salomaa received his Ph.D. in 1960 and has been Professor of Mathematics at the University of Turku since 1965. He has authored more than 400 scientific publications in major journals, as well as ten books, some of which have appeared also in French, German, Japanese, Romanian, Russian, Vietnamese and Chinese translations. Prof. Salomaa holds the degree of doctor honoris causa at six universities. He has been an invited speaker at numerous conferences in computer science and mathematics, and a program committee member or chairman for major computer science conferences, including STOC, ICALP, MFCS, FCT. Prof. Salomaa was EATCS President 1979-85 and is currently the editor-in-chief of the EATCS Monograph Series published by Springer-Verlag, as well as an editor of eight international journals of Computer Science. Festival conferences "Salodays in Theoretical Computer Science" (Bucharest 1992), "Salodays in Auckland" (Auckland, NZ, 1994) and "Important Results and Trends in Theoretical Computer Science" (Graz, 1994) have been arranged in his honor. Prof. Salomaa is a member of the Academy of Sciences of Finland, the Swedish Academy of Sciences of Finland and Academia Europaea 92. Currently, he is a Research Professor at the Academy of Finland and the head of the Academy's research project on mathematical structures in computer science.

Events and Languages

Early computer science

It is customary to begin a paper with an abstract or a summary. The purpose of this article is to tell about some *people* and *ideas* in the early history of computer science, roughly up to 1970. The article has a very subjective flavor: I will tell only about events and persons I have in some sense come into contact with. Since my point of view is over formal languages and automata, the happenings and phenomena reported in the sequel will be from these and related areas of theoretical computer science, rather than from the theory of algorithms and complexity. The relative importance of the latter areas has increased very much since the 1970s.

Although there have been many attempts of unification and clarification, it is still not at all clear what should be called computer science and what is theoretical computer science. Opinions and especially points of emphasis vary remarkably from place to place, country to country. Of course the setup was even less clear and more obscure in the 60s when important ideas were emerging or just about to emerge. However, by the end of the 60s well-established computer science curricula existed already at numerous universities. An indication of the ideological unclarity is the Finnish word for "computer science", "tietojenkäsittelyoppi", still prevalent in Finland. Literally translated the word means "discipline of data processing". I opposed this term very much in the 60s but without success. I was in the 60s also in London, Canada, for a long period. Here are some excerpts from my writing at the University of Western Ontario, in connection with a proposed Ph.D. program in computer science:

> Like any young and growing discipline, Computer Science is far from being saturated and, therefore, one should be prepared to add new areas to the program. When launching the program, one should consider simultaneously with the areas of study also the availability of staff members in these areas. ... *Formal language* and *automata theory* is in a rather central position in every Ph.D. program in Computer Science I know of and, thus, I don't see too much harm in this respect in a conflict with other universities in the province. ... The main part of *computer combinatorics* would consist of the theory of graphs and networks but also other combinatorial problems with applications in computers could be included. ... *Applied areas* could include some of the following: information retrieval, symbolic computation, artificial intelligence, theorem proving by computer, pattern recognition, learning theory (CAI), simulation, organization and design of systems for sequential, concurrent and time-shared processing both on line and off line, design of system components for assembly and compilation, computational linguistics. Any list like this is open-ended and in selecting the areas the availability of staff should be kept in mind. In all of the suggested areas, cooperation with other departments is possible and sometimes even necessary. Some such departments could be: mathematics (recursive function theory, combinatorics), applied mathematics (numerical analysis), linguistics (natural language aspects of formal languages), electrical engineering, library science.

That is my authentic writing from 1968, maybe still preserved in some files at the University of Western Ontario. I would still agree with some of it but obviously many things in it have become obsolete. However, I do not want to dwell on this topic longer; no final answers can be given anyway.

Formal languages: Are they formal and are they languages?

Grzegorz Rozenberg – I will talk more about him later – published a paper "Decision problems for quasi-uniform events" which had been presented to the Bulletin of the Polish Academy of Sciences by Andrzej Mostowski on August 7, 1967. In this paper we read:

> Let $\Sigma = \{\sigma_1, \sigma_2, \ldots, \sigma_n\}$ be a finite nonempty *alphabet* fixed for all our considerations. The elements of Σ are called *symbols*. Finite sequences of symbols are called *tapes*. The length of any tape x is defined to be the number of letters in the tape and is denoted by $\lg(x)$. The set of all tapes over Σ (including the empty sequence λ) is denoted by $T(\Sigma)$. We shall identify tapes of length 1 with symbols of Σ. If x and y are tapes, then xy denotes the concatenation of x and y. ... If P, Q are subsets of $T(\Sigma)$, then P, Q are called *events*. For arbitrary events P and Q, $P \cup Q$ is defined as the set union, the (complex) product PQ as $PQ = \{xy : x \in P, y \in Q\}$, and P^* as $\cup_{k=0}^{\infty} P^k$, where $P^0 = \{0\}$ and $P^{i+1} = P^i P$ $(i \geq 0)$. It is obvious that $T(\Sigma) = \Sigma^*$. ... An event P is *fundamental* if there exists a subset B of Σ such that $P = B^*$.

Everything is very well presented, maybe some of the terms sound strange nowadays. Indeed, (*formal*) *languages* were earlier mostly called *events*. I used the latter term in all of my first papers. The double meaning of the word "event" is reflected also in the title of this writing. Apparently, the usage of the word "event" in this sense stems from Kleene, [20]. I switched to "languages" in 1966–67. "Events" do not occur any more in my first book "Theory of Automata" which was written in 1966–67 but, due to my complete ignorance about the publishers, was delayed for more than two years by the publisher.

But does it make more sense to call subsets of Σ^* "formal languages" than "events"? Probably not. Most of the subsets lack any form whatsoever – how could they be called "formal"? And most of them do not serve any purpose in any kind of communication – how could they be called "languages"?

Let us go back to the quoted passage of Rozenberg, [13]. It is very mathematical in flavor. Moreover, the influence of the great Polish school of logicians becomes apparent in statements such as "We shall identify tapes of length 1 with symbols of Σ". In general, most of the early investigators of computer science, no matter how the term is understood, were mathematically trained and oriented. Thus it is very natural that typical mathematical topics, such as the effect of certain well-defined operations on languages, were studied early in the game. This tradition has continued, at least to some extent. Current investigators should not just continue the tradition for its own sake. They should rather make sure that their topic is of some practical importance, or else really significant mathematically.

I met a few times the aforementioned Andrzej Mostowski, a famous representative of the Polish school. He wore very elegant suits at conferences, quite unlike the casual so-and-so attire currently in use. He was also polite and considerate. He was chairing a session in 1962, where I gave a talk about infinite-valued truth functions. I was somewhat nervous, and apparently this was noticed by Mostowski. There were some simple questions after my talk but Mostowski did not want to embarrass me with his difficult question. Instead, he came to me much later and asked "I would like to know how you actually obtain the decidability".

In the early days of formal language theory, the motivation and the belief in the usefulness of the theory were quite strong. Let us look at some examples from well-known sources. They tell also what different people thought language theory was about. Noam Chomsky, [3], wrote in 1959:

> A language is a collection of sentences of finite length all constructed from a finite alphabet (or, where our concern is limited to syntax, a finite vocabulary) of symbols. Since any language L in which we are likely to be interested is an infinite set, we can investigate the structure of L only through the study of the finite devices (grammars) which are capable of enumerating its sentences. A grammar of L can be regarded as a function whose range is exactly L. Such devices have been called "sentence-generating grammars". A theory of language will contain, then, a specification of the class F of functions from which grammars for particular languages may be drawn. The weakest condition that can significantly be placed on grammars is that F is included in the class of general, unrestricted Turing machines. The strongest, most limiting condition that has been suggtested is that each grammar be a finite Markovian source (finite automaton). The latter condition is known to be too strong; if F is limited in this way, it will not contain a grammar for English. The former condition, on the other hand, has no interest. We learn nothing about a natural language from the fact that its sentences can be effectively displayed, i.e., that they constitute a recursively enumerable set. The reason for this is clear. Along with a specification of the class F of grammars, a theory of language must also indicate how, in general, relevant structural information can be obtained for a particular sentence generated by a particular grammar.

This can be viewed also as a motivation and background argument for the Chomsky hierarchy – of course there are other sources in this respect more comprehensive than [3]. By mid-60s the hierarchy was pretty well established; the following writing by Ginsburg, [7], emphasizes the interconnection between *ALGOL* and context-free languages. In fact, the latter were often called *ALGOL*-like languages.

> As is well-known, the context free languages are excellent approximations to the syntactic components of currently used programming

> languages (such as *ALGOL*). Pushdown automata (abbreviated *pda*) are devices used in parsing programming languages, for the most part in compiling. These two components are linked by the result that a set of words is a context free language if and only if it is accepted, i.e., recognized by some *pda*. To implement the construction (either by hardware or by software) and the usage of *pda*, it is important to have general classes of *pda* with particular properties. For example, it is convenient to discuss deterministic *pda* since they parse rapidly. In the same spirit, we investigate the class of *pda* having the property that the length of the pushdown tape alternatively increases and decreases at most a fixed bounded number of times during any sweep of the automaton. Such *pda* reject words faster than an arbitrary *pda*.

In baseball you sometimes face a situation: all or nothing. Bases loaded, bottom of the ninth inning, two out. An ε-small additional push on your high ball can bring it within the reach of a spectator (this happened to the Yankees in the 1996 playoffs: homerun) from the hands of an outfielder. Baseball being the most American game, such an all-or-nothing attitude has been visible in the writings of many Americans concerning formal languages and related areas. The quoted passage by Ginsburg is very optimistic: "excellent approximations of programming languages", "implement the construction either by hardware or by software", etc. Many similar examples of such a motivational "all"-mood, to be contrasted with the present "nothing"-mood, can be given from the 60s. I just mention papers by Greibach and Hopcroft, [10], and Greibach and Ginsburg, [9], the latter propagating in the Journal of the ACM the (nowadays esoteric) theory of multitape AFA (abstract families of automata).

By the end of the 60s, language theory had become already a very big field. A bibliography by Derick Wood, [21], from 1970 contains close to one thousand items, although it misses most of the Russian literature, which was quite extensive and also good in the early days. Some of the key points of language theory are brought home by following excerpts from well-known early authors. Each of the excerpts is of an introductory character. We begin with Michael Harrison, [11]:

> Formal language theory was first developed in the mid 1950s in an attempt to develop theories of natural language acquisition. It was soon realized that this theory (particularly the context-free portion) was quite relevant to the artificial languages that had originated in computer science. Since those days, the theory of formal languages has been developed extensively, and has several discernible trends, which include applications to the syntactic analysis of programming languages, program schemes, models of biological systems, and relationships with natural languages. In addition, basic research into the abstract properties of families of languages has continued.

Ron Book, [2], follows the same general line, emphasizing especially parsing and compiling:

> Formal language theory is concerned with problems of specification of sets of strings of symbols, i.e., languages, and properties of these sets. Mathematical models are constructed with the goal of faithfully representing properties of the languages involved and of devices which generate, accept or perform operations on languages. The languages studied arise from problems in abstract automata theory, in programming language and compiler theory, and in linguistics (particularly, mathematical models for natural languages). ... An extremely active area of formal language theory involves the construction of models for programming languages and for translator systems. Models are constructed in attempt to formalize the compiling process and to formalize various parsing methods, and resource requirements for parsing are considered.

The next two introductory excerpts are from Aho and Ullman, [1], and my own "Formal Languages", [18], respectively:

> The theory of languages is concerned with the description of languages, their recognition and processing. A language may contain an infinite number of strings, so at the least, one needs a finite description of the language. Particular types of finite descriptions will yield useful properties of the languages they define, especially when the class of languages they define is "small" (i.e., not every language of conceivable interest is defined). If C is the class of languages defined by a certain type of description, one would like to know whether membership in C is preserved under various operations. One would like to know that the languages in class C could be recognized quickly and simply, especially if one were attempting to develop a compiling system for a language or languages in C. Also useful are characterizations of languages in C, so one can tell easily if a given language is in class C. Finally, one wants algorithms, if they exist, to answer certain questions about the languages in C.

> A language, whether a natural language such as English or a programming language such as *ALGOL*, can be considered to be a set of sentences, that is, finite strings of elements of some basic vocabulary. The definition of a language given below is based on this notion and is, consequently, general enough to include both natural and programming languages. The syntactic specification of a language with finitely many sentences can be given, at least in principle, by listing the sentences. This is not possible for languages with infinitely many sentences. The main problem of formal language theory is to develop finite representations for infinite languages. Such a representation may be accomplished by a generative device or by a recognition device. By imposing restrictions on the devices, different language families are obtained.

Instead of trying to comment and develop further the many issues raised in the above quotations (the authors themselves might not agree any more with many points), I am taking a direct leap to the present. A summary of my current views is contained in the following excerpt from the Preface to the recent "Handbook of Formal Languages", [15]:

> The theory of formal languages constitutes the stem or backbone of the field of science now generally known as theoretical computer science. In a very true sense its role has been the same as that of philosophy with respect to science in general: it has nourished and often initiated a number of more specialized fields. In this sense formal language theory has been the origin of many other fields. However, the historical development can be viewed also from a different angle. The origins of formal language theory, as we know it today, come from different parts of human knowledge. This also explains the wide and diverse applicability of the theory.

Thereafter, various origins of language theory are discussed: mathematics, linguistics, diverse modeling situations. The reader is invited to read more about them in the Handbook itself!

Rational theories and irregular behavior

Regular languages (or events) are called *rational* by the French school. This two-fold terminology makes much sense. The languages come from strictly finite computing devices, so they should be very regular indeed. The term "regular" was apparently coined by Kleene. That regular languages are the same as rational ones can be argued on both linguistic and mathematical grounds. The Latin words "regula" and "ratio" are not so far apart as regards their meaning and usage. Regular languages emerge as supports of rational power series, where "rational" is understood in the mathematical meaning of rational functions.

Terminology aside, the theory of regular languages constitutes nowadays a comprehensive block of knowledge. The theory is pleasing in the sense that practically everything is decidable. However, the theory abounds with challenging problems. You are almost sure that your problem, or at least a closely related problem, has been settled somewhere in the early literature but cannot find the place. Let us take an example. Above, at the beginning of Section 2, I mentioned *quasi-uniform events* but did not tell what they are. They are catenations $R_0 R_1 \dots R_{2i}$ of regular languages R, where each language with an odd index is a singleton letter and each language with an even index is fundamental (that is, equals B^* for some subset B of the alphabet). Given a regular language, how do you decide whether or not it is a quasi-uniform event? The problem is apparently decidable but some details are not so clear. We are dealing with the possibility of a representation of a certain type, where some of the B^*-parts may reduce to the empty word. Think about it! You may also consult the already quoted paper by Rozenberg, [13], if you can get hold of it.

My own acquaintance with regular languages – they were exclusively called "events" that time – stems from a time ten years prior to the publication of the quoted paper [13] about quasi-uniform events. During the spring 1957 I was a student in John Myhill's seminar in Berkeley, where topics were chosen from the newly appeared red-cover Princeton book *Automata Studies*, [20]. For instance, the Kleene basic paper about "the representation of events in nerve nets" is in this book. Myhill gave also some lectures himself. He was very impressive, to say the least. But he was also out of this world; occasionally he was taken to a sanatorium. His behaviour could be very irregular. Once he did not show up at all. We started searching him and found him in a wrong room. He was lecturing, the board was already half-full, but he had apparently not noticed that there was no audience!

Myhill's speach was not easy to understand, at least not for me. It was referred to as the "Birmingham accent". My work in his seminar, carried out with Howard Jackson, was about self-reproducing automata. The automata worked in the plane, where necessary components, such as power supplies, were randomly scattered around. The final construction was quite detailed, with instructions for each specific configuration, a welding operation and so forth. Myhill was quite pleased and started to talk about a publication. But he had his good days and bad days. The day when we were supposed to talk about the details of the publication happened to be a bad one, and so nothing came out of the matter.

During one of his lectures Myhill formulated a theorem about regular events. Apparently nobody quite understood what the statement of the theorem was. At the end of the lecture, Myhill asked us to prove the converse of the theorem by the next time. He was then very upset when nobody had done it; apparently nobody knew what was to be shown. Later on I found out that the theorem and its converse constituted what is now known as the Myhill-Nerode Theorem.

It was not before 1963 that I regained my interest in regular languages. I just attacked some of the problems that were around in the literature – initially I had nobody in Turku to talk with about such problems. Those days the Russian literature about regular events was very good – perhaps better than the western literature. My book [17] contains the relevant references. Many of the problems investigated were closely connected with switching circuits and might now seem somewhat unusual, out of place or esoteric. I will mention some of the problems I was working with in this area in the first half of the 60s.

Take a regular language R. Following [17], I denote by $w(R)$ the smallest number of states in any deterministic automaton accepting R, and by $w_1(R)$ the smallest number of final states in any deterministic automaton accepting R. For instance, if R_k is the finite language consisting of the words $x_1, x_1x_2, x_1x_2x_3, \ldots, x_1x_2x_3\ldots x_k$, where the words $x_2, \ldots, x_k$ are nonempty, then a different final state is needed for each of the k words, implying that $w_1(R_k) = k$. A language R is termed *irreducible* if $w_1(R)$ equals the cardinality of the language. Clearly, all languages R_k are irreducible. It is a bit more difficult to see that there are no other irreducible regular languages, [17]. (There certainly are other irreducible nonregular languages. Then one starts to talk about automata with infinitely many states. In the 60s considerations were often extended to nonconstructive cases, for instance, in [6] and [17].) In the early papers dealing with $w(R)$ and $w_1(R)$ it was not at all clear that both numbers are actually achieved in the same minimal automaton. No trade-off is possible: $w_1(R)$ cannot be made smaller by allowing $w(R)$ to be larger than the minimal value. I developed a method, based on a maximal usage of the distributive law, to reduce a finite language into a normal form from which both of the numbers $w(R)$ and $w_1(R)$ can be readily seen. The details are explained in [17].

I did also some work concerning languages over the *one-letter alphabet*. Many of the people interested in finite automata were also working with *probabilistic automata*. This was certainly true of myself and, as a result, [17] contains some 40 pages about probabilistic automata. Nasu and Honda had given an example of a finite probabilistic automaton accepting a non-context-free language, although all the probabilities and the cut-point for the acceptance are rational numbers. Could the same be done for one-letter languages? The answer is positive but the automaton must have at least three states. These problems led me to considerations dealing with rational power series and characteristic values of positive matrices, very similar to the ones I needed a decade later for growth functions of L systems.

A problem of special interest concerned the characterization of equations between regular expressions. As long as you are dealing with union and catenation only, everything is clear: you just take the appropriate associative, commutative and distributive laws. The problems begin with the Kleene star. How do you characterize the relations arising from compositions of all three operations, especially in view of the fact that in the general case you need unboundedly many nested stars? Even some relatively simple equations, such as

$$(a \cup ab \cup ba)^* = (ba \cup a^*ab)^*a^*$$

might be hard to deduce from any set of basic equations you have in mind.

In fact, V. N. Redko from Ukraine showed in the early 60s that no finite set of equations can serve as a basis for all possible equations. The core of the argument dealt with equations of the form

$$a^* = (\lambda \cup a \cup a^2 \cup \ldots \cup a^{n-1})(a^n)^*,$$

valid for all n. (From the point of view of automata, the equation tells you how to make a loop longer.) Let n assume prime values in such equations. Then one can argue that infinitely many equations are needed to generate all of them because, for any prime p, equations with primes smaller than p are unable to produce the equation with p.

But could some rule stronger than the customary logic of equations do the trick? I was working with the following rule for "solving equations". Assume that α, β, γ are regular expressions and that the language denoted by β does not contain the empty word. (The latter condition can be expressed purely syntactically, following the recursive definition of a regular expression.) Then from the equation $\alpha = \alpha\beta \cup \gamma$ you may conclude the equation $\alpha = \gamma\beta^*$.

Now Redko's problem does not arise because the desired equation may be concluded from the equation

$$a^* = a^*a^n \cup (\lambda \cup a \cup a^2 \cup \ldots \cup a^{n-1}),$$

the latter, in turn, being a consequence of the basic expansion $a^* = a^*a \cup \lambda$, characterizing the Kleene star.

The validity of the rule is a straightforward matter. The assumption concerning the empty word is necessary because, otherwise, the solution is not necessarily unique. I was able to show in 1964 that, using the rule mentioned, all equations can be deduced from 11 simple equations. Essentially, the result is the same as that by John Brzozowski concerning the finiteness of the number of derivatives. Arguments similar in spirit were given also by V. Bodnarchuk, M. A. Spivak and S. Aanderaa. References and details can be found in [17]. Regular expression equations have later on turned out to be rather important and have popped up in various algebraic and semantic considerations. For instance, John Conway, Robin Milner, Vaughn Pratt and Dexter Kozen have made important contributions in the area. Even at the time of this writing, I am handling the refereeing of a paper, where a problem stemming from [17] is solved.

When I was working with axiom systems for regular expression equations, I did not yet know personally anybody in the area. I was corresponding with Bob McNaughton, Azaria Paz and Patrick Fischer about these questions. When I moved to Canada in 1966, I very soon met John Brzozowski, Neil Jones, Michael Harrison, Azaria Paz and many others.

Stars and Schützenberger

Since the foremost figure of the French school, Marco Schützenberger, passed away in 1996, I feel it very appropriate to discuss in a separate section some matters connected with him. This fits very well also as a continuation of the discussion concerning regular languages. I already said that if you see some fresh-looking challenging problem about regular languages, you can be almost sure that the problem has already been settled somewhere. If you ask a Frenchman, you very often get the answer that the solution is (at least implicitly) presented in such and such old paper by Schützenberger.

As a person Schützenberger was most remarkable and memorable. I always had the feeling that I lagged two steps behind in understanding his arguments and jokes. He was laughing very much, and often his laughter gave out a diabolic impression. Yet I got convinced that he was a very considerate person.

For the first time I met Schützenberger in March 1971 in Paris. We discussed only technical matters. Schützenberger acted as an "informant" for Samuel Eilenberg. He wanted to know all kinds of things about ω-words, most of which I could say nothing about.

Later on I met him at numerous conferences. Schützenberger was also an invited speaker at the Turku ICALP. He was in an especially good mood during a 1975 spring school about syntactic monoids in Vic sur Cer. Because of some reason, he did not attend the lectures in the room itself but was lying in the corridor and shouting his comments through the open door.

Schützenberger's lectures were by no means easy to follow, certainly not for me. Seymour Ginsburg put it as follows. "It doesn't make any difference whether Marco lectures in English or in French. I don't understand it anyway!" But it made a difference. Usually at a conference, if Schützenberger gave his talk in French, so did also all the other French participants. This happened at an IBM conference in Madrid. Again Seymour was ready to advise me. "The French give their talks in French, the Spanish in Spanish. Why don't you talk in Finnish?" That's what I did. But it was only a ten-minute portion of my lecture and, besides, quite an independent topic. Yet it was hard to get through at all, because the protests from the huge audience grew enormous.

Schützenberger could make his point. It was told that he submitted the same paper to an American conference in three consecutive years. The technical content remained the same, only the background material and, most importantly, the references were changed. At the third try, the paper was finally accepted. Schützenberger commented that the Catholic Church has much fairer methods and policy than American program committees, because the former tells openly who the saints are at any given moment of time.

It takes a lot of work to read through Schützenberger's papers – but it is usually worth the trouble. The argumentation is very compact yet reliable. The author does quite little to aid the reader. Most of the time, the latter has to rethink the proof. (Or maybe he hasn't if his thoughts do not lag two steps behind.) I will illustrate this with the basic notion from the solution of the *star height* problem. In doing so I will compare the original paper [5] with the exposition in my book [17], where an interested reader will find all the details.

In fact, this is one of the many star height problems that were around in the 60s – some of them remain still unsolved. The *star height* of a regular expression is the maximum number of nested stars in it. To get the star height of a regular language, you take the most economical regular expression for the language, that is, a regular expression with the least possible star height. Sometimes you can make rather surprising reductions. For instance, the star height of the regular language $(ab^*c)^*$ is 1 rather than 2, because the same language is denoted also by the regular expression $\lambda \cup a(b \cup ca)^*c$.

Are there languages over the binary alphabet $\{a, b\}$ with an arbitrary given star height? A positive answer was given in [5] already in the middle 60s. (In fact, there is also a somewhat older unpublished paper by McNaughton.) What would a language of star height q be, given $q \geq 1$?

Schützenberger defined such a language M_q as follows. Let h be the morphism of $\{a, b\}^*$ into the additive group $\{0, 1, \ldots, 2^q - 1\}$ of the integers modulo 2^q defined by $h(a) = -h(b) = 1$. Then $M_q = h^{-1}(0)$.

Got an idea how the words in M_q look like? Anyway, the definition is very clear. Moreover, the matter is referred to as "elementary", because the abstract of the paper reads: "An elementary answer is given to a question raised by Eggan." The title of the paper is most modest: "On a question of Eggan." Nothing fancy about star height, no bragging. Incidentally, Eggan might not be so well known anymore.

If you prefer to work with the inverse morphic image $h^{-1}(0)$, please do so. After quite much rethinking, I defined the language A_q recursively as follows:

$$A_1 = (ab)^*,$$
$$A_{i+1} = (a^{2^i} A_i b^{2^i} A_i)^*, \quad i \geq 1.$$

For instance,

$$A_3 = (a^4(a^2(ab)^*b^2(ab)^*)^*b^4(a^2(ab)^*b^2(ab)^*)^*)^*.$$

Actually, A_q is not the same language as M_q; the former is a subset of the latter. The definition of M_q is much more compact, perhaps more "elementary". On the other hand, the definition of A_q tells more explicitly how the words in the language look like. One also avoids the need of a proof that the star height of the language is at most q: this is obvious for A_q but not for M_q, because of the definitions. However, the other part of the proof is much more difficult, that is, to show that the star height of the language is at least q. In this respect the structure of A_q is still rich enough to allow the implementation of Schützenberger's extremely clever constructions.

Computable and beyond

The book by Martin Davis, [4], was the only one in existence about computability or automata in 1962 when I gave my first course about a topic that can be classified, according to the current terminology, as theoretical computer science. I was using the book in my course "Theory of Computability", attended by some 30 students who were in the 3rd, 4th or 5th year of their studies. As far as I remember, people were very enthusiastic and did not at all mind the detailed machine language constructions of the book which, undoubtedly, would seem rather tedious nowadays. A few of the students had done some programming with IBM 650; the first such machine was installed in Finland in 1958.

There were jokes about the "colorful and variegated pictures" of the book. (It has only one picture, quite big and labeled "Figure 1": a couple of empty squares of a tape of a Turing machine.) A topic of many serious discussions was: Could computability mean something different? It seems that I could not present very clearly ideas connected with Church's Thesis. How is the situation now? Do issues dealing with DNA computing and quantum computing open up again discussions about the nature of computability? Is there something beyond Turing?

Somewhat later, in 1963–64, I started a seminar in *automata theory* and gave also a course in *finite automata.* [8] and [12] were the books I followed. As I already mentioned before, Russian literature was quite strong in the early days. Also some interesting phenomena about it could be observed. With the advent of computers and the realization of the importance of logic in computing, the role of mathematical logic in Russia changed drasticly from an area of "bourgeois decadence" to a field of high esteem and prestige. Consequently, the same notions and theorems about many-valued truth functions and the axiomatization of many-valued logic, which previously were expressed in a hidden form by talking about functions over a finite domain and bases of equations, could after the change be stated quite openly. No such secretive phase could be observed in automata theory: the field had prestige in Russia from the very beginning. However, very soon it got stuck with endless discussions about switching circuits and gates.

There was quite a peculiar detail about Russian books. Very often a sheet of errors was attached. Each line told the wrong and correct versions but also the guilty party, for instance, the author, typist, printer or proof-reader. It was left to your imagination what happened to the guilty party in each case.

The courses I mentioned above were courses in mathematics. Computer science as a separate discipline was introduced in Finland about half a decade later. There was and still is an advantageous feature in the course structure, not only in Finnish universities but in many other European universities as well. The students can choose in their curricula quite many optional courses from an unspecified pool of special courses available. This means that no particular drill is needed in the introduction of an entirely new course. I could give at an early stage the courses mentioned above and, a few years later, various courses in formal languages, complexity and L systems. I will not go into these details, since this would bring us out of the time scope of this article.

Instead, I will go backwards in time and say something about my Ph.D. work completed in Turku in 1960. It dealt with the composition of *many-valued truth functions*, in particular, completeness criteria. Consider functions whose variables, finite in number, range over the finite set $\{1, \ldots, n\}$, $n \geq 1$, (intuitively, the set of *truth values*) and whose values belong to this same set. One can generate new functions from given ones by forming *compositions*. For instance,

$$f_1(x, y, z) = f(f(x, y), f(y, z))$$

is a ternary function generated by the binary function $f(x, y)$. (Only variables are allowed to be present in such composition sequences, not the constants $1, \ldots, n$.) A set S of functions is termed *complete* if all functions can be expressed as compositions of functions in S.

In the two-valued logic, $n = 2$, the *Sheffer stroke* (the truth function of "neither-nor") constitutes alone a complete set. This example can be readily extended to the n-valued case. Along the general line, Emil Post showed already in 1921 that it suffices to generate all binary functions $f(x, y)$; all other functions can be expressed as compositions of them. A step further was taken by Jerzy Slupecki in the late 30s. It suffices to generate all unary functions and one additional function that (i) depends essentially on at least two variables (that is, does not degenerate to a unary function) and (ii) assumes all the n values.

The main result in my Ph.D. Thesis was that one does not need the set of all unary functions; the set of permutations will suffice. Thus, an arbitrary set of three functions, where two of the functions constitute a basis of the symmetric group over n elements, and the third function satisfies the two Slupecki conditions (i) and (ii), is complete. This result, valid for all $n \geq 5$ but not for $n = 2, 3, 4$, appeared also in [16]. Later on I still continued along these lines, trying to replace in this criterion the symmetric group by smaller groups. For instance, any triply transitive group will do the trick. However, values of n that are powers of 2 are exceptional: quadruple transitivity is then needed.

Some "genealogies" have been published in computer science and other fields. This means family trees showing the ancestry in the doctoral sense: who was your Ph.D. supervisor, who was your supervisor's supervisor etc. The idea is deficient because there are persons, especially in Europe, without ancestry in this sense. I can surely count myself as such a person. A supervisor should suggest you the topic for your thesis or at least, if you had the topic yourself, advise you during the work. In my case there was no such person. I found the topic from the literature and did not approach anybody before the thesis was ready. Then many people helped me in the formal details and procedure in general. The only difficulty was to get the thesis accepted in mathematics – it would have been OK in philosophy – but with some delay and considerable effort it succeeded.

The difficulty mentioned was also an important reason why I did not start working in automata theory after returning to Finland from Berkeley in the late 50s. Automata theory was much too esoteric to qualify as mathematics, in particular, it could not be used as a thesis topic. Many-valued logic was already much better. The problems I considered were close to combinatorics and, to some extent, group theory. Although the latter were some kind of slum areas in a mathematical tradition strong in complex variable, they could still become acceptable.

My Berkeley background was decisive in my choice of a new area when I realized, around 1963, that I had exhausted my possibilities in many-valued logic. The same background also gave me confidence to work alone in mathematical logic, although I did no actual course work about many-valued logic in Berkeley. I will conclude this section with some further recollections about logic courses in Berkeley – I already told about John Myhill.

Alfred Tarski gave a course in the foundations of geometry. He said that he wants to make experiments in his courses. That's what he did. The originally given axiom system was modified in many ways as we went along. The whole material appeared as a book later on. Tarski was always dressed up elegantly, just as Mostowski, and chain-smoked Winston cigarettes. "Winston tastes good like a cigarette should", this was the slogan those days. Almost everybody smoked, including myself.

Tarski conducted also a seminar together with Leon Henkin. The topics, representing Tarski's interests that time, ranged from model completeness to logics with infinitely long formulas. Henkin's presentations were of utmost clarity; every sentence was lucid and crystalline. He also arranged social events – we went together to a concert of Louis Armstrong. Also Roger Lyndon participated in the seminar. He had just spent some time in England and was often consulted in matters concerning British pronunciation. Famous people such as György Pólya sometimes visited the seminar. "Seminar is a group of people gathered to work together" was Tarski's definition.

Quality and growth to 190 cm

After working in several areas of automata and language theory, I found *L systems* in the very early 70s. Then I also came into contact with Hermann Maurer, Derick Wood (MSW-group), Grzegorz Rozenberg, Karel Culik and Werner Kuich, who all were my partners in a fruitful and diverse cooperation later on. While details related to this cooperation are beyond the scope of this article, I find it very appropriate to conclude with some glimpses about the early stages of *L* systems.

I got acquainted with *L* systems on the very day I met Grzegorz "Bolgani" Rozenberg for the first time, in May 1971. This heralded the beginning of a cooperation that has continued uninterrupted and strong, being perhaps presently more active than ever. It is no serious obstacle that we both have nowadays our impediments: in fact, we are still called "a most efficient pair of invalids". As we have gone along, some difficult problems have piled up. But they will be settled "in Jerusalem", as we use to express it.

When I got an invitation to give talks at the University of Utrecht in the spring 1971 from G. Rozenberg, I had not heard the name before. He waited for me at the airport and arranged our meeting by an announcement through the loudspeakers. Our discussion became so hectic that Grzegorz, although usually an excellent driver, got completely lost. After about one hour of his driving and talking about parallel rewriting and developmental modeling, I remarked that we should definitely be already in Utrecht. "No, we are still in Amsterdam", was the answer.

The ideas about *L* systems seemed unusually fresh to me. I had thoughts about some of the problems, notably decidability, and presented them briefly in my talks the following day, otherwise dealing with regulated rewriting and probabilistic automata. Grzegorz had moved to Holland from Poland two years earlier, and when I first came to his seminar, it was already a country-wide event. There I met many of the early researchers in *L* systems, notably Aristid Lindenmayer himself.

Although told many times, the following story is still worth being told. It is about how Aristid, a biologist, got involved in formal languages. Aristid was passing a lecture room where a lecture about formal languages was being held. He stopped to listen and became alert. The lecturer mentioned $L(G)$ several times. "Algae, that's it!", was Aristid's immediate reaction. When Aristid constructed grammatical models for the development of simple filamentous organisms, algae constituted his first example.

I got so much carried away by the new ideas encountered during this visit to Utrecht that during the same summer I wrote to Grzegorz almost every day. I had difficulties in spelling his name (not to mention the impossibility of the pronunciation!) and so we gradually switched to Bolgani. That year 1971 I was running a summer program in formal languages in London, Ontario, with some 20 guest speakers including Alfred Aho, Seymour Ginsburg, Michael Harrison and Juris Hartmanis. The school was an idea of the department head, John Hart. I included something about *L* systems in the program: Gabor Herman was one of the speakers. I was also able to put still a section about *L* systems, written the same summer, in my book "Formal Languages", [18].

The number of papers about *L* systems used to grow exponentially, as powers of 2, during the first ten years. The reader is invited to consult [14] as regards the early history of the area, and [15] as regards the current point of view. I have not met elsewhere in a scientific community such a cozy family-like atmosphere as was prevalent in the *L* community during the early years. "*L*" was also everywhere identified as a symbol of quality, "*L* quality" being equivalent to "superb quality".

In questions concerning the biological background and significance, Aristid was always ready to help. He emphasized to me the importance of *determinism* in biological modeling. Consequently, from the very beginning, I was especially interested in deterministic *L* systems and their growth functions. The technique needed for the latter involved those parts of the Perron-Frobenius theory for matrices I had been using earlier in connection with finite probabilistic automata. Many of my first talks about *L* systems were focused on growth functions. There was a long Japanese report about my talk at a conference at MITRE near Washington. Only the following words, appearing in the middle of long passages of Japanese characters, were understandable to a western reader: Arto Salomaa, growth functions, 190 cm.

Never be clever for the sake of being clever

An article with an overall personal flavor would in my case be quite difficult to complete without discussing *sauna* at some stage. Apart from the fact that sauna as an institution is intrinsically and invariably connected with Finland, it is most important to me personally. In several connections I have mentioned grandchildren and sauna as my favorite pastime. The former are omitted from this article because family matters are not discussed. The latter, however, fits very well within the scope of "people and ideas". I have met most of the *people* I have worked with also in a sauna environment, an ideal atmosphere for the birth, development and exchange of *ideas*. In the MSW-group, my long-term cooperation with Hermann Maurer and Derick Wood, we used to speak of "three-sauna problems" instead of the "three-pipe problems" of Sherlock Holmes. When I was for a year in Waterloo with Karel Culik, the latter arranged at least weekly sauna sessions in his home. As regards Bolgani Rozenberg, sauna has always been a most important factor in our efficient friendly cooperation. These are only a few of my older sauna contacts – from the time span of this article.

Conditions are best for the birth, development and exchange of ideas in-between the actual sauna sessions, or, after sauna. Then the mood is relaxed in a very peculiar way. Apart from actual technical problems, many other important issues have been taken up. Such an issue concerns your publications. How should you present your results? How should you write a paper? If your result is a real break-through, something once in a century, it does not matter so much how you present it. People will dig it out also from a badly written paper. In other cases good writing is rather important.

The title of this section is a line from Glenn Gould's composition "So you want to write a fugue". The same principle applies for writing scientific articles. You should not try to be clever only for showing off. A Ph.D. thesis writer might want to impress the supervisor by the knowledge of some methods or of some new references. In general such an inclination is detrimental. You should mention something only if you really need and use it, not because you want to impress. In the 60s it was a key point in many scientific papers (for instance, in physics and chemistry) that the author had been able to use such and such a computer. This was even proposed as a merit in applications for positions in computer science. Well, never be clever for the sake of being clever.

Another point that has often been discussed in sauna sessions is the amount and extent of intuitive explanations that should be included in a paper. Most of theoretical computer science comes very close to traditional mathematics. Should also the papers be written in the traditional definition-theorem-proof style?

In most cases I would give a negative answer. A few clarifying sentences about a definition or a few intuitive remarks explaining what is really going on in a proof can really aid a reader. They can be crucial for the reader's decision whether or not he/she will study the proof. However, such remarks should definitely not replace the proof, especially if the proof is difficult or surprising. The paper should aid the reader to clarify all details he/she considers necessary. But sometimes the reader only wants to get some kind of an idea, and then intuitive remarks are very helpful. There cannot be any general rule concerning the relation between formal details and intuition. There are also masters of the most austere definition-theorem-proof style. Particular cases can be very different, and good taste is always important.

The last statement applies also to a number of other issues important in compiling a paper. How many details, and perhaps examples, should be included in each particular part? How much background should be included in the introduction? Should the results be generalized as much as possible? What about motivation: should a theoretician always try to invoke applications?

For such questions, there are very different particular cases and many aspects should be taken into consideration. I will not dwell longer in the topic of writing papers but will only add a couple of points. The intended audience is very important when you talk about applications. If you motivate some abstract, perhaps also strange, idea or notion by invoking such and such forthcoming applications, your writing is appropriate if it addresses some funding agency, but probably out of place if it addresses your colleagues. A generalization of a result does not as such always bring forth an improvement. If you extend a known result about regular sets (over free monoids) to regular sets over arbitrary monoids, there must be a reason, mathematical or otherwise, for doing so.

I have written about sauna elsewhere (for instance, see [19]), so I do not discuss here matters such as what is a sauna and what is a good sauna. Instead, I will let some of my friends and colleagues speak. During his visits, Hermann Maurer has written me so many poems that I could edit a book of sauna poetry by him. First a passage about sauna in general.

Salosauna, once again
heightens joy and heightens pain.
Underlines what maybe counts
what this life in truth amounts.
Through this sauna's windowpane
past some sunshine, wind and rain,
our hearts and eyes and ears
cut through all the passing years:
see the friendships that stay strong,
newborn faces, happy song.
Memories taste sad and sweet
as they rise in sauna's heat.

Then some lines of Hermann describing difficult problems:

Als Arbeitsaufenthalt ersonnen
hat es mit Sauna gleich begonnen.
Wir schwitzten, dachten, tranken viel
vergassen aber nicht das Ziel:
Erweiterung der Theorie
wir wollen ganz erforschen sie!
Wir haben also nachgedacht
ob das, was uns so Kummer macht:
die Dichte endlich Systeme
sich lösen lässt durch Theoreme
die ohne Sauna schwer zu finden
und Wert sind, dass wir sie verkünden.
Nun, das Problem, es scheint nicht leicht,
's hat eine Sitzung nicht gereicht,
sodass die Lösung wir mit Sorgen
verschieben mussten bis auf morgen.

Nobody can be more convincing than Werner Kuich:

Salosauna, Finnische Freunde, Ruhiges Rauhala
Allzulange entbehrt.
Kaarina, die kundige Köchin,
und Arto, den Allgewaltigen
sowie Salosauna,

grüsst Werner aus Wien
der deutsche Dichter.

Wer diese "letzten Sieben" überlebt hat,
der kann wahrlich sagen,
dass ihm nichts Saunamässiges mehr fremd ist.

Azaria Paz is also in a poetic mood:

Salosauna
What a sauna
With the flora and the fauna
Naveh shalom
And the sky above

And so is Bolgani Rozenberg ("löyly" is the Finnish word for sauna heat and "supikoira" is a raccoon dog):

When you come to Tarzan nest
You get sauna at its best
Where you can admire
Löyly and birch wood on fire
A lot of flora and fauna
Can be seen from Salosauna
But with Bolgani and nice weather
You can see two supikoiras together

Most of the time Bolgani is very practical:

> This time Bolgani was flown into sauna! It took some 50 minutes from the moment that the plane landed until the moment that Bolgani has entered Salosauna. We had three sittings. The EATCS Monograph matters were settled already in the first break between sittings – real Salosauna efficiency .... This was the most responsible day in Bolgani's life: I had to take care of fire going in three locations in the sauna building and in the living room of Salosauna Administration Building (= Rauhala). In heating Salosauna I've applied the ESP (Energy Saving Principle): the heat should be so good that for the next two days one can still use Salosauna in optimal conditions. It looks like I've succeeded. In fact, the family room in the sauna building was so hot that we were going from it into the sauna room just to cool down.

Derick Wood was always good in enumeration problems:

Fifty thousand buckets of water,
Thirty thousand logs,
Five thousand bottles,
Two thousand candles,

Two thousand matches,
One thousand newspapers,
One thousand birch twigs,
One radio, and
One salosauna
make one thousand salosauna sittings.

I conclude with the lines of Andy Szilard:

In the heat
when friends meet
it's a real treat
even though they burn their meat

as well as with those of Andy's former teacher of English, the late Ron Bates ("kiuas" is the Finnish word for sauna stove):

The kiuas is there,
The marriage of water and stone,
And fire, this is where
We come to be one.

References

[1] A. Aho and J. Ullman, The theory of languages. *Math. Systems Theory* 2 (1968) 97–126.
[2] R.V. Book, Problems in formal language theory. *Proc. 4th Princeton Conf. on Inform. Sciences and Systems* (1970) 253–256.
[3] N. Chomsky, On certain formal properties of grammars. *Information and Control* 2 (1959) 137–167.
[4] M. Davis, *Computability and Unsolvability*. McGraw-Hill (1958).
[5] F. Dejean and M.P. Schützenberger, On a question of Eggan. *Information and Control* 9 (1966) 23–25.
[6] S. Ginsburg, *An Introduction to Mathematical Machine Theory*. Addison-Wesley (1962).
[7] S. Ginsburg and E. Spanier, Finite-turn pushdown automata. *J. SIAM Control* 4 (1966) 429–453.
[8] V.M. Glushkov, Abstraktnaja teorija avtomatov. *Uspehi Mat. Nauk* 16 (1961) 3–62.
[9] S. Greibach and S. Ginsburg, Multitape AFA. *J. Assoc. Comput. Mach.* 19 (1972) 193–221.
[10] S. Greibach and J. Hopcroft, Scattered context grammars. *J. Comput. Syst. Sci.* 3 (1969) 233–247.
[11] M.A. Harrison, *Introduction to Formal Language Theory*. Addison-Wesley (1978).
[12] N.E. Kobrinskij and B.A. Trakhtenbrot, *Vvedenie b teoriju konechnykh avtomatov*. Gosud. izd. Fiz.-Mat. Lit., Moscow (1962).
[13] G. Rozenberg, Decision problems for quasi-uniform events. *Bull. Acad. Polon. Sci.* XV (1967) 745–752.
[14] G. Rozenberg and A. Salomaa (ed.), *The Book of L*. Springer-Verlag (1985).

[15] G. Rozenberg and A. Salomaa (ed.), *Handbook of Formal Languages*, I–III. Springer-Verlag (1997).

[16] A. Salomaa, A theorem concerning the composition of functions of several variables ranging over a finite set. *Journal of Symbolic Logic* 25 (1960) 203–208.

[17] A. Salomaa, *Theory of Automata.* Pergamon Press (1969).

[18] A. Salomaa, *Formal Languages.* Academic Press (1973).

[19] A. Salomaa, What computer scientists should know about sauna. *EATCS Bull.* 15 (1981) 8–21, and 35 (1988) 15–26.

[20] C.E. Shannon and J. McCarthy (ed.), *Automata Studies.* Princeton Univ. Press, Princeton (1956).

[21] D. Wood, Bibliography 23. Formal language theory and automata theory. *Computing Reviews* 11 (1970) 417–430.

Anatol Slissenko

Anatol Slissenko is a professor of Computer Science at the University Paris 12. Previously he held the positions of Head of the Lab for Theory of Algorithms at the Institute for Informatics of the Academy of Sciences of Russia and Head of the Chair of Computer Science of the University of St. Petersburg, Russia. Prior he was a researcher at the Leningrad Division of the Steklov Institute for Mathematics and Professor at the Leningrad Polytechnical Institute. For about 20 years (1973–1993) he organised the Leningrad Seminar on Computational Complexity at Steklov Institute for Mathematics which essentially influenced the theoretical computer science research in the former Soviet Union. Researchers as D. Grigoriev, A. Chistov, N. Vorobjov Jr. and many others were educated in his seminar. Professor Slissenko graduated from the Leningrad University with a diploma in mathematics (1963); he obtained his PhD degree from Leningrad Division of Steklov Institute, and his second Russian doctorate from Steklov Institute in Moscow. Professor Slissenko was an invited speaker at many conferences, including the International Congress of Mathematicians. He is best known for his real-time string-matching and palindrome recognition algorithms. He worked also in automatic theorem-proving, recursive analysis, algorithmics and theory of computation. His current research is focussed on verification of timed systems, and the problem of the shortest path amidst semialgebraic obstacles.

Leningrad/St. Petersburg (1961–1998): From Logic to Complexity and Further

This survey is a personal recollection of the development of research on computational complexity in St. Petersburg (named Leningrad from 1924 to 1991) in the period 1961–1998. This research was born in the womb of the Leningrad school of mathematical logic, hence a considerable part of this paper is dedicated to the latter. Information on the life, education and research in the Soviet Union is also given. More general remarks on some computer science problems can be found in the concluding part. This survey has been written at a stretch and governed by a flow of (sub)consciousness.

> "A retentive memory may be a good thing, but the ability to forget is the true token of greatness", Elbert Hubbard, *The notebook*, 1927.
>
> "The facts had happened and thus, disappeared. We can at best demand their images from our memory. But our memory is out of our control. Thus we are not responsible for our recollections".[1]

The creation of the Leningrad (St. Petersburg) school of mathematical logic, which gave rise to the research in computational complexity that I am writing about, is mainly due to professor Nikolai Aleksandrovich Shanin (1919–). I was his student and started my research in his seminars. Some features of the system of research and education might be of interest because of their high efficiency in some circumstances and within certain limits. I will not go into technical details (though basic references will be given) but I will write more on human aspects that will be soon completely forgotten. More information concerning the Soviet research related to the subject of this survey can be found in [Yan59] (logic), [Min91,MMO71] (theorem-proving), [Tra87,Tra84,Sli81a] (complexity), [Sli93] (computer science), [BGG97] (decision problems).

The Origins

> "Where is the beginning of that end which terminates the beginning?" Koz'ma Prutkov, *Fruits of Meditation*, 1854-1860. Translated from Russian.

Research in logic in Leningrad started after 1946 and was initiated and influenced mainly by A. A. Markov Jr. (1903-1979), a son of the famous mathematician A. A. Markov (1856–1922) who is widely known for Markov chains and various other classical results. Before the war of 1941–1945 A. A. Markov was a professor and the head of the chair of geometry at Leningrad State University (maybe, more precisely, at Leningrad National University, as it was subordinated to the Ministry of Superior Education; below it will be referred to as "University"). The famous Soviet geometer A. D. Aleksandrov was at the same chair. Just before the war the Leningrad Division of Steklov Mathematical Institute (LOMI) had been created; it played an exceptional role in the development of mathematics and not only in Leningrad. A. A. Markov has become also a part-time researcher at LOMI.

Somewhere in 1940 Markov started an educational seminar on mathematical logic just to learn it. His PhD student N. Shanin, the source of the information of this seminar, attended the seminar but he did not really participated in its work: he was doing research in general topology. The main interests of A. A. Markov,

---

[1] The epigraphs without references are either well known or are mine.

though rather diverse, were far from logic. He published papers on dynamic systems, differential equations, analysis, topology and even root separation. One extraordinary participant of this seminar, E. M. Livenson, was a polyglot. He knew the Bible by heart and used it to learn different languages: he took the Bible in an unknown language, after having learnt its basic grammar, and started to read and learn it. E. M. Livenson found in some reports of the London Royal Society a reference to Turing machines and gave a talk on this subject. His own research was on descriptive set theory on which he worked with L. V. Kantorovich, the future Nobel Prize winner in economy. However, at that time he himself developed an algorithm (based on a calculus with invertible rules) of inference search for the classical propositional logic (Gentzen's classical work [Gen34] was not known to him). At that seminar A. A. Markov reviewed the "Foundations of Mathematics" by Hilbert and Bernays [HB34,HB39]. The war which started in the summer of 1941, had interrupted this activity. E. M. Livenson, who lived in a suburb of Leningrad that was occupied by Germans, had disappeared during the war. The institutes of the Academy of Sciences and the University were evacuated to the East of Russia, as German troops started to besiege Leningrad, though they had never seized the city. During the war A. A. Markov was in Kazan.

Initiation of logic and complexity research of A. A. Markov Jr.

"The parish is like the priest. But where is the cause and where is the effect?"

From 1946 A. A. Markov entirely switched to the theory of algorithms and logic, and he was more and more interested in the foundations of mathematics where he analyzed the intuitionistic critical line initiated by E. Brauer and H. Weyl. Very soon these reflections led him to constructivism (see [NS64]).

His first results on undecidability were obtained in 1946, and started to appear from 1947; in particular, the famous undecidability of the equality of words in finitely defined semi-groups dates from that time. His lectures and seminars attracted his students and collaborators into the research on logic, theory of algorithms and constructivism. Among them the most important role in the development of the Leningrad school of logic and constructivism was played by his collaborator and disciple N. Shanin who was at that time a senior staff member of LOMI, and a part-time professor at the University (for more details on his scientific biography see [MMM+80,MMOS90,Kur59]).

The philosophy of intuitionism was vague. However, the logic was described precisely, and there were approaches to its interpretation (from the A. Kolmogorov's paper [Kol32] up to Kleene's realizability [Kle45]). A. A. Markov wished to develop a precise approach, taking into consideration E. Brauer and H. Weil criticism of the foundations of mathematics. This idea, quite normal for a mathematician, offered also some security from possible political accusations for dragging an "hostile bourgeois philosophy of intuitionism" into Soviet science; the word "constructivism" was itself risky. This idea has proved to be productive for the Leningrad research, independently of constructivism, because such a great scientist as A. A. Markov assured a profound and versatile knowledge of the theory of algorithms and logic. His own contribution includes Markov algorithms [Mar54,MN88] and Markov principle [Mar56,KV65] (giving a non trivial extension of intuitionistic logic within the constructive setting), not to speak about numerous excellent results that were not named after him.

A. A. Markov moved to Moscow at the end of 1955. He was the director of LOMI at that time. In Moscow he headed the chair of mathematical logic at the Faculty of Mechanics and Mathematics at Moscow State University, the highest rank university of USSR. However, by the time of his departure the foundations for further work have already been created. N. Shanin has written a profound paper on embedding operations, generalizing the results of Kolmogorov-Gödel, and some other papers related to constructivism. Moreover, he began to be interested in machine theorem proving that influenced considerably the Leningrad research in logic. Another person who is directly related to computational complexity was G. Tseitin (1936–), at that time a prodigy child who became a student of the University at the age of 16 (in 1952), but actually started to attend courses of A. A. Markov one year before. Two more known researchers, namely, I. Zaslavski (1932–) and E. Nechiporuk (1934–1970) were at the University. I. Zaslavski was a student in 1949–1954, and started to publish his papers from 1953; his first research was in constructive analysis [Zas55a,Zas55b]. Zaslavski was not accepted as PhD student (mainly because of being Jew) and worked in the Institute of Telephone Communications. He was arrested at the end of 1956 for his open protest against the Soviet invasion in Hungary, and was set free in November 1958. He was not permitted to reside in Leningrad and found a position in Erevan (Armenia) where he created a scientific school in the theory of algorithms and logic. Now he has six positions just to survive. As for Nechiporuk, he worked on the complexity of Boolean functions, closer to the research in Moscow, and after his classical results (see [Weg87]) he started a very general study of self-correcting circuits [Edi73]. He suffered from depressions that limited his interaction with people and had committed suicide while being treated in a hospital.

Among various brilliant results due to G. Tseitin, at least three are pioneering: first, an $n^2/\log^2 n$ lower bound on the complexity of Markov algorithms that was proved in 1957, secondly, compression and gap theorems for Markov algorithms proved in 1956 and mentioned in [Yan59] and, thirdly, axioms defining computational complexity and a version of the compression theorem for these axioms. The axioms were exactly the same that were later found by M. Blum [Blu67]. As it was traditional for Soviet mathematical logic this knowledge existed as folklore. The first of these results was at least mentioned in [Bar65]. As for the third one, I learnt it from G. Tseitin (before 1967; R. Freidson also remembers this); one more result on non efficiency of strong lower bounds was mentioned in my survey [Sli81a]. But now it is impossible to trace all this. When I asked G. Tseitin in April 1998 about this axiomatics, he was surprised and asked me what they are about. All these results, though well-known in Russia, had never been published, and the reason was mainly Tseitin's reluctance to write proofs. On the other hand, publishing papers in the USSR was never an easy business. Later G. Tseitin worked more in constructive (recursive) analysis where he proved his famous results on the continuity of constructive functions, and then started to move towards programming and artificial intelligence.

Education in the USSR

> "Nothing is more responsible for the good old days than a bad memory." Franklin P. Adams
>
> "At an exam a Soviet medical student is invited to describe some particular features of two skeletons shown to him. He is silent. No hints work. The examiner says: "Try to recollect what you were studying all these years!" The student exclaims: "Are they really the skeletons of Marx and Engels?!" Soviet joke.

My father, a military topographer, had a possibility to install himself in Leningrad in 1950, after more than 30 years in the service and having spent more than 20 years in expeditions in the Central Asia, Far East and Far North-East of the USSR. He had the rank of colonel, and he was a lecturer in the Military Academy of Communications in Leningrad. He retired in 1955. Such middle-class families were a typical source of university students, in particular, in mathematics. After spending a long time in military camps or small provincial towns, the life in Leningrad was a kind of paradise. After the war, highly demolished Leningrad which heroically resisted the famous 900 days besiege during the war, had a privileged status. Of course, such privileges could not solve desperate problems as the shortage of apartments. My father's family of three (or four when my grandmother was with us) had two small rooms in a four-room apartment without bathroom and hot water. That was considered as good living conditions. Other two rooms were occupied by two families with three persons each. We had to share a small kitchen with one cold water tap. Small kerosene-stoves were used to cook. But theaters, concert halls, libraries, excellent public lectures were superb sources to nourish young brains and souls, though large parts of the world and even Russian culture and science were absolutely invisible.

Schools were different from the point of view of education quality, though all of them were to follow the standard national programs supervised by the government bodies of district or regional level. Happily the school education was doubled by clubs and study groups; they were sufficiently numerous and were organized at houses of culture or pioneer houses (here the word "pioneer" refers to the Pioneers Organization formally controlled by the Communist Party but informally by nobody). My parents had found the only existing elite school with an intensive programme in English, the school No 213. The most paradoxical fact was that this school, though having been created mainly for the children of the Party nomenclature was, by that time, open to all children (via a competition). My parents hired an excellent teacher of English and after one or two years of additional lessons I passed the exams and was accepted. At school, starting from the fourth year, we had about 4-6 exams at the end of each year. The school

No 213 played a crucial role in liberating me from the communist orthodoxy, though pace was too slow.

As for mathematics, in addition to a relatively good national program at school all children had an opportunity to attend different study groups. The University study groups were open and headed by the best students of the Faculty as a kind voluntary public service (it was mandatory "to serve voluntary"). I have started in such a study group headed by V. G. Mazya, who became later a well-known mathematician in analysis (now in Sweden). In addition, I tried to read excellent books for school-children written by high-level mathematicians, for example, "Three Pearls of the Number Theory" by Khinchin. Many schools with various accelerated programmes for some disciplines like foreign languages, physics, mathematics, etc. appeared at some stage. It was harder to be a student there as the accelerated disciplines were taught as additional subjects on top of the national program.

To become a student of the University one had to pass an entrance exam, that is to go through a competition. Usually the Faculty of Mathematics and Mechanics had 3-4 candidates for one place in mathematics (there were about one hundred places for mathematics, one hundred for mechanics and astronomy). I got my baccalaureat in 1958 and became a student of the University the same year at the age of 17. That was a normal age to start. Universities offered only one type of diploma after usually 5 years of study; in some faculties, say, in medicine, the programme lasted longer. Students with good marks were paid scholarships, about 40 roubles (that was slightly less than the minimal salary) or even 60 roubles in case of highest marks. There were four marks: excellent, good, satisfactory, not satisfactory. Later, special scholarships were introduced for students with excellent marks during the first three years. I had such a scholarship that was called Lenin's scholarship; it amounted to 80 roubles. This was a sum that permitted to live in poverty. For the minimal salary one could only not die.

Universities were attractive for their liberalism and intellectual environment, and a rather free access to various lectures. Research positions in sciences or mathematics gave the freedom to work out of the Communist Party control. The extermination of disciplines like economy, philosophy and even genetic had been successfully accomplished earlier; physics and mathematics had been saved by nuclear weapons, avionics and other activities related to arms race. Surely, nuclear physics was considered incomparably superior than mathematics, and mathematical logic was treated as a marginal science in mathematics; its existence was somehow justified only by the appearance of computers. And again, nuclear and aero/hydrodynamic applications of computers were absolute priorities.

The late 50s and the early 60s were a relatively happy time of Khrustchev's era. We could see paintings of Picasso that were not exposed before. Ive Montand, a famous French chansonier, gave concerts in 1956. As a result, French became temporarily popular. He was a communist until he realized a part of the evident truth about the communism. When he had denounced communism, he was immediately officially forgotten. As well as Khrustchev after his dismissal. The next day after Khrustchev's dismissal, no Party official could answer the question "Who was Khrustchev?"

The Faculty was rich in old intelligentsia and outstanding mathematicians. Just to mention a few, one could take courses by L. V. Kantorovich who worked in various fields of mathematics and won a Nobel Prize for his pioneering contribution in mathematical methods in economy, number theory and statistics were represented by Yu. V. Linnik, geometry by A. D. Aleksandrov, algebra by D. K. Faddeyev, differential equations by V. I Smirnov and O. A. Ladyzhenskaya. Students were free to attend any course, and they were numerous. Later these opportunities disappeared, as well as some of these people of high culture and humanistic traditions. Those who came after them, though being good or excellent in research, were too exhausted by poverty and the struggle for existence, and had no access to the sources of culture available to previous generations.

As a first year student I had to start to take courses from the 1st of September according to the official rules of the Ministry. But for the Regional Committee of the Communist Party the laws written by the Communist Government were not compulsory. Thus, we went to gather potatoes in sovkhoz fields ("sovkhoz" is a state owned agricultural farm, similar to kolkhoz, but usually considered as more industrial and closer to communist forms of organization of labor). The weather was bad, the work was hard, only by hands, and as local people explained us, was useless because potatoes were to rest in the field in piles and would rot very soon. This communist absurdity was the core of the system. I gathered potatoes during the following thirty years until perestroika, and nothing changed.

We studied at the University 6 days a week, 6-8 hours a day with 10 minutes breaks every hour. No break for lunch. We spent 8 hours a week on Marxism-Leninism, one day for military preparation. I do not remember much concerning the latter, but as for the communism I learnt not only its practice but also the theory of this criminal ideology : I can say that practice corresponds to theory. In fact, this aggressive way of imposing the communist ideology on clever people had the opposite effect: it formed anti-communists with profound knowledge of all pathological destructiveness of communism.

Student seminars: free choice

> "It ain't the roads we take; it's inside of us that makes us turn out the way we do." O. Henry. *The roads we take.*

From the very beginning students had a variety of seminars where they participated as reviewers of papers that led them very quickly to open problems they could try to solve. The head of a seminar proposed interesting problems of reasonable difficulty. These seminars constituted supplementary units for first and second year students; later students had always optional seminars together with mandatory ones. Very often the results obtained by these younger students were really good and were published in journals. That was the case of Yu. Matiyasevich (1947–) who got his first results in my seminar on theory of algorithms, being a first year student. At the same time staff research seminars were also available to students. D. Grigoriev (1954–) started to participate in the seminar of complexity as a second year student, and arrived in his third year with his now classical results which will be mentioned below. By the third year a student was to chose a specialization, e. g. algebra, geometry, logic (that was at the chair of geometry), etc. and be attached to the corresponding chair. This choice was not very rigid, and later one could change the chair.

When I was a second year student N. Shanin started a course on mathematical logic with accent on proof theory. At the beginning the audience was enormous, I guess more that 200 people; at the end of the second year, and that was the duration of a non standard course, I was the only undergraduate student in the audience (there were several PhD students).

In 1959 the idea of automatic theorem proving was in the air, and publications started to appear. N. Shanin decided to study the question with his students S. Maslov (1939–1982), G. Mints(1939–) and G. Davydov (1939–) and some others who constituted a seminar on logic and theory of algorithms. Later V. Orevkov (1940–) and myself, A. Slissenko (1941–), joined the seminar, and all these people constituted a group on mathematical logic organized by N. Shanin within the Leningrad Division of Steklov Mathematical Institute (LOMI). We worked together for about 20 years.

N. Shanin's non standard idea was to develop a computer algorithm for searching *natural proofs*, firstly in the classical propositional logic. The starting point was the results of G. Gentzen and J. Herbrand. The idea of seeking a natural, human-oriented proof was not popular. As it became clear much later, it was strongly underestimated. But at that time the general belief was that, in spite of the undecidability of first order theories, the arriving computers—with power doubled every 2-3 years—would be able to prove interesting theorems. Further experiences and theoretical studies showed that the universality of these undecidable problems was destructive for general theorem proving algorithms. All the syntactic flexibility, as well as sophisticated programming solutions, used in modern theorem provers do not change the fundamental sources of their inefficiency. However, the inertia is high. I will come back to this point later.

To simplify the task of proof presentation, N. Shanin proposed to take as natural proofs the proofs in a sequent version of Gentzen's natural calculus which contains, in particular, modus ponens. But how to seek a proof? That was a problem that could not be solved by students, especially taking into consideration the goal to implement algorithms on computers.

Steklov Institute Group in Logic

> "The basic modus of Soviet logic: *A implies B, if there is no reasonable way to deduce B from A*." Experimental fact.

Several words on the organization of theoretical research in the USSR.

Research institutes in the USSR

> "**Theory.** Supposition explaining something, especially one based on principles independent of the phenomena etc. to be explained... . [Gk ... *theōros* spectator f. *thea* spectacle]". The Concise Oxford Dictionary, Fourth Edition. 1951

The best places for research in the former Soviet Union were the institutes of the Academy of Sciences of the USSR or of the associated republics, and universities research institutes. With the collapse of the Soviet Union the Academy was reorganized, and in fact, dismembered into Academies of the Independent States. The largest part has become the Academy of Sciences of Russia. No attempts to save the best part of research had been done, and in consequence, a considerable part of researchers (and one can guess not the worst part) has left for the West. Most institutes exist by letting parts of their buildings to companies making business.

Institutes had permanent positions only for the members of the Academy, though rules were vague. Non-tenured positions were for maximum five years; after each term researchers had to be re-elected by the scientific council. Not re-election was not too rare, especially for junior staff. This system worked well on the whole even within limited resources and the incompetent interference of the Communist Party. But it could not work well in the long term without sufficient resources, strong connections with applications and international competition. One can notice that elite research institutes with very small permanent staff of outstanding scientists exist in many countries, e. g. the Institute for Advanced Studies in the USA, Max-Planck Institute in Germany, I.H.E.S. in France. In the Soviet Union the non-permanent part was larger and more stable, and it was justified by the poverty of the population. All changes were highly difficult to implement.

People having positions in research institutes of the mentioned type were not obliged to teach and could, and had to, concentrate on research. Strictly speaking, there were political limitations on the subjects of research, but starting from the 60s, they were not considerable for sciences, mathematics or computer science. As always in the history of Russia, the competent layer of researchers was very thin and fragile. The number of people involved in applied science was not only inadequately small but also too limited in resources and freedom of action.

For theoretical research the main obstacle was the absence of contacts with the West. The information flow was limited, namely, the libraries were incomplete, reproduction and publication facilities were poor, and information exchange with the West was usually interrupted. On the other hand, the publishing houses "Nauka" ("Science") and "Mir" ("World") provided a good amount of books, the first one of Soviet authors and the second one translations. The amount of available translations from English was much higher than, say, in France nowadays.

As for any visible action, everything was forbidden except things specially approved by the Communist Party Committee. The crucial role in consolidating different research areas from different institutions was played by regular seminars, usually having had their meetings once a week and headed by competent leaders. Reviewing the literature was an important component of this activity. Seminars had no funds, and the only support from host institutions was a room for meetings. Nevertheless people were invited and were eager to give talks hosted by good seminars. The leading seminars were very prestigious, and to give a lecture in such a seminar was considered a sign of recognition. For PhD and second doctorate programmes a talk at an appropriate seminar was mandatory.

In Russia, as well as in USSR, there are two scientific *degrees*: candidate of sciences (physico-mathematical, technical, biological, etc.) and doctor of sciences. Both are validated by the state Higher Certification Committee which awards diplomas. The degree of candidate more or less corresponds to PhD, though, I guess, initially the level of candidate dissertations was, on the whole, higher than that of PhD theses in the West, at least in mathematics and physics. The second degree, to which I will refer to as "second doctorate", was much harder to obtain. In mathematics the most respectable second degree was the one obtained from the Steklov Institute for Mathematics in Moscow (among people mentioned in this survey, Yu. Matiyasevich, G. Makanin and myself got such degrees), or from Moscow or Leningrad Universities. The Steklov Institute was very slow, and many excellent people like D. Grigoriev got their degrees from universities to accelerate the process. Any dissertation had to be "defended" at a Scientific Council appointed by the Higher Certification Committee. The council appoints official opponents (2 for the candidate degree, one of them with second doctorate, and 3 with second doctorate for the doctor degree) and an "external" organization that must give its view on the dissertation. The report of the external organization is to be signed by a known expert in the field and endorsed by the head of the institution or its deputy. The defense procedure is governed by rather strict formal rules, is public, and at a certain moments anyone could pose a question or express on opinion on the subject. Critical remarks and hard questions are usual. The final vote of the Council is secret, and to get the degree two thirds of votes must be positive. But that is not all. The dissertation is to pass through the Certification Committee which could ask for additional assessment and change the decision of the Council. The system worked well for a rather long time but was gradually corrupted by the Communist Party. One of the sources of corruption was represented by classified dissertations (especially of generals) and dissertations of Party functionaries and their relatives.

Besides scientific degrees there exist also scientific *titles* "dotsent" (associate professor), "professor" (full professor) for university people, and "senior scientific collaborator" for researchers. Happily one could get a professor position without the title, but in this case the salary was smaller. In the USSR people were paid not for their work but for position and title, and these had often nothing to do with the qualifications of the person occupying it. I found something similar in Europe. Young brilliant brains normally drain from such systems.

Automatic theorem proving and proof theory

> "To err is human, but to really foul things up you need a computer." Well known.

The creation in 1961 of a group on logic and automatic theorem proving in LOMI was an outstanding organizational success of N. Shanin. The first three members of the his group were G. Davydov, S. Maslov, and G. Mints. One of the obstacles to the creation of the group was the (practically official) anti-Semitism of the administration of the Steklov Mathematical Institute in Moscow headed by academician I. M. Vinogradov, called by L. S. Pontryagin "one of the two greatest mathematicians of the century" (the second was L. S. Pontryagin himself). Steklov people called I. M. Vinogradov "Brother Vanya" (in Russian it is "Uncle Vanya"). He was an anticommunist respected by communist governments. He would criticize the Soviet regime in presence of reliable people; but if someone less reliable appeared in his office he started to attack Jews. His notorious and widely known anti-Semitism was operational mainly inside the Steklov Institute in Moscow.

Uncle Vanya's anti-Semitism gave him particular advantages to avoid any participation in the campaigns of persecutions of dissident scientists. For example, he refused to sign the official letter blaming the known human rights defender academician A. D. Sakharov because he noticed a Jew name among signatories; later he refused to sign any letter using various other arguments. Among the most brilliant jokes concerning his anti-Semitism is the following one. A KGB officer comes to Uncle Vanya and draws his attention to the dissident activity of R. Shafarevich, a member of the Institute. He speaks for a long time and demands "measures" to curb this activity. Uncle Vanya interrupts him and says: "I have studied this question in the deepest way. Shafarevich is not Jew." And that was this final judgement.

Thus, N. Shanin, supported by the then Director of LOMI (who formally was a Deputy Director of I. M. Vinogradov) G. I. Petrashen, has managed to take in LOMI three persons among which one was Jew even according to his passport (G. Mints) and another one had a Jew mother (S. Maslov). Next year V. Orevkov arrived, and in 1963 it was me, and this finishes the first phase of this history. There were other people hired for programming, but they left the group after rather short terms except Alla Sochilina who stayed with us for about 20 years.

The late 50s and early 60s were a time of unlimited believe in the possibilities of computers. Any algorithm will eventually work, that was a widespread point of view.

The principal ideas of the first theorem proving algorithm for classical propositional calculus were developed by N. Shanin. In addition to various considerations to accelerate the inference search, he designed an algorithm transforming a proof of logic sequent calculus to a natural proof.

We started programming in 1963 on a computer "Ural-4" that was situated in the city Penza in the institute that developed that computer. To debug and run programs we traveled there, and the trip took about two days by train.

The programming was in code. Of course, we created some environment but basically it did not change much. In parallel with design and implementation, all members of the group continued their theoretical research either in logic or in constructive analysis, the latter being my case. The program [SDM+83] has been accomplished very quickly and it was a success. Our productivity was more than 30 instructions per day, the design of algorithms included, compared to the average productivity of 2-3 instructions per day of programming only, without design. The program consisted of about 15,000 instructions, and about 30% concerning the transformer into natural proofs was written by myself.

Then the turn of predicate calculus arrived. Almost immediately S. Maslov invented his "inverse method" [Mas64]. Though finally it proved to be, in a way, equivalent to the resolution method [Rob64] that became known to us much later, Maslov's method worked directly for arbitrary formulas and had an essential flexibility in representing strategies of inference search. Maybe it still has some advantages (for lower level complexity classes), but this question was premature at that time, and nowadays it may be too late to study it. The implementation that we had done showed results that were, in fact, excellent. The computer proved that $\sqrt{2}$ was not a rational number. But our hopes to prove new, interesting theorems had not come true. The following decades, on the whole, did not change the situation from the point of view of really interesting theorems.

This disillusionment caused by the weakness of general provers motivated the search for more efficient automatic provers. So a part of the group passed to proof theory and decidable classes, and another part (consisting of myself)—to computational complexity.

The logic seminar flourished; among participants there were V. Lifschitz (1947–), M. Gelfond (1945–), V. Kreinovich (1952–) (all of them later became professors in the USA), E. Dantsin (1951–) (now a senior researcher in LOMI), N. Kossovski (1945–) (now head of chair of computer science at the St. Petersburg University), S. Soloviev (now a senior researcher at the Institute of Informatics of the Academy of Sciences of Russia), PhD students from other republics as N. Zamov from Tatarstan (now a professor at Kazan University) and many others. The Jewish emigration of the beginning of 70s diminished slightly its rows but not too much. The core stayed intact. The hard blows were Maslov's death in 1982 and Mints's departure to Estonia in 1985.

From logic to computational complexity

> "Fast algorithms are too complicated to implement. Simple algorithms are too slow to run. And any program for any algorithm is erroneous." Known to researchers with vast knowledge.

Two open problems stimulated my switch to computational complexity: first, the fact that theorem proving algorithms were unable to prove really interesting theorems, and, secondly, the fact that the constructive approach to mathematics was, clearly, practically non-constructive. Other members of the group for logic had started to analyze how to get really practical algorithms for theorem proving. The idea of natural sciences approach has appeared, but how to put it into operation having the mentality of mathematical logician? A. A. Markov, who was educated as physicist, and was considered by many as a person with a strong naturalist way of thinking, had shown two examples: to analyze complexity and semantics from the point of view understandability. As for complexity, A. A. Markov had published a paper [Mar57] on Boolean complexity already in 1957. Then in an elegant [Mar64] he used complexity considerations to prove the undecidability of some problems; that was, in fact, the method that was much later associated with Kolmogorov complexity-theoretic approach to undecidability (recall that Kolmogorov paper [Kol65] appeared only in 1965).

The second approach, that is to look at semantics, could be more productive from a practical point of view if realized in the spirit of artificial intelligence. A. A. Markov and N. Shanin continued their profound study of semantics of constructivism which interacted well with the Western research on the semantics of intuitionistic logic, e. g. [Kle60]. Much later N. Shanin declared himself as a "finitist" [Sha87]. There were no direct practical consequences of this approach to theorem proving or other domains of algorithmics, as the complexity of admissible constructions was too high. However, this research influenced the research of some PhD students on characterizations of complexity classes.

Another path was undertaken by S. Maslov though his fate was dramatic. Having obtained his second doctorate in 1972 and being well known in the field for his various results—not only in theorem proving but also in the theory of algorithms, logic (Maslov class) [BGG97]—he had not been promoted to the senior staff by the Council of Steklov Institute of Mathematics in Moscow (MIAN). I guess that the main reason was the traditional anti-Semitism of MIAN leaders, but some role was caused by Maslov's unconventional vision on problems of automation of reasoning. In his arguments he used considerations from various fields of knowledge like semiotics, psychology, linguistics that were always considered as inferior and unworthy of genuine mathematicians by the more traditional mathematicians of MIAN. He left LOMI in 1974 and was a professor and senior researcher in various institutions in Leningrad. He was prevented from going abroad, though he attended the Congress on Logic, Methodology and Philosophy of Science in Amsterdam in 1968. He was monitored by KGB because of his political views and contacts with known dissidents. His interesting ideas on using general deductive systems to model various phenomena of social developments are presented in [Mas87]; another heuristic idea of how to solve hard problems is discussed in the collection [KM97] with a preface by V. Kreinovich [Kre97]. S. Maslov had tragically died in a car accident in the summer of 1982, presumably having fallen asleep while driving. He was not only brilliant, he was a pole of interaction and activity for different people of various domains of research.

Promotion was delayed or impossible (as in the case of Maslov) for all people who entered the field of mathematical logic and theory of algorithms as students of A. A. Markov or N. Shanin. G. Mints, who is a professor of logic at Stanford University from 1991 (he hold the position held before him by Barwise and

Kreisel), simply had no practical possibility to obtain his second doctorate in the 70s though he published very actively in theorem proving, proof theory and semantics (e. g., see [MMO71,Min72b,Min72a]). The worse came when he tried to immigrate in 1979. He left LOMI not to "throw a shadow" on the institute and being prohibited to leave the country, worked as programmer in various institutions (in particular, in the Institute of Meat and Milk Industry in 1984-85) and translated books in logic. He was, in a way, saved by E. Tyugu (now a professor in Stockholm) in 1985 who invited him to Tallinn (the capital of Estonia) as a senior staff member. Their joint work had started earlier and was profitable for both. They found a fascinating representation of Tyugu's program PRIZ in terms of intuitionistic logic [MT82]. Only after perestroika, in 1989 G. Mints had received his second degree (he published by that time more than 150 papers and was well-known in the world since long time). Interestingly, he wrote on complexity as well [Min92].

V. Orevkov was highly appreciated by A. A. Markov for his clever constructive (and, thus, continuous) mapping of the circle into itself without fixed points [Ore64]. He published various papers on recursive analysis, theorem proving, proof theory, and started his research on the complexity of proofs that finally produced this well-known paper [Ore93]. By the time he got his second doctorate in 1990 he had published more that 70 papers.

Exceptionally brilliant, Yu. Matiyasevich appeared in my student seminar on the theory of algorithms in 1965 being a first year student at the University. His first presentation in the seminar was on Kolmogorov algorithms. When studying Post normal systems I posed a question on the possibility to minimize the number of rules and premises in Post normal systems for a concrete problem that I do not remember. In a very short time he proved a general theorem that one axiom and one rule with one premise is always sufficient. Later, these results constituted a part of this PhD thesis. Presumably influenced by G. Tseitin, he moved to undecidability of finitely defined semi-groups where he drastically decreased the number of relations, a fact that impressed J. Cohen at the International Congress of Mathematicians (ICM) in Moscow in 1966. Having touched the string-matching problem he devised in 1969 a real-time algorithm for the case when the text and the pattern arrive simultaneously. The known linear-time algorithm by Morris-Pratt appeared in 1970 [MP97]. Matiyasevich's algorithm was unknown in the West until I wrote about it to Galil in the mid 70s. Very soon Yu. Matiyasevich got to know other members of our logic group. S. Maslov was interested in 1966 in Hilbert's tenth problem which had become from 1967 the main point of Yu. Matiyasevich's interest. He became a PhD student in 1969, and in January 1970 Hilbert's tenth problem has been solved. Clearly, he had to get his second doctorate directly for this result. But, taking into consideration the general attitude towards our group, he firstly got his PhD degree for a part of results obtained during his early student years, and after that he got the second degree and was promoted to a senior staff position. Now Yu. Matiyasevich is the head of laboratory of mathematical logic of LOMI, and consequently the head of the seminar on mathematical logic which is now slowly recovering, though being very far from its best days. In 1997 he was elected a corresponding member of the Academy of Sciences of Russia—better late than never.

One unforgettable event was the International Congress of Mathematicians (ICM) in 1966 in Moscow. All of us gave task. The classics like S. Kleene and A. Church were among the participants. J. McCarthy visited Leningrad after the Congress, and we discussed the automatic theorem proving problem. He predicted that all of us would visit the USA: it proved to be true. V. Livshits even worked with him after his immigration at the beginning of the 70s.

Computational Complexity

> "Everyone considers as the best way the one he is inclined to follow." Koz'ma Prutkov, *Fruits of Meditation*, 1854-1860. Free translation from Russian.

All of us wished to find some rigorous foundations for further research on algorithms. The complexity results of G. Tseitin were known and, quite naturally, young enthusiasts, myself and R. Freidson (1942–), asked him to head a seminar on computational complexity. That happened in March of 1967. By that time I had almost finished my theorem proving activity and got my PhD degree for results in constructive analysis.

Birth of the complexity seminar

> "Sometimes the zeal overcomes the reason." Koz'ma Prutkov, *Fruits of Meditation*, 1854-1860. Free translation from Russian.

The research we started was aimed at lower bounds of computational complexity. G. Tseitin improved his mentioned lower bound for Markov algorithms that invert the words, from $n^2/\log^2 n$ to n^2, and intensified his attacks on propositional validity. He found that many combinatorial problems were polytime equivalent either to satisfiability or validity. His pioneering paper on lower bounds of complexity of propositional proofs [Tse68] (that was an attempt to approach $coNP \neq P$ hypothesis) appeared at that time. It was not an easy job to persuade him to write down his results, that was my main effort as editor. By 1971 G. Tseitin started to loose interest in the seminar, and mainly myself and R. Freidson supervised it.

R. Freidson generalized the classical method for finite automata to real-time algorithms [Fre70] leading to, for example, the impossibility of real-time string-matching on Turing machines of any dimension (here we speak about the hardest case, when the text arrives before the pattern).

I attacked the problem of palindrome recognition on multi-head Turing machines that was very popular at the end of the 60s, and it was conjectured that the real-time recognition of palindromes was impossible. By 1969 it was clear that the conjecture was wrong. The proof based on a direct analysis of periodicities has appeared in [Sli73] that is the longest paper in the history of computer science with a proof of one simply formulated theorem. The drastic simplification of this result by Z. Galil [Gal76], who used basic ideas of [Sli73] and the result [FP74]—this was not known to me while writing [Sli73]—made him famous. As for me, having found no support, I had not even tried to get the second doctorate for this result, though it evidently deserved it. However, I got the first prize of LOMI and was awarded a month extra salary. More essentially, I was lucky to become known in the United States. The paper [Sli73] has been translated into English twice, the first time by Bob Daley in 1974, and the second time on a regular basis (as mentioned in the reference). I received a very warm letter from Albert Meyer whose later support for our research on complexity is hard to overestimate. For several years he was sending us not only important technical reports, but entire volumes of STOC and FOCS symposia which influenced the theoretical research. As for palindrome recognition, using Galil's idea of applying the results of [FP74], I wrote (perhaps) the shortest existing proof of real-time recognizability of palindromes [Sli77].

Somewhere in 1969 I met B. Trakhtenbrot who was an influential figure in the logic and complexity community, and slightly later, L. Levin, brilliant and eccentric. Levin impressed me by his optimal algorithm for recognizing propositional tautologies. The idea is simple: we gradually apply all algorithms to the input, and having got an output, check it for a model. Any algorithm for the propositional tautology will be not better than the functioning of this algorithm up to some multiplicative constant (exponentially depending on the size of the algorithm). This result explained me the importance of a careful interpretation of mathematical results. The notion of optimal algorithm used here is more than dubious as the multiplicative constant may radically change the meaning of the result.

L. Levin presented his candidate dissertation to the anti-Semitic Novosibirsk Council though I warned him. He had been turned down, it was somewhere in 1973. Thus, he was forced to immigrate to the USA, and later he expressed thanks for this turn of events.

The complexity seminar continued to work though its head became less and less visible. Due to recollections of Dima Grigoriev, the last blow was the paper [SM73] by L. Stockmeyer and A. Meyer. G. Tseitin presented this paper himself. All the ideas were well-known to him since long time. And says Dima "he probably understood that the West was strong by its mass research, and exceptional individuals, as himself, would have no chances." And he disappeared. Somewhere at that time N. Shanin was, in a way, pushed out of the Faculty, and thus the first phase of the development of the seminar was over. Happily, Dima Grigoriev had already started to work at the seminar. He was a second year student in the winter of 1973, and he also attacked the lower bound problem.

In September of 1973, being a third year student, D. Grigoriev introduced the notion of matrix rigidity, later also introduced by L. Valiant (the latter did it for Hamming metric while D. Grigoriev used L_1-metric), and applied it to get some lower bounds, now well-known; see the recent review [KR98]. The same year, during a soirée on the occasion of passing an exam on quantum mechanics, D. Grigoriev got an idea to use the uncertainty principle in complexity, and produced Grigoriev's method on time-space trade-off [Sav98]. Using this method he proved an n^2 lower bound for the time-space product for polynomial multiplication, and an n^3 lower bound for matrix multiplication. These results were, in a way, underestimated, and I, as his supervisor, has not done special efforts for a fast publication. They appeared only in 1976 in the Proceedings of LOMI Seminars [Gri76]. Happily, Dima was sufficiently active, and very soon started to find his way to publish his results, a much more efficient approach than rely on me. He got his PhD degree in 1979 having published already many excellent results.

One "dramatic" episode happened at that time at the seminar (as I learnt later from Albert Meyer, similar episodes happened several times in the USA). One of our participants, a really good mathematician, announced a sensational result that $P \subseteq SPACE(\log^2 n)$. Neither G. Tseitin, nor D. Grigoriev nor myself noticed an error that was nontrivial, though I had some unpleasant inner feeling during the talk. We proposed to publish it in the current volume of Proceedings of Seminars of Steklov Institute edited by Yu. Matiyasevich and me, and the text had been included in the volume under preparation. In a couple of days D. Grigoriev found an error, and ran to LOMI where he found Yu. Matiyasevich holding a telegram from G. Tseitin where the latter also signaling the same gap. I had to do the unpleasant job of deleting the text from the issue.

Hard time for the complexity seminar

> "It's desperately bad now. But I am an optimist. Surely, it will become worse." Dima Grigoriev.
>
> "If you wish to be happy, be so." Koz'ma Prutkov, *Fruits of Meditation*, 1854-1860. Translation from Russian.

The Faculty of Mathematics and Mechanics started to move from Leningrad, N. Shanin was pushed out of the Faculty, G. Tseitin had not appeared at the seminar since 1973. Gradually I became the actual head of the seminar on complexity in 1973, and from 1975 the time has become hard. N. Kossovsky was among other participants of the old seminar. Though he did not attend this new seminar, his PhD students A. Beltiukov and S. Pakhomov did, and together with D. Grigoriev and myself we composed a very strong working group. The seminar also participated in joint meetings with the seminar in logic (we were all members of the latter). Verification of proofs during seminars was an usual activity. One outstanding case was the famous Makanin's result on the decidability of equations in free semi-groups. That was his dissertation for the second doctorate, and Yu. Matiyasevich, as a representative of LOMI, had to write one of the reports. Makanin's paper had been accepted to Izvestia of the Academy of Sciences of the USSR, that was one of the most prestigious journals in mathematics in the Soviet Union and with a high international reputation. Makanin made several two-hours presentations at the joint logic-complexity seminar, and during the his fourth talk, at the 8th hour (counted from the beginning) we found a considerable gap in his proof. He had been shocked, and left for Moscow in a state of profound depression, as his paper was in printing. But he has managed to find a patch that was added as a footnote in his paper. That was not the only case when gaps were revealed during the talks, and some of gaps were fatal for final results.

It became clearer and clearer that the problem of lower bounds was much more difficult that one could have thought at the beginning (the same applies to the $P = ?NP$ problem), so one reasonable way to advance in lower bounds was to consider more limited models that RAM or Boolean circuits. Thus, the shift towards computer algebra was natural. However, other, more traditional activities went on, e. g. A. Beltuikov introduced the model [Bel79] known as Beltuikov' machine [Clo97] seeking exact descriptions of various complexity classes. Beltuikov's outstanding power was that he was able to review an entire volume of STOC or FOCS proceedings during one or two seminar meetings. S. Pakhomov also worked on complexity classes, that was regarded at that time as a possible approach to $P = ?NP$ problem. After having got his PhD, he moved to neurophysiology and worked with the well-known professor N. Bekhtereva. He gave very impressing talks on some algorithmic models of perception.

After 1974 I attacked string-matching on RAMs, hoping to either find a real-time algorithm or prove its impossibility. In 1976 it was clear that a real-time algorithm can be constructed. Here it is important to stress that we deal with the problem when the text arrives before the pattern, and occurrences of the pattern must be detected in real-time while arriving from the latter. As periodicities played a crucial role in the solution, I decided to look at the possibility of finding all of them in real-time in a natural succinct form (otherwise it is impossible as the number of periodicities in a word can be quadratic). That was an erroneous decision. The real-time string-matching algorithm is two or three times simpler than the algorithm finding all periodicities [Sli81b], and though the algorithm in [Sli81b] solves in real-time many known "string-matching" problems (for example, finding maximal repetitions, the longest common substrings etc.) it is reasonable to give string-matching algorithms separately. The paper [Sli81b] is too voluminous and hard to read. An attempt of [Kos94] to simplify the construction contains no proof and repeats general considerations on approximating the suffix tree that can be found in [Sli81b]. Thus up to now, my text, admitting its awful quality of presentation, remains the only source on the subject. This result determined an invitation to the Symposium on Mathematical Foundations of Computer Science in 1979 in Czecho-Slovakia; this time my reputation was much better, and I was permitted to go. This was my first trip abroad. An important point was that Czecho-Slovakia was a "socialist" country, and the Communist Party norm was to start foreign trips with socialist countries. In fact, some people from the USA proposed to send me an invitation (that covered all the expenses, otherwise there was no point of discussion), but when I tried to estimate my chances to get permission Party officials told me something like: "Who are you to speak with you about such a trip?" And indeed, who was I?

In 1981 I received an invitation to attend a meeting in Oberwolfach and finally I was permitted to go. It was in February, and in the train in Germany I met L. Valiant who easily observed my Russian origin looking at my clothes. Unfortunately, permissions to go abroad were not stable. The most regrettable for me was the refusal of the Regional Party Committee to permit me go to a meeting on string-matching algorithms in Italy in the 80s, though Z. Galil had found money to pay all the expenses, including traveling.

The main problem of the seminar on complexity was the lack of young researchers and students; connections with the University were weak, and life was difficult. The salary was low, and the apartment problem seemed unsolvable (I succeeded to get a permission to buy a two pieces apartment in 1973 and stayed since then there with my family of four). We were wasting a lot of time to solve all these problems of basic living. My personal situation was almost desperate in 1975-1976. I was paying for the apartment and my wife, who had to stay often with our little daughter, got no salary. I remained in the junior research staff (palindromes did not help to be promoted). Happily, the logic community supported me, namely, I got a chance to translate the "Design and Analysis of Computer Algorithms" by Aho-Hopcroft-Ullman (with Yu. Matyasevich as editor); then the tough policy of Steklov Institute was weakened, and in 1977 I was promoted to a senior position before formally getting the second degree. There were other forms of support, some opportunities to teach at the Electrotechnical Institute (arranged by R. Freidson who was an associate professor there).

But the research in complexity remained productive. In addition to the results mentioned above I can add the result of D. Grigoriev concerning the relation between real-time calculations on Turing machines of higher dimensions and Kolmogorov algorithms [Gri77] that permitted to fix an imprecision in the paper [Hen66] that was unnoticed since its publication in 1966 and its Russian translation in 1970.

Towards recovery

> "The zeal overcomes everything." Koz'ma Prutkov, *Fruits of Meditation*, 1854-1860. Translated from Russian.

But we had opportunities to move around the country. The community of people doing algorithmics, complexity and logic was not numerous, personal contacts were rather all-embracing. However, joint research with people living in other cities was very difficult. In the 70s people from abroad could be invited by the Academy of Sciences, and we had talks, for example, by G. Kreisel, A. Tarski who were invited due to the efforts of G. Mints and N. Shanin. G. Kreisel, who stayed in Leningrad for a rather long time, impressed me by his vast knowledge in various fields of science and culture. The visit of A. Tarski was very short, I remember that he spoke Russian and was remarkably clear in his presentation.

The external signs of improvement of the state of complexity in the USSR are associated in my memory with the travel to MFCS'79. There I met among many other well-known researchers, J. Hartmanis and K. Mehlhorn. This year 1979 was fruitful for meeting with people. In September we had an International Symposium at Urgench, Uzbekistan, dedicated to al-Khwarezmi (al-Khuwārizmi) who was born in the region. The symposium was organized by the late A. Ershov and D. Knuth, and was attended by numerous computer scientists from the West. In the proceedings [EK81] one can find historical and conceptual papers and photos of many people (captions of Buda and Barzdin must be interchanged), in particular, of some people mentioned in this paper. On their way to Urgench, V. Strassen and E. Specker visited Leningrad before the symposium and we met them in LOMI. All these meetings gave more possibilities to be invited to the West.

On the whole, foreign visitors remained rare and were viewed like Martians. Their visits in the country were not always smooth though they were permanently accompanied by their Soviet colleagues. One curious case happened in the same 1979 with J. Hartmanis. In the summer of 1979, after having visited his motherland Latvia, he decided to make a short stop in Leningrad on his way to one of Scandinavian countries. His Soviet visa expired the day of his arrival. I had some family problems, and asked Dima Grigoriev to take care of him in what concerns his departure. But J. Hartmanis missed his plane and got only a ticket for the next day. When Dima tried to install him in hotels, he got everywhere refusals because of his visa: from the midnight his stay in the Soviet Union became illegal. Thus, they came back to the airport that had a special room for foreigners, and Dima requested the frontier guard on duty whether J. Hartmanis will be permitted to leave tomorrow if he slept somewhere in the city. The guard said that he personally could do it, but tomorrow there would be another guy whose behaviour could not be predicted. To avoid risks, J. Hartmanis was to sleep on a bench at the airport under the continuous supervision of guards. He said later that Dima saved him.

Urgench symposium produced my joint paper with G. Adel'son-Vel'ski [AVS81]. It concerned different approaches to solve NP-hard problems. Among others we mentioned the method developed by S. Maslov. The method [KM97] concerned the propositional satisfiability and was based on the philosophy of 'free choice'. Technically, the main ingredient was a kind of gradient method, but the general vision was wider. The technical part, clearly explained in [Fre96], is the following. Consider a CNF-formula F in n variables $a_1, \ldots, a_n$. As usually, we call variables or their negations *literals*. With every literal a of F we associate a non-negative number $\xi(a)$ called the *defference* of a. The vector $\xi = \langle \xi(a_1), \ldots, \xi(a_n), \xi(\bar{a}_1), \ldots, \xi(\bar{a}_n) \rangle$ is called an *obstacle*. An obstacle ξ defines a Boolean vector $X = \langle x_1, \ldots, x_n \rangle$ if

$$(x_i = 1) \Rightarrow (\xi(a_i) = 0),\ (x_i = 0) \Rightarrow (\xi(\bar{a}_i) = 0),\ \ i = 1, \ldots, n.$$

An obstacle ξ is said to be *correct* if it defines a unique Boolean vector. Let F be of the form $\bigwedge_k D_k$, where D_k are clauses. An *iterative method* for F is given by an operator $K_{R,L}$ of the type $R_+^{2n} \to R_+^{2n}$ which is defined by $2n$ formulas of the form

$$\xi'(a) = R \cdot \xi(a) + L \cdot \sum_{\bar{a} \in D_k} \min_{b \in D_k,\, b \neq \bar{a}} \xi(b). \tag{1}$$

Here a, b are literals from F, and R, L are non-negative constants. The operator $K_{R,L}$ is a piecewise linear continuous homogeneous operator. S. Maslov proved [Mas81,Mas83] that for any K and R the space of obstacles defining the satisfiable Boolean vector for F is an eigenspace of $K_{R,L}$, and the iterations of $K_{R,L}$ converge to a model of satisfiable Tseitin's formulas [Tse68] and 2-CNF-formulas. It follows from the eigenspace property that seeking a model can be reduced to finding linear eigenspaces of $K_{R,L}$.

After Maslov's death in 1982, R. Freidson organized a seminar that invested a considerable effort in analyzing Maslov's method. The seminar worked until 1991 at Leningrad Electrotechnical Institute and gave a good possibility for young researchers, including students, interested in algorithmics of hard problems to work together. In particular, some participants of the logic seminar were there, namely, E. Dantsin, G. Davydov and V. Kreinovich, mentioned above. The seminar had a good reputation, and many known researchers gave talks there.

Grandeur and Decadence

> "Pride goes before destruction, and an haughty spirit before a fall." Proverbs XVI, 18.

In the fall of 1980 I got at last my second doctorate from the Steklov Institute for Mathematics in Moscow for the results on real-time algorithms. This fact was also a tacit recognition of the importance of complexity and algorithmics research by the Academy of Sciences. This action was supported by L. Faddeev, the Director of LOMI, who was elected to the Academy some years ago and was becoming a more and more influential figure at the Department of Mathematics; I guess that his support was essential. There was a minor filth from O. Lupanov as the Dean of the Faculty of Mechanics and Mathematics of the Moscow University. The Faculty had been chosen by the Council for the external report on the dissertation. This paper was signed by A. Kolmogorov. But to be valid this paper had to be endorsed by the Dean, and this endorsement was always automatic especially if signed by a scientist of the caliber of A. Kolmogorov. But O. Lupanov refused to do it, and disappeared. And it was on the eve of the meeting of the Scientific Council where I had to present the dissertation. Happily A. Semenov found a deputy dean who endorsed the paper without discussion.

An official state certificate for my degree arrived very soon, and I could satisfy the bureaucratic demands as professor and organizer of conferences. The same year, with support of LOMI and together with D. Grigoriev, we organized the first Soviet Workshop in Computational Complexity in Leningrad. It was attended by classics like G. Adel'son-Vel'ski (one of the two authors of AVL-trees) and young researchers like L. Khachian—who had already published his polytime algorithm for linear programming. L. Babai was an illegal participant (to make him legal we had to obtain the status of an international workshop which was much harder). He was experienced in illegal visits to Leningrad as he came there before (from Moscow) without official permission. His visit was productive for the graph isomorphism problem, and determined the paper [BGM82] —this was unusual for our practice as, at that time, D. Grigoriev never met his co-author D. Mount. An instructive publication [GS82] written by some participants of this workshop which appeared rather quickly played an essential role in the development of complexity research in the Soviet Union. We had more workshops, in 1983 and 1985; the 1983 one was in Grodno (Belorussia) and gathered together a large amount of researchers. An exceptionally brilliant A. Razborov was among the participants.

In 1981 V. Strassen proposed me as a member of the Panel for Mathematical Aspects of Computer Science of the International Congress of Mathematicians to be held in 1982 in Warsaw (actually, it took place in 1983 because of the political situation in Poland, and the threat of a Soviet invasion). Such a panel consists of about 9 members representing world-wide the field and serves as a consulting body on invited speakers for the Executive Committee that makes the final decision. Now the role of panels is higher than at that time. In the Soviet Union all interaction with ICM was via the National Committee headed by the Director the Steklov Mathematical Institute in Moscow who was considered by the officials as the Chief Mathematician of the USSR. (This structure and mentality were universal, there was a Chief Composer, a Chief Writer etc.) Thus I had to ask permission from I. M. Vinogradov to participate in the Panel. However, first I wrote to the Head of the Panel, that was R. Karp, that I was ready to serve, and then sent a letter to Brother Vanya demanding permission. There had been no answer. On the other hand, this invitation to the Panel, the high appreciation of my results by A. Kolmogorov and the support of L. Faddeev led to the invitation as a 45-minute speaker of the Section of Mathematical Logic of ICM at Warsaw.

As a Panel member I was rather efficient as I knew the situation well, and most part of speakers proposed by myself were accepted. I was also a Panel member for the next ICM (1986, Berkley) for which I proposed as 45-minute section invited speakers D. Grigoriev and R. Krichevsky, and both were accepted; R. Krichevsky was not permitted to go. He was at the Institute of Mathematics of the Siberian Division of the Academy of Sciences, his invitation was a surprise to officials, but from their point of view he had a defect—he was Jew. But other reasons could have been considered; for example, A. Razborov was not permitted to attend the same ICM as an invited section speaker because of some groundless revenge of V. Yablonsky's group that was executed by O. Lupanov as the Dean of the Faculty at Moscow University where A. Razborov was a PhD student. The Dean refused to sign some standard paper concerning A. Razborov. By now all has gone, A. Razborov, who had won a prestigious Nevanlinna Prize (an analogue of Fields Medal) of the International Union of Mathematicians in 1990, is famous all over the world, except maybe France, and Yablonsky people are happy to collaborate with the National French Institute INRIA. By some reason, that I do not know exactly, I was also in the Panel for Mathematical Logic for the ICM of 1990, but there I could not be as efficient as at the Computer Science Section.

Booming of complexity and laboratory for theory of algorithms

"The higher flight — the deeper fall."

In the winter of 1982 L. Faddeev proposed me to organize a laboratory in the Leningrad Institute of Informatics of the Academy of Sciences of the USSR that was created not long ago before and was thought to become the center of computer science research in Leningrad. If I could foresee the perestroika, I would have refused. But the communist regime seemed eternal, and as the head of a lab I could got a 20% increase in salary and a possibility to promote the complexity research in Leningrad. I was always poor, and for a very long time was the world poorest researcher in theoretical computer science for my or higher scientific level. Now maybe I share this title with Yu. Matiyasevich and A. Razborov, but if we take into consideration that I have two children, Yu. Matiyasevich one and A. Razborov no one for the moment, and other living circumstances, I may still hold the title.

The director of that institute was a professor from a military academy who retired in the rank of colonel, and started as a head of computer center of the well-known Physico-Technical Institute of the Academy of Sciences of the USSR. This Center constituted the core of the Institute. The starting activity of the Institute for Informatics (which changed its name several times) was to supply computer resources to the numerous institutes of the Academy of Sciences in Leningrad. It was rather successfully until the invasion of personal computers. But the ambitious (that was good) director wished to make career in the Academy by any means (that was not good). He decided to rely on the support of the Regional Party Committee and not on the quality of research. The style of the Party can be diagnosed as a maniacal syndrome, groundless promises to solve any problem in an incredibly short time. Even to build a supercomputer that will surpass all the existing by the end of the year.

Nevertheless, I moved there, and my first collaborators were A. Chistov (1954–) and S. Evdokimov (1950–), several months later S. Baranov (1950–). The latter was a participant of the initial seminar on complexity, and then switched on programming and got his PhD degree for compiler construction. It seemed quite natural to unite the efforts of people from algorithmics, computer algebra and programming to make a system of symbolic computation for personal computers more advanced then the existing systems that were available, some of them, like SAC-2 of G. Collins and R. Loos, even as source. One way to program it was to use the Forth language perfectly handed by S. Baranov. Yu. Matiyasevich who was doing some computer aided research in number theory was also ready to participate. All we needed were three personal computers and a couple of young programmers. Yu. Matiyasevich and I had written a proposal that was personally handled to the Chairman of the State Committee on

Science and Technology academician G. Marchuk, who later became the President of the Academy of Sciences of the USSR. I have an impression that he did not understand what it was about, and we got no support. The favorable moment has been lost. An attempt to use a Soviet made version of an IBM 360 series computer revealed an error in hardware (that was found by A. Chistov) at a rare moment when the computer functioned. Normally it was out of work. Later S. Baranov had done a direct Forth implementation of SAC-2 system that showed excellent results on small computers, but small computers were not the main trend of software industry — the Moore law governed the market of computers. For S. Baranov it was profitable as this work finally permitted him to get his second doctorate; he become a professor and even get the title of professor. Besides that he headed some Motorola lab in St. Petersburg.

A. Chistov and D. Grigoriev concentrated on computer algebra. In 1982 the fundamental paper [LLL82] has appeared and I proposed them to study it and to look at the case of several variables. Here we were in time. That gave rise to now well-known results of these authors on factorization of polynomials and search of connected components. At the beginning of this activity an unpleasant accident happened. In the first joint preprint on the multivariate case they announced a fast algorithm of factorization for the case of non-separable fields. During their presentation at the complexity seminar I expressed doubts in the validity of this result, but they made a reference to a theorem of Hilbert that implied the result. I could not overcome Hilbert. The minor detail was that this theorem of Hilbert was invalid, and that was known but not to them. So their debut was not quite smooth. Later they were 45-minutes invited speakers at the ICMs in 1986 at Berkeley (D. Grigoriev) and 1990 at Kyoto (A. Chistov).

From 1982 the seminar on complexity started to flourish. The worst was over, new people appeared even in 1979, like N. Vorobjov Jr. (1967–) in 1979 and I. Ponomarenko (1957–) in 1981. Later they joined LOMI when D. Grigoriev became the head of laboratory for algorithmic methods in 1988 (now headed by A. Vershik under some other name). Then came D. Burago (1964–) in 1986, A. Burago (1968–) in 1989, S. Fomin (1958–) in 1991 who were members of my laboratory, and people from other places, namely, A. Barvinok (1963–), D. Ugolev. The results of that time are numerous, of high level and many of them are well-known. They gave better or much better complexity bounds for many problems in computer algebra, in particular, for the factorization of algebraic varieties over algebraically closed fields (A. Chistov, D. Grigoriev [CG82,CG83]), deciding the theory of algebraically closed fields (D. Grigoriev [Gri86]), finding real solutions of polynomial equations (D. Grigoriev, N .Vorobjov [GVJ85]), deciding Tarski algebra (D. Grigoriev [Gri88]), similar problems in differential algebra (D. Grigoriev [Gri89]), consistency of systems of polynomial in exponent inequalities (H. Vorobjov [VJ88]), polytime subclasses of graph isomorphism (I. Ponomarenko [Pon92]), factoring polynomials over finite fields (S. Evdokimov [Evd88]), counting integer points in convex polyhedral and other applications of statistical sums to algorithmics (A. Barvinok [Bar92]), lower bounds for pointer machines (D. Burago [Bur92]), etc. These results constituted several volumes of "Zapiski nauchnikh Seminarov LOMI. Series: Theory of Computational Complexity" edited at the beginning by D. Grigoriev and myself, and later by D. Grigoriev alone. They are all translated into English either as issues of *J. of Soviet Mathematics* or *J. of Mathematical Sciences.*

At the same time people from pure mathematics like D. Burago and S. Fomin continued to work mainly in their particular fields. D. Burago later proved the famous Hopf conjecture (with S. Ivanov) and S. Fomin had a good opportunity to work in combinatorics where he is now well-known.

At the beginning of the 80s I had got two results that I like. The first of them says that for any class of graphs determined by a context-free graph grammar of a certain type the travelling salesman problem can be solved in polytime [Sli82]. This result had been noticed with some enthusiasm, as it gave a general tool of describing polytime subclasses of hard problems, the subject of my permanent interest. However, no essential generalizations have been found, though it contributed to the theory of graph grammars. In fact, this property is similar to the bounded tree-width of Robertson-Seymour which neither have been generalized to get interesting polytime classes of hard problems.

The second result consists of a system of notions to evaluate the quality of knowledge representation. It has some interesting applied consequences and leads to the thesis *"The development of knowledge tends to minimize the entropy of its representation".* This thesis was not explicitly formulated in [Sli91], thus, I give some ideas here. First, we introduce a general notion of inference of logic formulas (without loss of generality we may consider the first order syntax). Such an inference is determined by a system S of inference rules, each of them being treated as an algorithm transforming proofs, that are lists of formulas labeled as assumptions or conclusions. This general notion of inference defines the relative complexity of a set of formulas Φ with respect to a set of formulas (assumptions) Γ as the shortest inference of all formulas of Φ from Γ. Denote this complexity by $d_S(\Phi|\Gamma)$. Given two proof systems S and V and a natural number ξ we define the ξ-entropy $H_S(\Phi|V,\xi)$ of a set of formulas Φ as the minimum of sizes of proof systems U that are consistent with S and such that $d_{V \cup U}(F) \leq \xi$ for all $F \in \Phi$. The non relative ξ-entropy of Φ with respect to S is defined as $H_S(\Phi,\xi) = H_S(\Phi|\emptyset,\xi)$. This is an analogue of ε-entropy of metric spaces. The system S is a way to represent the knowledge given by Φ. Each domain of knowledge has its favourite range of ξ, that is of "acceptable" lengths of proofs. People of this field do their best to minimize the size of the representation S to justify all "interesting" consequences Φ by reasonably short proofs. In other words, they seek an U mentioned above and replace the initial S by U. One practical consequence is that to construct an efficient proof search computer system one have to base it on user friendly tools of manipulation of knowledge and "almost on-line" strategies. My last Russian PhD student V. Tarasov [Tar96] had implemented this idea for symbolic search of limits, and the results were very promising.

The complexity seminar had no participants from the University. Its stable core was represented by the Institute for Informatics, LOMI and the Institute of Evolutionary Physiology and Biochemistry of the Academy of Sciences of the USSR. The Laboratory of the Physiology of Child gave shelter to A. Barvinok who has the warmest feelings about this laboratory. Another member, the most unusual for the seminar was D. Ugolev, a researcher on the physiology and quality of food. He has a non-mathematical type of mind, but nevertheless he obtained several results concerning algorithmic models, e. g. [USP93], including some complexity estimations. He had many visitors both from the Soviet Union and from abroad. Our frequent speaker was A. Razborov.

Formal promotion also went on smoothly, people got their first and second doctorates in time. Traveling abroad was feasible but for short terms and very uncertain.

Theory of complexity versus computer science

> "One of the greatest sins is to work too much". From an interview of the Director of the Bank "Société Générale" (France), about 1996.
>
> "One of the symptoms of approaching nervous breakdown is the belief that one's work is terribly important. If I were a medical man, I should prescribe a holiday to any patient who considered his work important." Bertrand Russell, *The Autobiography of Bertrand Russell*, Vol. 2, 1968.

In 1981 I was invited as a part-time professor of the chair of computational mathematics of the Faculty of Physics and Mechanics of the Leningrad Polytechnical Institute. The Faculty had long standing traditions in applied nuclear physics and optimal control. The chair was headed by Professor V. Troitsky who wished to make the computer science education more apt to current demands. So I initiated some reforms introducing or improving courses on basic theory, on networks and distributed algorithms, operating systems and data bases. The attitude towards me was very favorable, and my activity was well accepted by colleagues and Faculty. I was there from 1981 until 1987. This experience was new and interesting, as students were engineering oriented and were interested in other type of research problems than more mathematically oriented people. But we were blocked by the lack of computers.

Somewhere in 1984 the Regional Party Committee, which was the highest and absolute local power, had all of a sudden realized a dangerous lack in technology with respect to the West. The decision was not original. They decided to announce a program that would get rid of the back-log in five years. The programme was called "Intensification-90". It was clear from the beginning that the programme will be on the whole reduced to declarations and demonstrations, but all researchers in computer science were pressed to participate as automation was one of the corner stones of the programme. Some reasonable actions to educate people or improve the education had been done. The programme was considered as terribly important. Tons of methodical papers had been written by the people from the Institute for Informatics. One of our fervent activists had got infarct (now he lives in the USA).

People from industry proposed me to write a curriculum for computer science. I started this work that was supported by G. Tseitin and S. Lavrov (1923–)

who headed the chair of programming at the University and was a known researcher and practitioner in programming. It gave [LST85]. The curriculum was too idealistic and too big to be implemented, but it gave a good point to discuss the situation. It was on the whole supported by A. Ershov, the most dynamic figure in Soviet programming, who organized a public discussion in Moscow. The public reaction was positive; definitely negative was only a representative of the Ministry of the Superior Education who was clearly irritated by the fact that such a programme came not from the Ministry. But all he could show was his definite incompetence in the subject. The programme was read by people responsible for computer science education in the USSR, they criticized, but used it.

Later, in 1987, I was invited to head the chair of computer science at the University, that I accepted and started to function in 1988. With some efforts and support from the Faculty, computer science was separated in an independent discipline with 250 students. I modernized the curriculum, though it was hard to find people to teach such topics as distributed computations and parallelism. The first one was taught by Yu. Matiyasevich, the second one by M. Gordin who worked in probability and later joined LOMI. All that stopped with my departure to France.

Though my teaching and reforming of the educational curriculum were time consuming, they were not useless. My organizational work in the Institute of Informatics implied more dramatic consequences for me. I had no time to continue my research in theory, and what was worse, I had no time to write papers on applied topics where I had done a lot during 1983–1990. The activity gave rise to new laboratories, in technology of programming, signal processing and artificial intelligence, but nothing gave any profit to me. My living conditions remained bad. The administration of the Institute for Informatics actively and explicitly did not like me. In 1990 the colonel director was replace by major-general from the same military academy—the progress took place.

As for more applied aspects of computer science the work usually was blocked because of lack of computers. It was not a problem of money. The Party created enormous and useless institutes with thousands of people. They wasted more than 3 billion roubles (about $600 million if we take the black market exchange rate of dollar) on the absurd programme "Intensification-90", but to support our advanced applied research with 2-3 PCs there were no money. We have done prototypes that showed good results, and got many promises to support that were never realized. Nobody was interested in real progress. It was clear that the system was dead. And it had fallen down instantaneously with Gorbachev.

Some people (not I) had hopes about an improving computer science research when in 1983 the Department of Informatics, Computing Machinery and Automation was created in the Academy of Sciences of the USSR. For me it was clear that the train has gone away. The computing machinery has been destroyed by the stupid Party decision to copy IBM 360 series that was clearly impossible, and had proved to be impossible. The people related to automation, who were elected to the Department, were mostly very far from research, they were from defense industry. Computer science was represented by veteran programmers, people doing numerical methods and Party functionaries. The Party leading role was personified in too many members of that department. One of them, M. L. Aleksandrov, from Leningrad, had been elected as corresponding member. According to the definition of the position of corresponding member, he had to obtain scientific achievements of first-rate importance. His personality was advertised in the most respectable central newspapers like "Izvestia" which published a whole page article on his outstanding qualities. As for his scientific results, I doubt that one can find any traces, though he knew the word "Fortran". After perestroika he made some money by selling computers, then he had stolen one million German marks in cash from a too credulous West German company, and disappeared. People say that Interpol is looking after him since that time. Really outstanding result! The department of informatics must be proud!

Agony and exodus

> "We wished to do it better. But, as always, it turned out for the worse." Victor Chernomyrdin, Russian ex-prime-minister, on the economic reforms, about 1996.

After 1989 the seminar was very strong but the level of living was going down. Communications were becoming better, though the mail was slow and monitored, and the e-mail was of low capacity and unreliable. The perestroika lifted the "iron curtain", but the power inside it continued to be in the hands of communist nomenclature united with more traditional criminals. The nomenclature explicitly switched on making money. The USSR collapsed. The situation in former Soviet republics became bad, even awful. Armenia was without electricity, heating and normal food for about four years. A part of Soviet research had been saved mainly by Soros Foundation with its highly efficient individual support with minimal formalities. Mathematicians got some support from the American Mathematical Society. Later the Western Europe also contributed to this and happily continues to do it. Not long ago the Russian Parliament (Duma)—which is dominated by communists and which did nothing for Russian science (professor's salary stays at about $90 per month)—accused Soros for distroying Russian Science by buying the brains. Ridiculous and insulting.

In 1990 the situation became simply bad, there were food shortages. On the other hand, traveling abroad became almost free. Nevertheless I tried to do research, and an invitation to Buenos-Aires arranged by J. Heintz resulted in a paper on shortest paths amidst semi-algebraic obstacles in the plane. We had found that in that context one can do things in polytime [HKSS91] as compared with higher dimensions where the problem is at least NP-hard.

During that time of permanent stress, the necessity to earn money to support the family had completely exhausted me. My health started to worsen, the brain was paralyzed by the hopeless situation. I was not an exception, the number of suicides and sudden deaths went up. I started to try to find a position somewhere abroad, but in vain. Happily somewhere in 1990 I met Michel de Rougemont who was in St. Petersburg. He invited me to the University Paris-11 in 1991, and then helped me to get a French habilitation as professor from the French Ministry of Education. After some time spent at the University of Poitiers, I obtained a professor position at the University Paris 12. The adaptation was easy. In Poitiers and Paris 12 the environment was friendly and cooperative. There were no essential problems of adaptation. But it took about three years to restore my capacity to think. I was strongly supported by members of my seminar, in particular, D. Burago, and also by my French colleagues.

Many of the mentioned people are scattered all over the world, mainly in the USA: A. Barvinok is at University of Michigan, Ann Arbor, A. Burago at Microsoft, Seattle, D. Burago and D. Grigoriev at Pennsylvania State University, S. Fomin at M.I.T., A. Slissenko at University Paris 12, France, N. Vorobjov, at University of Bath, England, M. Gelfond and V. Kreinovich at University of Texas at El Paso, V. Livschits at University of Texas, Austin, G. Mints at Stanford University.

On the current situation in Russia

"Blessed is he who expects no gratitude, for he shall not be disappointed".

The state of the research in Russia is desperate. Those who have no permanent or semi-permanent positions in the West and continue to do research are supported by invitations to the West, different Western programs, or are earning money out of research. Russian programmes of support are miserable. For example, G. Tseitin, whose 60s anniversary was marked only by forcing him to retire, worked as programmer in a bank, and now has several contracts with Motorola in St. Petersburg. Nevertheless there are PhD students at LOMI, some of them coming from my last student seminar.

Concluding Remarks

"...
Et je m'en vais
Au vent mauvais
Qui m'emporte
Deçà, delà,
Pareil à la
Feuille morte".
P. Verlain, *Chanson d'automne*

On education and research systems

"Only in the State Service one perceives the Truth." Koz'ma Prutkov, *Fruits of Meditation*, 1854-1860. Free translation from Russian.

In the Soviet system of research and education there were some strong points that deserve to be studied. They concern mainly activities relevant to research. First, the system of school education. Numerous study groups and specialized schools with intensified teaching of selected disciplines give school-children an important freedom of choice to realize their inclinations without missing the standard national minimum. Various olympiads and competitions add some vigor to these specializations. In the university system a reasonable premise is the fixed number of places in universities assured by the State, positions that are to be won via competitions (surely, the system was always more or less corrupted). There are not permanent positions, neither in education nor in research. This permits to get rid of people who do not keep their promises.

The university education and research system was not efficient in the long term because of its centralization and lack of resources from industry. Nevertheless, the system was less centralized than, say, in France, and gave to universities and research institutes enough power to assure high efficiency—at least in theory. The basic principle that a good research in many fields (mathematics, theoretical physics, computer science, etc.) is done by strong individuals or small strong groups of 2-3 persons was partially satisfied by the system. The weak point was the absence of diverse and competitive sources of direct support of research. The latter point is strong in the USA where any researcher can apply to NSF grants directly and where the industry thinks competitively and invests in research. These incentives were absent in the USSR and are absent, in France, and consequences are evident. This defect is aggravated in all West European countries by protectionism. An USA university can invite a brilliant foreigner directly on a permanent position. In Europe this is not the case. E. g. in France everyone must go through the procedure defined by the Ministry (whose essential activity is to change every year the dates and minor bureaucratic procedures, informing the public about all these changes in the last moment). In addition to protectionism, one ought to face the local protectionism which does its best to avoid brilliant people in order not to worry local people incapable to compete at an international level. An international rating of researchers and research groups working in one particular subject, as in chess, could be an important indicator. Why not take it into consideration?

On the notion of algorithm and computational problem

> "Nobody can embrace non-embraceable." Koz'ma Prutkov, *Fruits of Meditation*, 1854-1860. Free translation from Russian.
>
> "The number of posed problems, as well as of obtained results, goes to infinity. Thus, the value of an average one goes to zero." Slissenko's Law.

The failure to get non trivial lower bounds for concrete problems for general enough notions of algorithms (RAM, Boolean circuit) leads to a conjecture that the notions of algorithm and problem are too wide. If we compare any diagonal construction (in a way, all of them are similar) and any algorithm that is of practical importance we can easily see that their computations have completely different structures. In diagonal constructions no "geometry" is visible: they are highly "discontinuous". In practical algorithms, on the contrary, one can intuitively feel something like piecewisely smooth geometry. Such feelings have not been formalized until now, though algebraic decision trees may give some analogy. For the latter there are non trivial lower bounds. If one limits oneself to algorithms whose computations are "piecewise smooth" one can hope to get sufficiently persuasive lower bounds, though not for all possible algorithms, but for algorithms that we can really construct. That is as in mathematics, the general notion of function is not productive, even that of continuous function is too general. Most part of mathematical theories deal with sufficiently smooth operators and varieties. In case of algorithms the "smoothness" is to be understood rather largely, to include, for example, finite automata. To go on further in this direction one may try to look at sets of computations of different algorithms solving the same problem trying to explain their different efficiencies in terms of the structure of their sets of computations. This kind of work, which is in the spirit of natural science, may help.

There is another general question related to the improvement of the value of theory. If we look at known hard problems one can notice that they are over-generalized. But the situation is not too straightforward. Sometimes, as in cryptography, we seek really hard problems with some particular properties. These problems may be of any nature, in particular diagonal constructions are not prohibited (though for the moment they are not productive). Let us consider problems that arise from an activity where we do our best to economize efforts and resources. As an example, compare formulas describing routine properties concerning program specifications, say of timed systems, and formulas used in the proof of undecidability of the corresponding logic. They are quite different *semantically*, that is, the set of models of specification formulas has many structural restrictions that sometimes can be formulated rigorously and can imply decidability. One can conjecture that a more profound analysis of such restrictions may lead even to feasibility of the validity.

This situation of simplicity of concrete instances is more clear for Boolean functions. The known concrete Boolean functions are not as complicated as a random Boolean function. Why we never arrive at really hard instances from natural problems? The philosophical conjecture is that instances are generated on the basis of many laws of nature that impose some smoothness in structure; however, this smoothness is in the semantics and is not visible on the level of the syntax.

Some time ago strong efforts could have noticed in attempts to prove the logical independence of, say $P \neq NP$ conjecture (in the sense that neither it nor its negation is provable in arithmetic). Such a result may be of purely mathematical interest, but not of computer science interest. The thesis is that any "computationally" reasonable problem can be and must be expressible in the form $\forall W M(W)$, where W is a word over some alphabet, and $M(W)$ is a formula with bounded quantifiers built from easily decidable predicates. For such formulas logical independence implies its validity, so is not of much interest for computer science. It is not always the case with known problems, but why $P \neq NP$ conjecture is well formulated? Let's look at it.

Remarks on logical variations of P vs NP

> "An open problem, however important it seemed at its birth, finally grows old and dies."

Here are some remarks on the logical form of SAT $\notin \boldsymbol{P}$. For formulas of the form $\forall X F(X)$, where $F(X)$ is bounded quantifier, independence from formal arithmetic implies its validity: if $\forall X F(X)$ is not valid, then for some X_0 the formula $\neg F(X_0)$ must be valid, and arithmetic is complete with respect to quantifier bounded closed formulas. Thus $\exists X \neg F(X)$ is provable that contradicts to the independence.

Consider SAT based version of $\boldsymbol{P} \neq \boldsymbol{NP}$:

$$\forall\alpha(\alpha \text{ is an algorithm recognizing SAT} \rightarrow \text{time complexity of } \alpha \tag{2}$$

$$\text{is superpolynomial}) \tag{3}$$

To fix some notation rewrite it as

$$\forall\alpha(C(\alpha) \rightarrow L(\alpha)), \tag{4}$$

or

$$\forall\alpha(\forall x C'(\alpha, x) \rightarrow \forall k \exists y L'(\alpha, k, y)) \tag{5}$$

where

- $C(\alpha) =_{df} \forall x C'(\alpha, x)$,
- $\forall x C'(\alpha, x) =_{df} \forall x \exists m \exists v (T(\alpha, x, v, m) \,\&\, (v = 0 \leftrightarrow x \in SAT))$,
- T is the Kleene predicate

$$T(\alpha, x, v, m) \leftrightarrow \text{“}\alpha \text{ finishes its work on } x \text{ after exactly } m \text{ steps and}$$

$$\text{outputs } v\text{”},$$

- $L(\alpha) =_{df} \forall k \exists y L'(\alpha, y, k)$,
- $\forall k \exists y L'(\alpha, k, y) =_{df} \forall k \exists y (time_\alpha(y) > |y|^k)$,
- $time_\alpha(y)$ is time complexity of α for argument y.

Clearly, the predicate T is polytime, and the existential quantifiers in $C'(\alpha, x)$ can be bounded: $\exists m \leq 2^{\mathcal{O}(|x|)}$ $\exists v \in \{0, 1\}$; the predicate $time_\alpha(y) > |y|^k$ is polytime; the formula $x \in SAT$ is bounded quantifier.

Thus all unbounded quantifiers are explicitly given in the representation (5). One can transform this formula into the prenex form

$$\forall\alpha\forall k \exists x \exists y (C'(\alpha) \rightarrow L'(\alpha, k, y)) \tag{6}$$

or any other. But there are no visible arguments to bound the quantifier on x, as it is actually a universal quantifier in the premise assertion of correctness of α.

Restrict ourselves to algorithms α whose correctness is provable in a formal system $\mathcal{F}$. To define such algorithms let us introduce the following notation:

$$C_{\mathcal{F}}(\alpha) =_{df} \exists D(D \text{ is a proof in } \mathcal{F} \text{ of } C(\alpha))$$
$$=_{df} \exists D\, PROOF_{\mathcal{F}}(D, \alpha).$$

As a restricted version of (5) we take:

$$\forall \alpha (C_{\mathcal{F}}(\alpha) \to \forall k \exists y L'(\alpha, k, y)). \tag{7}$$

Now the prenex form

$$\forall \alpha \forall D \forall k \exists y (PROOF_{\mathcal{F}}(D, \alpha) \to L'(\alpha, k, y)). \tag{8}$$

of (7) is more tractable.

Restricting the existential quantifier $\exists y$ in (8) may preserve the "physical meaning" of the initial assertion. Thus, the formula

$$\forall \alpha \forall D \forall k \exists y \leq \varphi(\alpha, D, k)(PROOF_{\mathcal{F}}(D, \alpha) \to L'(\alpha, k, y)) \tag{9}$$

with φ fast growing but tractable in the formal system, say hyperhyper exponential, can be considered as a satisfactory version of $\boldsymbol{P} \neq \boldsymbol{NP}$, though it is weaker than the initial formulation (4). This version is of the form for which independence implies validity.

So, there could be two sources of independence which do not imply validity:

(1) algorithms with not provable correctness;

(2) no sequence $\{y_k\}$ providing high complexity, i. e. such that $time_\alpha(y_k) > |y_k|^k$, is representable in a provable way, e. g. every such a sequence is very sparse: $length(y_k)$ grows faster than any provable function.

On computer aided theorem proving versus verification

It does not work but it is dear, and I will keep it on.

The problem of program verification, which exists since long time as a research domain, was not a favourite child of computer science. As a new-comer I could remark that one reason lies in using languages and settings not related to real software engineering and weak theoretical level. Even now many people working in the field do not know enough about industrial programming. Usually verification people invent their language to write program specification motivated by their inner considerations not taking into account the real practice, and give no theoretical analysis. What specifications are considered?

The standard idealized description of software engineering process [Som92] starts from an informal conception phase that must give a requirements definition. It always exists in practice though, as a rule, is vague and incomplete.

Usually program specification is being extracted from the requirements definition without analyzing it, and the later verification at best touches a small part of the requirements definition. On the other hand, it is well known that the most dangerous and difficult to detect errors are either in the requirements definition or in the process of extraction the first program specification that is later refined to an executable program. (Our first experience with this area revealed an error in the algorithm of the founders of the widely analyzed benchmark problem, the Generalized Railroad Crossing, see [BS97].) These phases of program development are verified by model checking in a very restricted and incomplete way. So we have a large gap between requirements definition and program specification, considerable gaps on the way to program in a standard programming language compilable to a executable code, and in addition, a hard to patch gap between the compilable program and the executable program that is of final interest. To prove anything about the latter one must have a formal semantics of the programming language and operating environment. Is it realistic to have all that? I guess no. And even if we have it, is it realistic to verify all? Clearly, not. At least not now. Hence, testing will be crucial for a very long time, and industry will rely more on testing than on verification, though testing is a particular case of the latter. But testing, as it exits, cannot prove the validity of programs. Thus, we have to develop formal methods.

As for testing, practically there is no theory on the subject. By theory I mean provable positive properties of particular feasible testing procedures, that is, something saying that if a given program and tests have these or those properties, then the result of testing guarantees that the program is correct (with respect to a given specification) with this or that probability. The mentioned probability is hard to define in a way interesting for applications if we do not consider only toy examples as linear functions or matrix multiplication for which even a low level formal verification is feasible.

The domain of formal verification looks as a good application field to interactive provers oriented on semantical considerations controlling the inference search. Unfortunately, such provers do not exist, as far as I know. The known one are syntactically controlled as early provers of the 60s. But this can be remedied. A flexible notion to specify programs, which permits to easily change the level of abstraction and is close to logic, are Gurevich machines [Gur95]. Their basic idea was known: to represent a current global state as an interpretation of the appropriate vocabulary. Yu. Gurevich transformed this idea into a system of notions and showed how to apply them. This notion seems of value for problems of verification oriented to logic. It is far from being a tool. But a tool similar in spirit, though more limited from the point of view of logic does exists, that is J.-R. Abrial's **B** [Abr96]. Maybe it is not so hopeless, the verification?

"Nothing destroys the memory so much as writing memoirs."

Acknowledgments. The present survey is due to the iron grip of B. A. Trakhtenbrot. C. Calude helped me to start it. My colleagues and friends N. A. Shanin, R. Freidson, D. Grigoriev, G. . Mints, V. Orevkov, G. Tseitin, I. Zaslavski,

H. Maranjan, Yu. Matiyasevich, V. Kreinovich, M. Gelfond, V. Livshits, A. Beltiukov, A. Barvinok, D. Burago, N. Vorobjov Jr., S. Evdokimov, I. Ponomarenko provided me with different materials, more than I could use. I am thankful to all of them. The contents is the responsability of my subconsciousness.

References

[Abr96] J.-R. Abrial. *The B-book: Assigning programs to meanings.* Cambridge University Press, 1996.

[AVS81] G.M. Adel'son-Vel'ski and A. Slisenko. What can we do with problems of exhaustive search? In A. Ershov and D. Knuth, editors, *Algorithms in Modern Mathematics. Proc. Intern. Symp., Urgench, Uzbek SSR, September 16-22, 1979*, pages 315–342. Springer Verlag, 1981. Lect. Notes in Comput. Sci, vol. 122.

[Bar65] J. Barzdins. Complexity of recognizing the symmetry on turing machines. In A. A. Lyapunov, editor, *Problems of Cybernetics*, pages 245–248. Nauka, Moscow, 1965. (Russian.).

[Bar92] A. Barvinok. Computing the volume, counting integral points, and exponential sums. In *Proc. of the 8th Annu. ACM Symp. on Computational Geometry*, pages 161–170, 1992.

[Bel79] A. Beltiukov. Machine description and the hierarchy of initial grzegorczyk classes. In Yu. Matiyasevich and A. Slissenko, editors, *Zapiski Nauchnykh Seminarov LOMI. Series: Studies in Constructive Mathematics and Mathematical Logic, 8.*, volume 88, pages 30–46, 1979. (Russian. English: *J. of Soviet Mathematics*, 20(1982).).

[BGG97] E. Börger, E. Grädel, and Yu. Gurevich. *The Classical Decision Problem.* Springer-Verlag, 1997.

[BGM82] L. Babai, D. Grigoriev, and D. Mount. Isomorphism of graphs with bounded eigenvalue multiplicity. In *Proc. of the 14th ACM Symp. Theory of Comput.*, pages 310–324, 1982.

[Blu67] M. Blum. A machine-independent theory of the complexity of recursive functions. *J. Assoc. Comput. Mach.*, 14(2):322–336, 1967.

[BS97] D. Beauquier and A. Slissenko. On semantics of algorithms with continuous time. Technical Report 97–15, University Paris 12, Department of Informatics, 1997. Revised version.

[Bur92] D. Burago. A lower bound for the split-find problem for a pointer machines. *J. of the St. Petersburg Math. Soc.*, 159:23–38, 1992. Amer. Math. Soc. Trans. (2).

[CG82] A. Chistov and D. Grigoriev. Polynomial–time factoring of the multivariable polynomials over a global field. Technical Report E-5-82, Leningrad Division of Steklov Mathematical Institute, Academy of Sciences of the USSR, 1982. 39 p.

[CG83] A. Chistov and D. Grigoriev. Subexponential–time solving systems of algebraic equations. i, ii. Technical Report E-9-83, E-10-83, Leningrad Division of Steklov Mathematical Institute, Academy of Sciences of the USSR, 1983. 119 p.

[Clo97] P. Clote. Non-deterministic stack register machines. *Theor. Comput. Sci.*, 178(1–2):37–76, 1997.

[Edi73] Editorial. On the work of e. i. nechiporuk. In A. A. Lyapunov, editor, *Problems of Cybernetics, vol. 26*, pages 9–18. Nauka, 1973. (Russian).

[EK81] A. Ershov and D. Knuth, editors. *Algorithms in Modern Mathematics. Proc. Intern. Symp., Urgench, Uzbek SSR, September 16-22, 1979.* Springer-Verlag, 1981. Lect. Notes Comput. Sci., vol. 122.

[Evd88] S. Evdokimov. Factoring a solvable polynomial over a finite field and the Generalized Riemann Hypothesis. In D. Grigoriev, editor, *Zapiski Nauchnykh Seminarov LOMI. Series: Theory of Computational Complexity. 5*, volume 176, pages 104–117, 1988. Russian. English: *J. of Soviet Mathematics.*

[FP74] M. Fischer and M. Paterson. String-matching and other products. In *Complexity of Computations. (SIAM-AMS Proc., vol.7)*, pages 113–125, Providence, R. I., 1974.

[Fre70] R. Freidson. A characterization of the complexity of recursive predicates. In *Proc. Steklov Inst. of Mathematics*, volume 113, pages 78–101, 1970. Russian. English: *Proc. Steklov Inst. Math.*, 113(1970):91-117, Amer. Math. Soc.

[Fre96] R. Freidson. Algorithmics of np-hard problems. In *Amer. Math. Soc. Trans. (2) Vol. 178*, pages 1–4. AMS, Providence, R. I., 1996.

[Gal76] Z. Galil. Real-time algorithms for string-matching and palindrome recognition. In *Proc. 8th Annu. ACM Symp. on Theory of Computing*, pages 161–173, 1976.

[Gen34] G. Gentzen. Untersuchungen über das logische schließen. 1, 2. *Mathematische Zeitschrift*, 39:176–210, 405–443, 1934.

[Gri76] D. Grigoriev. An application of separability and independence notions for proving lower bounds of circuit complexity. In Yu. Matiyasevich and A. Slissenko, editors, *Zapiski Nauchnykh Seminarov LOMI. Series: Studies in Constructive Mathematics and Mathematical Logic*, volume 60, pages 38–48, 1976. Russian. English: *J. of Soviet Mathematics*, 14:5(1980).

[Gri77] D. Grigoriev. Embedding theorems for Turing machines of different dimensions and Kolmogorov's algorithms. *Doklady Akad. Nauk SSSR*, 234(1):15–18, 1977. Russian. English: *Soviet Math. Dokl.*, 18:3(1977):588–592.

[Gri86] D. Grigoriev. Complexity of deciding the first–order theory of algebraically closed fields. *Izvestia Akad. Nauk SSSR*, 50(5):1106–1120, 1986. Russian. English: *Math. USSR Izvestia*, 29:2(1987):459–475.

[Gri88] D. Grigoriev. Complexity of deciding tarski algebra. *J. Symp. Comput.*, 5:65–108, 1988.

[Gri89] D. Grigoriev. Complexity of quantifier elimination in the theory of ordinary differential equations. *Lect. Notes in Comput. Sci*, 378:11–25, 1989.

[GS82] D. Grigoriev and A. Slissenko, editors. *Theory of Computational Complexity 1.* Zapiski Nauchnykh Seminarov LOMI. Vol.118. Nauka, Leningrad, 1982. Russian. English: *J. of Soviet Mathematics*, 29(1985).

[Gur95] Y. Gurevich. Evolving algebra 1993: Lipari guide. In E. Börger, editor, *Specification and Validation Methods*, pages 9–93. Oxford University Press, 1995.

[GVJ85] D. Grigoriev and N. Vorobjov Jr. Finding real solutions of systems of algebraic inequalities in subexponential time. *Doklady Akad. Nauk SSSR*, 283(6):1561–1565, 1985. Russian. English: Soviet Math. Doklady, 32:1(1985):316–320.

[HB34] D. Hilbert and P. Bernays. *Grundlagen der Mathematik*, volume 1. Berlin, 1934.

[HB39] D. Hilbert and P. Bernays. *Grundlagen der Mathematik*, volume 2. Berlin, 1939.

[Hen66] F. Hennie. On-line Turing machine computations. *IEEE Trans. Electron. Computers*, EC–15(1):35–44, 1966.

[HKSS91] J. Heintz, T. Krick, A. Slissenko, and P. Solernó. Search for shortest path around semialgebraic obstacles in the plane. In D. Grigoriev, editor, *Zapiski Nauchnykh Seminarov LOMI. Series: Theory of Computational Complexity, 5*, volume 192, pages 163–173, 1991. Russian. English: *J. of Math. Sciences*, 70:4(1994):1944–1949.

[Kle45] S. Kleene. On the interpretation of intuitionistic number theory. *J. Symbolic Logic*, 10:109–124, 1945.

[Kle60] S. Kleene. Realizability and Shanin's algorithm for the constructive deciphering of mathematical sentences. *Logique et Analyse. Nouvelle série*, 3(11–12):154–165, 1960.

[KM97] V. Krenovich and G. Mints, editors. *Problems of Reducing the Exhaustive Search.* AMS Translations, series 2, vol. 178. American Mathematical Society, 1997.

[Kol32] A. N. Kolmogorov. Zur deutung der intuitionistischen logik. *Mathematische Zeitschrift.*, 35(1):58–65, 1932.

[Kol65] A. N. Kolmogorov. Three approaches to defining the notion 'quantity of information'. *Problems of Information Transmission*, 1(1):3–28, 1965. (Russian.).

[Kos94] S. R. Kosaraju. Real-time suffix-tree construction. In *Proc. 26th ACM Annu. ACM Symp. on Theory of Comput.*, 1994.

[KR98] B. Kashin and A. Razborov. Improved lower bounds on the rigidity of hadamard matrices, 1998. To appear. Russian. Translated into English on a regular basis.

[Kre97] V. Kreinovich. S. Maslov iterative method: 15 years later (freedom of choice, neural networks, numerical optimization, uncertainty reasoning, and chemical computing). In V. Krenovich and G. Mints, editors, *Problems of Reducing the Exhaustive Search*, AMS Translations, series 2, vol. 178, pages 175–189. American Mathematical Society, 1997.

[Kur59] A. G. et al Kurosh, editor. *Mathematics in the USSR for 40 years, 1917–1957*, volume 1–2. Fizmatgiz, Moscow, 1959. (Russian).

[KV65] S. K. Kleene and R. E. Vesley. *The Foundations of Intuitionistic Mathematics, Especially in Relation to Recursive Functions.* North-Holland, 1965.

[LLL82] A. Lenstra, H. Lenstra, and L. Lovasz. Factoring polynomials with rational coefficients. Technical Report Preprint 195/82, Math. Centrum Amsterdam IW, 1982.

[LST85] S. Lavrov, A. Slissenko, and G. Tseitin. Curriculum for informatics and system programming. project. *Microprocessor Devices and Systems*, 4:20–28, 1985. (Russian.).

[Mar54] A. A. Markov. *The Theory of Algorithms.* Proceeding of Steklov Math. Inst., vol.42. AN SSSR, 1954. (Russian).

[Mar56] A. A. Markov. On one principle of constructive mathematical logic. In *Proceeding of the 3d All-Union Mathematical Congress*, volume 2, pages 146–147. Fizmatgiz, 1956. (Russian).

[Mar57] A. A. Markov. On inversive complexity of systems of functions. *Doklady AN SSSR*, 116:917–919, 1957.

[Mar64] A. A. Markov. On normal algorithms computing boolean functions. *Doklady AN SSSR*, 146(5):1017–1020, 1964.

[Mas64] S. Maslov. Inverse method of establishing the deducibility in the classical predicate calculus. *Doklady AN SSSR*, 159(1):17–20, 1964. Russian. English: *Soviet Math. Doklady.*

[Mas81] S. Maslov. Iterative methods in intractable problems as a model of intuitive methods, 1981. (Russian).

[Mas83] S. Maslov. Asymmetry of cognitive mechanisms and its applications. In *Semiotics and Information Science, vol. 20*, pages 3–31. VINITI, Moscow, 1983. (Russian).

[Mas87] S. Maslov. *Theory of Deductive Systems and its Applications.* MIT Press, Cambridge, 1987.

[Min72a] G. E. Mints. Exact estimates of the provability of transfinite induction in the initial segments of arithmetic. *J. Soviet Mathematics*, 1:85–91, 1972.

[Min72b] G. E. Mints. Quantifier-free and one-quantifier systems. *J. Soviet Mathematics*, 1:71–84, 1972.

[Min91] G. E. Mints. Proof theory in the USSR 1925-1969. *J. Symb. Logic*, 52(2):385–424, 1991.

[Min92] G. E. Mints. Complexity of subclasses of the intuitionistic propositional calculus. *BIT*, 32(1):64–69, 1992.

[MMM$^+$80] S. Maslov, Yu. Matiyasevich, G. Mints, V. Orevkov, and A. Slissenko. Nikolai Aleksandrovich Shanin (On his sixtieth anniversary). *Uspekhi Matematicheskikh Nauk*, 35(2(212)):241–245, 1980. (Russian.).

[MMO71] S. Maslov, G. Mints, and V. Orevkov. Mechanical proof-search and the theory of logical deduction the USSR. *Revue Intern. de Philosophie*, 98(4):575–584, 1971.

[MMOS90] Yu. Matiyasevich, G. Mints, V. Orevkov, and A. Slissenko. Nikolai Aleksandrovich Shanin (On his sixtieth anniversary). *Uspekhi Matematicheskikh Nauk*, 45(1(271)):241–245, 1990. Russian. English: *Russian Math. Surveys*, 45:1(1990):239–240.

[MN88] A. A. Markov and N. M. Nagorny. *The Theory of Algorithms.* Mathematics and its Applications. Soviet series, 23. Dordrecht etc.: Kluwer Academic Publishers, 1988. The 2nd edition appeared in 1996 in Russian. It contains A. A. Markov's biography.

[MP97] J. Morris and V. Pratt. A linear pattern-matching algorithm. Technical Report TR-40, Computing Center, University of California, Berkeley, 1997.

[MT82] G. Mints and E. Tyugu. Justification of the structural synthesis of programs. *Sci. Comput. Progr.*, 2:215–240, 1982.

[NS64] N. M. Nagornyi and N. A. Shanin. Andrei Andreyevich Markov (To his sixtieth anniversary). *Uspekhi Matematicheskikh Nauk*, 19(3(117)):207–223, 1964.

[Ore64] V. Orevkov. On constructive mapping on circle into itself. In *Proc. Steklov Inst. of Mathematics*, volume 172, pages 437–461, 1964. Russian. English: *Proc. Steklov Inst. of Mathematics*, 100(1972):69–100, AMS Transl. (2).

[Ore93] V. P. Orevkov. *Complexity of proofs and their transformations in axiomatic theories.* AMS Translations of Mathematical Monographs, vol. 128, 153p. Amer. Math. Soc., 1993.

[Pon92] I. Ponomarenko. Polynomial-time algorithms for recognizing and isomorphism testing of cyclic tournaments. *Acta Appl. Math.*, 29:139–160, 1992.

[Rob64] J. Robinson. On automatic deduction. *Rice Univ. Studies*, 50(1):69–89, 1964.

[Sav98] J. Savage. *Models of Computations.* Addison-Wesley, 1998.

[SDM+83] N. Shanin, G. Davydov, S. Maslov, G. Mints, V. Orevkov, and A. Slissenko. An algorithm for machine search of a natural logical deduction in a propositional calculus. In *The Automation of Reasoning I. Classical Papers on Computational Logic 1957–1966*, pages 424–483. Springer Verlag, 1983. The Russian original appeared as a booklet, published by Nauka (1965, 39pp.).

[Sha87] N. Shanin. On finitary development of mathematical analysis on the base of euler's notion of function. In *The 8th Intern. Congress on Logic, Methodology and Philosophy of Science. Abstracts.*, pages 60–63. Moscow, 1987. vol. 1.

[Sli73] A. Slisenko. Recognizing a symmetry predicate by multihead turing machines with input. In V. Orevkov and N. Shanin, editors, *Proc. Steklov Inst. of Mathematics*, volume 129, pages 30–202, 1973. Russian. English: *Proc. Steklov Inst. of Mathematics*, 129(1976):25–208.

[Sli77] A. Slisenko. A simplified proof of real-time recognizability of palindromes on turing machines. In G. Mints and V. Orevkov, editors, *Zapiski Nauchnykh Seminarov LOMI. Series: Theoretical Applications of Methods of Mathematical Logic, 2.*, volume 68, pages 123–139, 1977. Russian. English: *J. of Soviet Mathematics*, 15:1(1981):68–77.

[Sli81a] A. Slissenko. Complexity problems of theory of computation. *Russian Mathematical Surveys*, 36(6):23–125, 1981.

[Sli81b] A. Slissenko. Detection of periodicities and string-matching in real time. In G. Mints and V. Orevkov, editors, *Zapiski Nauchnykh Seminarov LOMI. Series: Theoretical Applications of Methods of Mathematical Logic, 3.*, volume 105, pages 62–173, 1981. Russian. English: *J. of Soviet Mathematics*, 22:3(1983):1316–1386.

[Sli82] A. Slissenko. Context-free grammars as a tool for describing polynomial-time subclasses of hard problems. *Inform. Process. Lett.*, 14(2):52–56, 1982.

[Sli91] A. Slissenko. On measures of information quality of knowledge processing systems. *Information Sciences: An International Journal*, 57–58:389–402, 1991.

[Sli93] A. Slissenko. A view on recent years of research in theoretical computer science in the former Soviet Union. *RAIRO, Technique et science informatique*, 12(1):9–28, 1993.

[SM73] L. Stockmeyer and A. Meyer. Word problems requiring exponential time: preliminary report. In *Proc. 5th Annu. ACM Symp. on Theory of Comput.*, pages 1–9, Austin, Texas, 1973.

[Som92] I. Sommerville. *Software Engineering.* International Computer Science Series. Addison-Wesley, 1992. 4th edition.

[Tar96] V. Tarasov. *Inference control in expert systems based of explicit meta-rules of inference search.* PhD thesis, St. Petersburg Institute for Informatics and Automation of the Academy of Sciences of Russia, 1996. 132 p.

[Tra84] B. Trakhtenbrot. A survey of Russian approach to perebor (brute-force search) algorithms. *Annals of the History of Computing*, 6(4):384–400, 1984.

[Tra87] B. Trakhtenbrot. Selected developments in Soviet mathematical cybernetics (finite automata, combinatorial complexity algorithmic complexity). Technical Report 73/87, Delphic Associates Incorporated, 1987.

[Tse68] G. Tseitin. On the complexity of proofs in propositional calculus. In A. Slisenko, editor, *Zapiski Nauchnykh Seminarov LOMI, vol. 8*, pages 234–259, 1968. Russian. English: *Studies in Constructive Mathematics and Mathematical Logic*, Plenum Press.

[USP93] D. Ugolev, T. Sokolova, and I. Ponomarenko. On an algorithmic model of non-destructive chemical analyzer for complex biological and ecological systems. *Vestnik S.-Peterburgskogo Universiteta. Series 4.*, 4:49–60, 1993. Russian.

[VJ88] N. Vorobjov Jr. Deciding consistency of systems of polynomial in exponent inequalities in subexponential time. In D. Grigoriev, editor, *Zapiski Nauchnykh Seminarov LOMI. Series: Theory of Computational Complexity. 5*, volume 176, pages 7–40, 1988. Russian. English: *J. of Soviet Mathematics*, 59:3(1992):1322–1360.

[Weg87] I. Wegener. *The Complexity of Boolean Functions.* Wiley-Teubner, 1987.

[Yan59] S. A. Yanovskaya. Mathematical logic and foundations of mathematics. In A. G. et al Kurosh, editor, *Mathematics in the USSR over the forty years 1917-1957*, pages 13–120. Fizmatgiz, Moscow, 1959. (Russian).

[Zas55a] I. Zaslavsky. Invalidity of some theorems of classical analysis in constructive analysis. *Uspekhi Metem. Nauk*, 10(4(66)):209–210, 1955. Russian.

[Zas55b] I. Zaslavsky. Some criteria of compactness of metric and normed spaces. *Doklady Acad. Nauk SSSR*, 103(6):953–956, 1955. Russian.

Boris Trakhtenbrot

Professor Boris Trakhtenbrot was born on 20 February 1921 in Brichevo (now Moldova). Graduated from the Chernovtsy (Ukraine) University (1947). Ph.D. studies in mathematical logic and computability at the Institute of Mathematics of the Ukrainian Academy of Sciences in Kiev (1947-1950); Ph.D. degree in 1950 and the (soviet) Doctor of Sciences degree in 1962. From 1950-1960 he held positions at the Pedagogical and Polytechnical Institutes of Penza (Russia). From 1960 until 1980, he conducted research at the Mathematical Institute of the USSR Academy of Sciences' Siberian Branch, and lectured at the Novosibirsk University. Since 1981, after the emigration to Israel, he is professor of computer science at Tel-Aviv University. At the same time he visited and collborated with many Western Universities and Research Centers. Retired 1991.

Professor B. Trakhtenbrot enjoyed a variety of research interests in mathematical logic, computability, automata, complexity, semantics, concurrency. He published about one hundred papers and four books (the dates refer to Russian editions): *Introduction to the Theory of Finite Automata* (1962, co-authored with N. E. Kobrinski), *Algorithms and Computing Machines* (1960, and extended version in 1974), *Complexity of Algorithms and Computations* (1967), *Finite Automata: Behavior and Synthesis* (1970, co-authored with J. Barzdin).

He married Berta I. Rabinovich in 1947. They have two sons, Mark and Yossef, and five grandchildren.

From Logic to Theoretical Computer Science

Foreword

In October 1997, whilst touching up this text, exactly 50 years had past since I was accepted for graduate studies under P. S. Novikov. I started then study and research in logic and computability, which developed, as time will show, into research in Theoretical Computer Science (TCS).

After my emigration from the Soviet Union (December 1980) I was encouraged by colleagues to experience the genre of memoirs. That is how appeared [T84, T85], and more recently [T97], conceived as contributions to the history of TCS in the SU. The present paper is intended as a more intimate perspective on my research and teaching experience. That is mainly an account of how my interests shifted from classical logic and computability to TCS, notably to Automata and Computational Complexity. Most of these reminiscences, recounting especially the scientific, ideological and human environment of those years (roughly, 1945-67), were presented earlier at a Symposium (June 1991) on the occasion of my retirement. Occasionally, I will quote from [T84, T85, T97], or will refer to them.

Before starting the main narrative I would like to recall some important circumstances which characterized those years.

First of all, the postwar period was a time of ground-breaking scientific developments in Computability, Information Theory, Computers. That is widely known and need no comments. The subjects were young and so were their founders. It is amazing that at that time the giants Church, Kleene, Turing, von Neumann, were only in their thirties and forties!

Now, about the specific background in the Soviet Union.

The genealogical tree of TCS in the SU contains three major branches leading from A. N. Kolmogorov, A. A. Markov and P. S. Novikov. In those troublesome times these famous mathematicians also had the reputation of men with high moral and democratic principles. Their scientific interests, authority and philosophies, for several generations, influenced the development of mathematical logic, computability, and subsequently TCS in the SU.

Whereas Markov and Kolmogorov contributed directly to TCS, Novikov's involvement occurred through his strong influence on his disciples and collaborators. The most prominent of them – A. A. Lyapunov (1911-1974) – became a widely recognized leader of "Theoretical Cybernetics" – the term which covered at that time most of what is considered today to belong to TCS.

As a matter of fact, for many offsprings of those three branches, including myself, the perception of TCS was as of some kind of applied logic, whose conceptual sources belong to the theoretical core of mathematical logic. The affiliation with logic was evident at the All-Union Mathematical Congress (Moscow, 1956), where theoretical cybernetics was included in the section of mathematical logic. Other examples: the books on Automata [KT62,G62] appeared in the series "Mathematical Logic and Foundations of Mathematics"; also my first papers on Computational Complexity were published in Anatoly I. Maltsev's journal "Algebra and Logic".

The early steps in TCS coincided with attacks of the official establishment on various scientific trends and their developers. In particular, cybernetics was labeled a "pseudo-science", and mathematical logic – a "bourgeois idealistic distortion". That was the last stage of the Stalin era with persecution and victimization of "idealists", "cosmopolites", etc. The survival and the long overdue recognition of Mathematical Logic and Cybernetics is in many respects indebted to Lyapunov, Markov, Novikov, Kolmogorov and S. A. Yanovskaya. But even after that, academic controversies often prompted such bureaucratic repression as the prevention of publications and the denial of degrees. Difficulties with publications also happened because of the exactingness and selfcriticism of the authors and/or their mentors, or because the community was far from prepared to appreciate them. I told about that in [T84] and [T85].

Above, the emphasis was on the Soviet side; now, some remarks on the international context in which research in TCS was conducted in the SU.

The chronology of events reveals that quite a number of ideas and results in TCS appeared in the SU parallel to, independent of, and sometimes prior to, similar developments in the West. This parallelism is easy to explain by the fact that these were natural ideas occurring at the right time. In particular, that is how comprehensive theories of automata and of computational complexity emerged in the 50's-60's; I will elaborate on this subject in the next sections. But for a variety of reasons, even in those cases where identical or similar results were obtained independently, the initial motivation, the assessment of the results and their impact on the development and developers of TCS did not necessarily coincide. In particular, in the SU specific interest in complexity theory was aroused by discussions on the essence of brute force algorithms (*perebor* - in Russian). However, despite this difference in emphasis from the motivating concerns of the American researchers, after a few years these approaches virtually converged.

In the past, the priority of Russian and Soviet science was constantly propounded in Soviet official circles and media. This unrestrained boasting was cause for ironic comments in the West and for self-irony at home. But, as a matter of fact, the West was often unaware of developments in the SU, and some of them went almost entirely unnoticed. To some extent this was a consequence of the isolation imposed by language barriers and socio-political forces. In particular, travels abroad were a rare privilege, especially to the "capitalist" countries. My first trip abroad, for example, took place in 1967, but visits to the West became possible only in 1981 after my emigration to Israel.

Against this unfavorable background it is worth mentioning also the encouraging events and phenomena, which eased the isolation.

The International Mathematical Congress in Moscow (1966) was attended by the founders of our subject, namely, Church, Kleene, Curry, Tarski and other celebrities. It was an unforgettable and moving experience to have first-hand contact with these legendary characters. Later, Andrey P. Ershov (1931-1988) managed to organize a series of International Symposia on "Theoretical Programming", attended also by people from the West. For many years, A. Meyer used to regularly send me proceedings of the main TCS symposia, a way to somehow compensate for the meetings my colleagues and myself were prevented to attend. All this reinforced our sense of belonging to the international TCS community.

Early Days

I was born in Brichevo, a village in Northern Bessarabia (now Moldova). Though my birth place has nothing to do with my career or with other events I am going to write about, let me begin with the following quotation: "Brichevka a Jewish agricultural settlement, founded in 1836. According to the general (1897) census of the population - 1644 inhabitants, 140 houses ..." (From Vol. 5 of "The Jewish Encyclopedia", St. Petersburg, 1912. Translated from Russian).

Among the first settlers were Eli and Sarah Helman, the grandparents of my maternal grandfather. World War 2 brought about the collapse of Brichevo (or Brichevka). The great majority of the population did not manage to flee and were deported to the notorious Transnistria camps; only a small number survived and they dispersed over countries and continents. For years I used "Brichevo" as a reliable password: easy for me to remember, apparently impossible for outsiders to guess, and still a way to retain the memory of a vanished community.

After completing of elementary school in Brichevo I attended high school in the neighboring towns of Belts and Soroka, where I was fortunate to have very good teachers of mathematics. My success in learning, and especially in mathematics, was echoed by the benevolence of the teachers and the indulgence of my fellow pupils. The latter was even more important to me, since it to some degree compensated for the discomfort and awkwardness caused by my poor vision.

In 1940 I enrolled in the Faculty of Physics and Mathematics of the newly-established Moldavian Pedagogical Institute in Kishinev. The curriculum covered a standard spectrum of teachers' training topics. In particular, mathematical courses presented basics in Calculus, Linear Algebra and Algebra of polynomials, Analytical Geometry, Projective Geometry, Foundations of Geometry (including Lobachevski Geometry), Elements of Set Theory and Number Theory.

On June 22, 1941, Kishinev (in particular the close neighbourhood of our campus) was bombed by German air forces. In early July, I managed to escape from the burning city. Because of vision problems I was released from military service and, after many mishaps, arrived as a refugee in Chkalov (now Orenburg) on the Ural River. Here, I enrolled in the local pedagogic institute. A year later we moved to Buguruslan in the Chkalov region, where the Kishinev Institute was evacuated to in order to train personnel for the forthcoming return home as soon as our region would be liberated. Almost all lecturers were former high school teachers – skilled people whose interests lay in the pedagogic aspects of mathematics and physics. (There were no recipients of academic degrees among them, but one of the instructors in the Chkalov institute bore the impressive name Platon Filosofov). Nikolai S. Titov, a former Ph. D. student of the Moscow University, who happened to flee to Buguruslan, lectured on Set Theory. I was deeply impressed by the beauty and novelty of this theory. Unfortunately, this was only a transient episode in those hard and anxious days. Actually, during the war years 1941-1944, my studies were irregular, being combined with employment in a felt boot factory, a storehouse and, finally, in the Kuybyshev-Buguruslan Gas Trust.

In August 1944 the institute was evacuated to Kishinev and I returned to my native region for a position in the Belts college to train elementary school teachers. Only a year later I took my final examinations and qualified as a high school mathematics teacher. That was my mathematical and professional background in September 1945 when (already at the age of 24 and a half) I decided to take a chance and seriously study mathematics.

Chernovtsy

I enrolled at the University of Chernovtsy (Ukraine) to achieve the equivalent of a master's degree in mathematics. In that first postwar year the university was involved in the difficult process of restoration. Since my prior education covered only some vague mathematical-pedagogical curriculum with examinations partially passed without having attended lectures, I did not know much to start off with. But there were only a few students and the enrollment policy of the administration was quite liberal. There were also only a few academic staff in our Faculty of Physics and Mathematics and soon I became associated with Alexander A. Bobrov, a prominent character on the general background. A. A. (b. 1912), who completed his Ph. D. thesis in 1938 under Kolmogorov, gave an original course in probabilities. The distinguishing quality was not so much in the content of the course as in his style (completely new to me) of teaching and of involving the audience. A. A. did not seem to be strongly committed to his previously prepared lectures; during class he would try to examine new ideas and to improvise alternative proofs. As such trials did not always succeed he would not hesitate to there and then loudly criticize himself and appeal to the audience for collaboration. This challenging style was even more striking in a seminar he held on Hausdorff's famous book on Set Theory, with the participation of both students and academic staff. Due to the "Bobrovian" atmosphere dominating the seminar, I started to relish the idea of research in this fascinating area. A. A. also helped me secure a job in the new founded departmental scientific library. My primary task was to take stock of the heaps of books and journals extracted earlier from basements and temporary shelters, and to organize them into some bibliographical service. I remember reverently holding volumes of the "Journal für reine und angewandte Mathematik" with authentic papers and pictures of Weierstrass and other celebrities. As I later understood the mathematical library was exclusively complete, and, as a matter of fact, disposed of all the important journals before WW2. As there was only a handful of graduate students it soon turned out that my library was not in much demand – in truth, for days there were no visitors at all; so most of the time I shared the roles of supplier and user of the library services. Through self study I mastered a significant amount of literature and reached some scientific maturity. I soon identified "Fundamenta Mathematicae" to be the journal closest to my interests in descriptive set theory. All the volumes starting with the first issue dated 1921 were on my table and I would greedily peruse them.

After considering some esoteric species of ordered sets I turned to the study of delta-sigma operations, a topic promoted by Andrei N. Kolmogorov and also tackled in "Fundamenta". At this stage Bobrov decided that it was the right time to bring me together with the appropriate experts and why not with Kolmogorov himself! In the winter of 1946 Kolmogorov was expected to visit Boris V. Gnedenko at the Lvov University. So far so good, except that at the last moment Kolmogorov canceled his visit. Gnedenko did his best to compensate for that annoying failure. He showed me exclusive consideration, invited me to lunch at his home and attentively inquired about all my circumstances. It was the first time that I had talked to a full professor and I felt somehow shy in his presence and in the splendor of his dwelling. B. V. listened to me patiently and, I guess, was impressed not as much by my achievements (which were quite modest, and after all, outside the field of his main interests) but by my enthusiastic affection for Descriptive Set Theory. Anyway he explained to me that for the time-being Kolmogorov had other research preferences and it would be very useful to contact Piotr S. Novikov and Alexei A. Lyapunov who, unlike Kolmogorov and other descendants of the famous Lusin set-theoretical school, were mostly still active in the field.

During this period I met Berta I. Rabinovich, who was to become my wife.

In the summer of 1946 I visited Moscow for the first time. Because it was vacation time and since no prior appointments had been set up, it was very difficult to get hold of people. Nevertheless I managed to see Kolmogorov for a short period at the university and to give him my notes on delta-sigma operations. He was in a great hurry, so we agreed to meet again in a couple of weeks on my way back home; unfortunately this did not work out. Novikov was also unreachable being somewhere in the countryside. I was more fortunate with A. A. Lyapunov in whose house I spent a wonderful evening of scientific discussions alternated with tea-drinking with the whole family. A. A. easily came to know my case and presented me with a deeper picture of the Moscow set-theoretical community with a stress on the current research done by Novikov and by himself. He offered to inform Novikov in detail about my case and suggested that I visit Moscow at a more appropriate time for further discussions.

My second trip to Moscow was scheduled for May 1947 on the very eve of my graduation from the Chernovtsy University, when, beyond pure mathematics, the question of my forthcoming (if any) Ph. D studies was on the agenda. All in all I had to stay in Moscow for at least a couple of weeks and that required appropriate logistics – a very nontrivial task at that time, in particular, because of the shortage of food and the troublesome train connections. Alas, at the first connection of the Lvov railroad station, local pickpockets managed to cut out the pocket with all my money. Despite this most regrettable incident, the trip ultimately turned out to be quite successful. The meetings with Novikov were very instructive and warm. And again, as in the case of the Lyapunovs, the atmosphere in the Novikov family was friendly and hospitable. Occasionally Novikov's wife, Ludmila V. Keldysh, a prominent researcher in set theory in her own right, as well as A. A. Lyapunov, would also participate in the conversations. Counterbalancing my interests and efforts towards descriptive set theory, Novikov called my attention to new developments I was not aware of in provincial Chernovtsy. He pointed to the path leading from a handful of hard set theoretical problems to modern concepts of mathematical logic and computability theory. He also offered his support and guidance should I agree to follow this path. I accepted Novikov's generous proposal although with a sense of regret about my past dreams about descriptive set theory.

Novikov held a permanent position at the Steklov Mathematical Institute of the USSR Academy. At that time departments of mathematical logic did not yet exist in the USSR but Novikov together with Sofia A. Yanovskaya had just started a research seminar "Mathematical Logic and Philosophical problems of Mathematics" in the Moscow University unofficially called The Bolshoy (great) Seminar. So, it was agreed that wherever other options might arise, Novikov would undertake my supervision and would do his best to overcome bureaucratic barriers.

Ph. D. Studies

In October 1947 I began my Ph. D. studies at the Kiev Mathematical Institute of the Ukrainian Academy of Sciences. The director of the institute, Mikhail A. Lavrentiev, approved my petition to specialize in mathematical logic under P. S. Novikov and agreed to grant me long-term scientific visits to Moscow where I would stay with my advisor. In Moscow the Bolshoy Seminar was then the main medium in which research and related activities in that area were conducted. In particular, it was the forum where mathematical logicians from the first post-war generation (mostly students of P. S. Novikov, S. A. Yanovskaya and A. N. Kolmogorov) joined the community, reported on their ongoing research, and gained primary approval of their theses; and that is also what happened to me.

The atmosphere dominating the meetings of the seminar was democratic and informal. Everybody, including the students, felt and behaved at ease without strong regulations and formal respect for rank. I was happy to acquire these habits and later to promote them at my own seminars.

Actually the seminar was the successor of the first seminar in the USSR for mathematical logic, which was founded by Ivan I. Zhegalkin (1869-1947). After Zhegalkin's death it became affiliated with the Department of History of Mathematical Sciences of the Moscow University, whose founder and head was

Yanovskaya. Its exceptional role in the development of mathematical logic in the USSR is a topic of its own and I will touch on it only very briefly.

The seminar usually engaged in a very broad spectrum of subjects from mathematical logic and its applications as well as from foundations and philosophy of mathematics. Here are some of the topics pursued by the senior participants: Novikov – consistency of set-theoretical principles; Yanovskaya – philosophy of mathematics and Marx's manuscripts; Dmitri A. Bochvar (a prominent chemist in his main research area) – logic and set-theoretical paradoxes; Victor I. Shestakov (professor of physics) – application of logic to the synthesis and analysis of circuits.

Among the junior participants of the seminar I kept in close contact with the three Alexanders. Alexander A. Zykov, also a Ph. D. student of Novikov, was at that time investigating the spectra of first order formulas. A. A. called my attention to Zhegalkin's decidability problem, which became the main topic of my Ph. D. thesis. He also initiated the correspondence with me sending lengthy letters to Kiev with scientific Moscow news. This epistolary communication, followed later by correspondence with Kuznetsov and Yablonski, was a precious support in that remote time.

Alexander V. Kuznetsov (1927-87) was the secretary of the Bolshoy Seminar and conducted regular and accurate records of all meetings, discussions and problems. For years he was an invaluable source of information. For health reasons A. V. did not even complete high school studies. As an autodidact in extremely difficult conditions, he became one of the most prominent soviet logicians. I had the good fortune to stay and to collaborate with him.

Alexander S. Esenin-Volpin was a Ph. D. student in topology under P. A. Alexandrov, but he early on became involved in logic and foundations of mathematics. A. S. became most widely known as an active fighter for human rights, and already in the late forties the KGB was keeping an eye on him. In the summer of 1949 we met in Chernovtsy, where he had secured a position after defending his thesis. Shortly thereafter he disappeared from Chernovtsy and we later learned that he had been deported to Karaganda (Kazakhstan). A couple of years later I received a letter from him through his mother. I anxiously opened the letter, fearful of what I was about to learn. The very beginning of the letter was characteristic of Esenin-Volpin's eccentric character – "Dear Boris, let f be a function ...".

During the years of my Ph. D. studies (1947-50) I actively (though not regularly) participated in the seminar meetings. Also the results which made up my thesis "The decidability problem for finite classes and finiteness definitions in set theory" were discussed there. S. A. Yanovskaya offered the official support of the Department in the future defense at the Kiev Institute of Mathematics; the other referees were A. N. Kolmogorov, A. A. Lyapunov and B. V. Gnedenko.

My thesis included the finite version of Church's Theorem about the undecidability of first order logic: the problem of whether a first order formula is valid in all finite models is, like the general validity problem, undecidable, but in a technically different way. The novelty was in the formalization of the algorithm concept. Namely, I realized that, in addition to the process of formal inference, the effective process of (finite) model checking could also be used as a universal approach to the formalization of the algorithm concept. This observation anticipated my future concern with constructive processes on finite models. Other results of the thesis which are seemingly less known, deal with the connection between deductive incompleteness and recursive inseparability.

In 1949 I proved the existence of pairs of recursively enumerable sets which are not separable by recursive sets. I subsequently learned that P. S. Novikov had already proved this, but, as usual, had not taken the trouble to publish what he considered to be quite a simple fact. (Note, that in 1951, Kleene who independently discovered this fact, published it as "a symmetric form of Gödel's theorem".) In the thesis I showed that the recursive inseparability phenomenon implies that no reasonably defined set theory can answer the question of whether two different finiteness definitions are equivalent. This incompleteness result was also announced in my short note presented by A. N. Kolmogorov to the "Doklady", but after Novikov's cool reaction to "inseparability", I refrained from explicitly mentioning that I had used these very techniques. Clearly, A. N. had forgotten that these techniques were in fact developed in the full text of my thesis and he later proposed the problem to his student Vladimir A. Uspenski. Here is a quotation from "History of Mathematics" [His70], p. 446: "A. N. Kolmogorov pointed to the possible connection between the deductive incompleteness of some formal systems and the concept of recursive inseparability (investigated also by Trakhtenbrot). V. A. Uspenski established (1953) results, which confirm this idea ..."

Those early years were a period of fierce struggle for the legitimacy and survival of mathematical logic in the USSR. Therefore the broad scope of the agendas on the Bolshoy seminar was beneficial not only for the scientific contacts between representatives of different trends, but also, in the face of ideological attacks, to consolidate an effective defense line and to avoid isolation and discredit of mathematical logic. For us, the junior participants of the seminar, it was also a time when we watched the tactics our mentors adopted to face or to prevent ideological attacks. Their polemics were not free of abundant quotations from official sources, controlled self-criticism and violent attacks on real and imaginary rivals.

It was disturbing then (and even more painful now) to read S. A. Yanovskaya's notorious prefaces to the 1947-1948 translations of Hilbert and Ackermann's "Principles of Mathematical Logic" and Tarski's "Introduction to Logic and the Methodology of Deductive Sciences" in which Russel was blamed as a warmonger and Tarski, as a militant bourgeois. Alas, such were the rules of the game and S. A. was not alone in that game. I remember the hostile criticism of Tarski's book by A. N. Kolmogorov (apparently at a meeting of the Moscow Mathematical Society): "Translating Tarski was a mistake, but translating Hilbert was the correct decision" he concluded. This was an attempt to grant some satisfaction to the attacking philosophers in order to at least save the translation of Hilbert-Ackermann's book. I should also mention that S. A.

was vulnerable – she was Jewish – a fact of which I was unaware for a long time. I learned about it in the summer of 1949 during Novikov's visit to Kiev. He told me then with indignation about official pressure on him "to dissociate from S. A. and other cosmopolitans".

However difficult the situation was, we - the students of that time - were not directly involved in the battle which we considered to be only a confrontation of titanics. As it turned out this impression was wrong.

Toward TCS

In December 1950 after the defense of my thesis, I moved to Penza, about 700 km SE of Moscow, for a position at the Belinski Pedagogical Institute.

At the beginning it was difficult for me to appropriately pattern my behaviour to the provincial atmosphere so different from the informal, democratic surroundings of the places I came from. These circumstances unfavorable influenced my relationships with some of the staff and students (in particular because of the constant pressure and quest for high marks). Because of this, though I like teaching, at the beginning, I did not derive satisfaction from it. [1] The situation was aggravated after a talk on mathematical logic I delivered to my fellow mathematicians. The aim of the talk entitled "The method of symbolic calculi in mathematics", was to explain the need and the use of exact definitions for the intuitive concepts "algorithm" and "deductive system". I was then accused of being "an idealist of Carnap-species". In that era of Stalin paranoia such accusations were extremely dangerous. At diverse stages of the ensuing developments, P. S. Novikov and A. A. Lyapunov (Steklov Mathematical Institute) and to some degree A. N. Kolmogorov and Alexander G. Kurosh (Moscow Mathematical Society) were all involved in my defense, and S. A. Yanovskaya put my case on the agenda of the Bolshoy Seminar. This story was told in [T97].

My health was undermined by permanent tension, fear and overwork (often more than 20 hours teaching weekly). It goes without saying that for about two years I was unable to dedicate enough time to research. It was in those circumstances that only the selfish care and support of my wife Berta saved me from collapse. I should also mention the beneficial and calming effect of the charming middle-Russian landscape which surrounded our dwelling. Cycling and skiing in the nearby forest compensated somewhat for our squalid housing. (Actually, until our move to Novosibirsk in 1961, we shared a communal flat, without water and heating facilities, with another family.)

But despite all those troubles I remember this period mainly for its happy ending. In the summer of 1992, 40 years after this story took place, Berta and I revisited those regions. The visit to Penza was especially nostalgic. Most of the

[1] Of course, I also had good students and one of them, Ilya Plamennov, was admitted through my recommendation to Ph. D. studies at the Moscow University. Later he became involved in classified research and was awarded the most prestigious Lenin Prize (1962).

participants of those events had already passed away. Only the recollections and of course the beautiful landscape remained.

Returning to the "Idealism" affair, the supportive messages I received from Moscow stressed the urgent need for a lucid exposition of the fundamentals of symbolic calculi and algorithms for a broad mathematical community. They insisted on the preparation of a survey paper on the topic, which "should be based on the positions of Marxism-Leninism and contain criticism of the foreign scientists-idealists". There was also an appeal to me to undertake this work which would demonstrate my philosophical ideological loyalty. Nevertheless I did not feel competent to engage in work which covered both a mathematical subject and official philosophical demands. These demands were permanently growing and changing; they could bewilder people far more experienced than myself. So it seemed reasonable to postpone the project until more favorable circumstances would allow separation of logic from official philosophy. Indeed, such a change in attitude took place gradually, in particular due to the growing and exciting awareness of computers.

In 1956 the journal *Mathematics in School* published my tutorial paper "Algorithms and automated problem solving". Its later revisions and extensions appeared as books which circulated widely in the USSR and abroad [T57]. (Throughout the years I was flattered to learn from many people, including prominent logicians and computer scientists, that this tutorial monograph was their own first reading on the topic as students and it greatly impressed them.)

Meanwhile I started a series of special courses and seminars over and above the official curriculum, for a group of strong students. These studies covered topics in logic, set theory and cybernetics, and were enthusiastically supported by the participants. Most of them were later employed in the Penza Computer Industry where Bashir I. Rameev, the designer of the "Ural" computers, was a prominent figure. Later, several moved with me to Novosibirsk. They all continued to attend the seminar after graduating from their studies. We would gather somewhere in the institute after a full day of work in Rameev's laboratories (the opposite end of town), inspired and happy to find ourselves together. Here is a typical scene – a late winter's night, frosty and snowy, and we are closing our meeting. It is time to disperse into the lonely darkness, and Valentina Molchanova, a most devoted participant of our seminar, has still to cross the frozen river on her long walk home.

The publication of my tutorial on algorithms and the above mentioned work with students increased my pedagogical visibility to such a degree that I was instructed by the Education Ministry, to compile the program of a course "Algorithms and Computers" for the pedagogical institutes. Moreover, the Ministry organized an all-Russian workshop in Penza, dedicated to this topic, with the participation of P. S. Novikov, A. I. Maltsev, and other important guests from Moscow.

In Penza there was a lack of scientific literature, not to mention normal contacts with well established scientific bodies. This obvious disadvantage was partially compensated by sporadic trips to Moscow for scientific contacts (and food supply), as well by correspondence with Kuznetsov, Sergey V. Yablonsky and Lyapunov.

I continued the work on recursive nonseparability and incompleteness of formal theories, started in the Ph. D. thesis. At the same time, I was attracted by Post's problem of whether all undecidable axiomatic systems are of the same degree of undecidability. This super-problem in Computability and Logic, with a specific flavor of descriptive set theory, was for a long time on the agenda of the Bolshoy Seminar. It inspired also my work on classification of recursive operators and reducibilities. Later, A. V. Kuznetsov joined me and we extended the investigation to partial recursive operators in the Baire space. These issues, reflected our growing interest in relativized algorithms (algorithms with oracles) and in set-descriptive aspects of computable operators. I worked then on a survey on this subject, but the (uncompleted) manuscript was never published. Nevertheless, the accumulated experience helped me later in the work on relativized computational complexity.

In 1956 Post's problem was solved independently by Albert A. Muchnik - a young student of P. S. Novikov - and by the American Richard Friedberg. Their solutions were very similar and involved the invention of the priority method of computability theory. At that point it became clear to me that I had exhausted my efforts and ambitions in this area, and, that I am willing to switch to what nowadays would be classified as "Theoretical Computer Science". From the early 50's this research was enthusiastically promoted by A. A. Lyapunov and S. V. Yablonski under the general rubric "Theoretical Cybernetics"; it covered switching theory, minimization of boolean functions, coding, automata, program schemes, etc. Their seminars at the Moscow University attracted many students and scholars,and soon became important centers of research in these new and exciting topics. I was happy to join the cybernetics community through correspondence and trips to Moscow. The general atmosphere within this fresh and energetic community was very friendly, and I benefited much from it. Many "theoretical cybernetists" started with a background in mathematical logic, computability and descriptive set theory and were considerably influenced by these traditions. So, no wonder that despite my new research interests in switching and automata theory, I considered myself (as did many others) to be a logician. My formal "conversion" to cybernetics happened on 9 January 1960 when Sergey L. Sobolev invited me to move to the Novosibirsk Akademgorodok and to join there the cybernetics department of the new Mathematical Institute.

Topics in combinational complexity were largely developed by the Yablonski school, which attributed exceptional significance to asymptotic laws governing synthesis of optimal control systems. The impetus for these works was provided by Shannon's seminal work on synthesis of circuits. However, the results of S. V. Yablonski, Oleg B. Lupanov and their followers surpassed all that was done in the West at that time as can be seen from Lupanov's survey [L65]. But focusing on asymptotic evaluations caused the oversight of other problems for which estimates up to a constant factor are still important.

A perebor algorithm, or perebor for short, is Russian for what is called in English a "brute force" or "exhaustive search" method. Work on the synthesis and minimization of boolean functions led to the realization of the role of perebor as a trivial optimization algorithm, followed by Yablonski's hypothesis of its nonelimination. In 1959 he published a theorem which he considered to be a proof of the hypothesis [Y59]. However his interpretation of his results was not universally convincing, – a presage of future controversies in the TCS community. I told this story in detail in [T84], and will touch it briefly in the next section.

In the winter of 1954, I was asked to translate into Russian a paper by A. Burks and J. Wright, two authors I didn't know earlier. Unexpectedly, this episode strongly influenced my "Cybernetical" tastes and provided the impetus to research in automata theory. A curious detail is that in [BW53], the authors don't even mention the term "automaton", and focus on Logical Nets as a mathematical model of physical circuits. Afterwards, "Logical Nets" would also appear in the titles of my papers in Automata Theory, even though the emphasis was not so much on circuitry, as on operators, languages and logical specifications.

The use of propositional logic, promoted independently by V. I. Shestakov and C. Shannon, turned out to be fruitful for combinational synthesis, because it suffices to precisely specify the behaviour of memoryless circuits. However, for the expression of temporal constraints one needs other, appropriate, specification tools, which would allow to handle synthesis at two stages: At the first, behavioral stage, an automaton is deemed constructed once we have finite tables defining its next-state and output functions, or, equivalently, its canonical equations. This serves as raw material for the next stage, namely for structural synthesis, in which the actual structure (circuit) of the automaton is designed. (Note, that in [AS56] Kleene does not yet clearly differentiate between the stages of behavioural and structural synthesis.) After some exercises in structural synthesis I focused on behavioral synthesis and began to collaborate with Nathan E. Kobrinsky, who at that time held a position in the Penza Polytechnical Institute. Our book "Introduction to the Theory of Finite Automata" [KT62] was conceived as a concord of pragmatics (N. E. 's contribution) and theory (summary of my results). The basic text was written in 1958, but the book was typeset in 1961, and distributed only in early 1962, when both of us had already left Penza.

Automata

Languages and Operators

The concept of a finite automaton has been in use since the 1930s to describe the growing automata now known as Turing machines. Paradoxically, though finite automata are conceptually simpler than Turing machines, they were not systematically studied until the fifties, if we discount the early work of McCulloch and Pitts. A considerable part of the collection "Automata Studies" [AS56] was already devoted to finite automata. Its prompt translation into Russian, marked the beginning of heightened interest by Soviet researchers in this field. In particular, the translation included a valuable appendix of Yuri T. Medvedev (one of the translators), which simplified and improved Kleene's results, and anticipated some of Rabin and Scott's techniques for nondeterministic automata.

As in the West, the initial period was characterized by absence of uniformity, confusion in terminology, and repetition of basically the same investigations with some slight variants. The subject appeared extremely attractive to many Soviet mathematicians, due to a fascination with automata terminology with which people associated their special personal expectations and interests. Automata professionals who came from other fields readily transferred their experience and expertise from algebra, mathematical logic, and even physiology to the theory of finite automata, or developed finite-automata techniques for other problems.

Kleene's regular expressions made evident that automata can be regarded as certain special algebraic systems, and that it is possible to study them from an algebraic point of view. The principal exponents of these ideas in the SU were Victor M. Glushkov and his disciples, especially Alexander A. Letichevski, Vladimir N. Red'ko, Vladimir G. Bodnarchuk. They advocated also the use of regular expressions as a primary specification language for the synthesis of automata. Later, adherents of this trend in the SU and abroad developed a rich algebraic oriented theory of languages and automata (see [RS97]).

Counterbalancing this "algebra of languages" philosophy, I followed a "logic of operators" view on the subject, suggested by A. Burks and J. Wright. In [BW53] they focused on the input-output behaviour of logical nets, i. e. on operators that convert input words in output words of the same length, and infinite input sequences into infinite output sequences. [2] Apparently, they were the first to study infinite behaviour of automata with output, and to (implicitly) characterize input-output operators in terms of retrospection and memory. Furthermore, they considered Logical Nets as the basic form of interaction between input-output agents.

To summarize, Burks and Wright suggested the following ideas I adopted and developed in my further work on the subject:

1. Priority of semantical considerations over (premature) decisions concerning specification formalisms.
2. Relevance of infinite behaviour; hence, ω-sequences as an alternative to finite words.
3. The basic role of operators as an alternative to languages.

According to those ideas, I focused on two set-theoretical approaches to the characterization of favorite operators and ω -languages (i. e. sets of ω-sequences).

[2] Compare with D. Scott's argumentation in [S67]: "The author (along with many other people) has come recently to the conclusion that the functions computed by the various machines are more important - or at least more basic - than the sets accepted by these devices. The sets are still interesting and useful, but the functions are needed to understand the sets. In fact by putting the functions first, the relationship between various classes of sets becomes much clear. This is already done in recursive function theory and we shall see that the same plan carriers over the general theory".

The first is in terms of memory; hence, operators and languages with finite memory. The second one, follows the spirit of descriptive set theory (DST), and selects operators and ω-languages by appropriate metrical properties and set-theoretical operations. (Note that the set of all ω-sequences over a given alphabet can be handled as a metrical space with suitably chosen metrics.)

My first reaction to the work of Burks and Wright was [T57], submitted in 1956, even before the collection "Automata Studies" was available. A footnote added in proof mentions: "the author learned about Moore's paper in [AS], whose Russian translation is under print".

The paper [T57] deals with operators, and distinguishes between properties related to retrospection, which is nothing but a strong form of continuity, and those related to finite memory. In [T62] a class of finite-memory ω-languages is defined which is proved to contain exactly those ω-languages, that are definable in second order monadic arithmetic. Independently Buchi found for them a characterization in terms of the famous "Büchi automata". In the paper [T58] I started my main subject – synthesis of automata, developed later in the books [KT62] and [TB70].

Experiments and Formal Specifications

Usually, verbal descriptions are not appropriate for the specification of input-output automata. Here are two alternative approaches.

Specification by examples. This amounts to assembling a table which indicates for each input word x, belonging to some given set M, the corresponding output word z. Further, the synthesis of the automaton is conceived as an interpolation, based on that table. This approach was very popular among soviet practitioners, and suggested the idea of algorithms for automata-identification. Such an algorithm should comprise effective instructions as to: 1) what questions of the type "what is the output of the black box for input x?" should be asked; 2) how the answers to these questions should be used to ask other questions, and 3) how to construct an automaton which is consistent with the results of the experiment.

In his theory of experiments [AS56] Moore proved that the behavior of an automaton with k states can be identified (restored) by a multiple experiment of length $2k-1$. Independently, I established in [T57] the same result, and used it in [KT62] to identify automata, with an a priori upper bound of memory. I conjectured also in [T57] that the restorability degree of "almost" all automata is of order $\log k$, i. e. essentially smaller than $2k-1$. This conjecture was proved by Barzdins and Korshunov [TB70]. Barzdin developed also frequency identification algorithms [TB70] which produce correct results with a guaranteed frequency, even when there is no apriori upper bound of the memory. The complexity estimation for such algorithms relies on the proof of the $\log k$ conjecture. Later Barzdin and his group in Riga significantly developed these ideas into a comprehensible theory of inductive learning.

Formal Specifications. The second approach, initiated by S. C. Kleene in [AS56] amounts to designing special specification formalisms, which suitably use logical connectives. However the use of only propositional connectives runs into difficulties, because they cannot express temporal relationships.

Actually, Kleene's paper in [AS56] contains already some hints as to the advisability and possibility of using formulas of the predicate calculus as temporal specifications. Moreover, Church attributes to Kleene the following *Characterization Problem* (Quotation from [Ch62]): "Characterize regular events directly in terms of their expression in a formalized language of ordinary kind, such as the usual formulations of first or second order arithmetic. "

Towards logical specifications

The years 1956-61 marked a turning point in the field and Church reported about that on the 1962-International Mathematical Congress. Here is a quotation from [Ch62] "This is a summary of recent work in the application of mathematical logic to finite automata, and especially of mathematical logic beyond the propositional calculus".

Church's lecture provides a meticulous chronology of events (dated when possible up to months) and a benevolent comparison of his and his student J. Friedman's results with work done by Büchi, Elgot and myself. Nevertheless, in the surveyed period (1956-62) the flow of events was at times too fast and thus omission prone. That is why his conclusion: "all overlaps to some extent, though more in point of view and method than in specific content" needs some reexamination. Actually, the reference to Büchi's paper [Bu62] as well as the discussion of my papers [T58], [T61] were added only "in proof" to the revised edition of the lecture (1964). My other Russian papers [T61b], [T62] were still unknown to Church at that time.

Independently, myself, [T58] and somewhat later A. Church [Ch59], developed languages based on the second order logic of monadic predicates with natural argument. Subsequently another variant was published by R. Büchi [Bu60] In those works the following restrictions were assumed: [T58]: restricted first order quantification; [Ch59]: no second order quantification; [Bu60]: restriction to predicates that are true only on a finite set of natural numbers.

All these languages are particular cases of a single language, widely known now as S1S - Second Order Monadic Logic with One Successor, in which all the restrictions above are removed.

Various arguments can be given in favor of choosing one language or another, or developing a new language. Nevertheless, two requirements seem to be quite natural: The first one (expressiveness) represents the interest of the client, making easier for him the formulation of his intention. The second requirement reflects the viewpoint of the designer; there must be a (fairly simple?) algorithm for the synthesis problem in the language.

These two requirements are contradictory. The more comprehensive and expressive the language, the more universal and so more complex is the algorithm. Moreover, if the language is too comprehensive the required algorithm may not exist at all. It turned out that the choice of S1S supports the demand of expressiveness and still guarantees a synthesis algorithm. Indeed, one can show, that all other known specification formalisms can be embedded naturally into S1S. However, this process is in general irreversible.

Synthesis

Church's lecture focuses on four problems, namely: 1) simplification, 2) synthesis, 3) decision, 4) Kleene's Characterization Problem. [3]

Problem 2, better known as the Church-synthesis problem, amounts roughly to the following: Given a S1S-formula $A(x, y)$: a) Does there exist an automaton M with input x and output y, whose behaviour satisfies $A(x, y)$? b) If the answer is "yes", then construct such an automaton. Solutions are algorithms which provide the correct answers and/or constructions.

Problem 4 presumes the invention of a logical formalism L (actually a rich sublanguage of S1S), which expresses exactly the operators (or events) definable by finite automata, and is equipped with two translation algorithms: (i) from formulas to automata (Kleene-synthesis) and (ii) from automata to formulas (Kleene-analysis).

According with the above classification, [KT62] deals with Kleene-synthesis and Kleene-analysis. Actually, in [KT62] we used the following three formalisms to specify input-output operators:[4] 1) at the highest level – formulas of S1S); 2) at the intermediate level – finite input-output automata represented by their canonical equations; 3) at the lower level – logical nets.

Correspondingly, we dealt there with both behavioral synthesis (from 1 to 2) and with structured synthesis (from 2 to 3).

Büchi was the first to use automata theory to logic and proved [Bu62] that S1S is decidable. These achievements, notwithstanding the general Church-synthesis problem for specifications in full S1S, remained open, not counting a few special classes of S1S-formulas, for which the problem was solved by Church and myself (see [Ch57] and [T61a]). The game theoretic interpretation of Church-synthesis is due to Mc. Naughton [Mc65]. R. Büchi and L. Landweber used this interpretation to solve the general Church-synthesis problem. Note that the original proof in [Mc65] was erroneous. Unfortunately I did not detect this error, which was reproduced in the Russian edition of [TB70], and corrected later by L. Landweber in the English translation.

Part 1 of the book [TB70] constitutes a revised version of my lectures at Novosibirsk University during the spring semester of 1966; it summarizes the results of Church, Büchi-Landweber, Mc. Naughton and myself, as explained above. Part two, written by Barzdin, covers his results on automaton identification.

About The Trinity

The choice of the three formalisms in [KT62] is the result of two decisions. The first identifies three levels of specifications; one can refer to them respectively as the declarative, executable and interactive levels. The second chooses

[3] Of course there is also the problem of *efficiency*: estimate and improve the complexity of the algorithms and/or the succinctness of the results they provide.

[4] *Note that in [KT62] regular expressions are not considered!*

for each of these levels a favorite formalism. In [KT62] those were, respectively, S1S-Formulas, Automata and Logical Nets; these three are collectively called "The Trinity" in [T95]. The first decision is more fundamental, and is recognizable also in computational paradigms beyond finite automata. The second decision is flexible even for finite automata; for example, the Trinity does not include Regular Expression (in [KT62], they are not even mentioned!) After Pnueli's seminal work, Linear Temporal Logics (LTL) became very popular as a declarative formalism. But note that various versions of LTL are in fact just the friendly syntactical sugar of S1S-fragments, and that the most extended one, called ETL, has the same expressive power as the whole S1S. In this sense one can argue that S1S is the genuine temporal logic, and that the Trinity has a basic status. Moreover, recent computational paradigms are likely to revive interest in the original Trinity and its appropriate metamorphosis.

Complexity

Entering the field

In 1960, I moved to the Akademgorodok, the Academic Center near Novosibirsk, where, through the initiative and guidance of Lyapunov, the Department of Theoretical Cybernetics was established within the Mathematical Institute.

I continued to work on automata theory which I had begun at Penza, at first, focusing mainly on the relationship between automata and logic, but also doing some work in structural synthesis [T64]. At that time automata theory was quite popular, and that is what brought me my first Ph. D. students in Novosibirsk: M. Kratko, Y. Barzdin, V. Nepomnyashchy.

However, this initial interest was increasingly set aside in favor of computational complexity, an exciting fusion of combinatorial methods, inherited from switching theory, with the conceptual arsenal of the theory of algorithms. These ideas had occurred to me earlier in 1956 when I coined the term "signalizing function" which is now commonly known as computational complexity measure. (But note that "signalizing" persisted for a long time in Russian complexity papers and in translations from Russian, puzzling English-speaking readers.) In [T56] the question was about arithmetic functions f specified by recursive schemes R. I considered there the signalizing function that for a given scheme R and nonnegative x, returns the maximal integer used in the computation of $f(x)$ according to R. As it turned out, G. S. Tseytin, then a 19-year old student of A. A. Markov at Leningrad University, began in 1956 to study time complexity of Markov's normal algorithms. He proved nontrivial lower and upper bounds for some concrete tasks, and discovered the existence of arbitrarily complex 0-1 valued functions (Rabin's 1960 results became available in the SU in 1963). Unfortunately, these seminal results were not published by Tseitin; later, they were reported briefly (and without proofs) by S. A. Yanovskaya in the survey [Ya59].

Because of my former background, my interest in switching theory, automata, etc. it never meant a break with mathematical logic and computability. In fact, the sixties marked a return to those topics via research in complexity of computations.

I profited from the arrival of Janis M. Barzdins and Rusins V. Freivalds in Novosibirsk as my postgraduate students. These two, both graduates of the Latvian University in Riga, engaged actively and enthusiastically in the subject. Alexey V. Gladkiy and his group in mathematical linguistics became also interested in complexity problems, concerning grammars and formal languages. Soon other people joined us, mainly students of the Novosibirsk University. My seminar "Algorithms and Automata" was the forum for the new complexity subjects, and hosted often visitors from other places. This is how research in computational complexity started in Novosibirsk; a new young generation arose, and I had the good fortune to work with these people over a lengthy period.

Subsequently I joined forces with A. V. Gladkiy in a new department of our Mathematical Institute, officially called the Department of Automata Theory and Mathematical Linguistics. Its staff in different periods included our former students Mikhail L. Degtyar, Mars K. Valiev, Vladimir Yu. Sazonov, Aleksey D. Korshunov, Alexander Ya. Dikovski, Miroslav I. Kratko and Valeriy N. Agafonov (1943-1997).

The basic computer model we used was the Turing machine with a variety of complexity measures; for example, besides time and space, also the number of times the head of the machine changes its direction. Along with deterministic machines we considered also nondeterministic machines, machines with oracles, and probabilistic machines.

It is not surprising that we were attracted by the same problems as our colleagues in the West, notably - as J. Hartmanis and R. Stearns. Independently and in parallel we worked out a series of similar concepts and techniques: complexity measures, crossing sequences, diagonalization, gaps, speed-up, relative complexity, to cite the most important ones.

Blum's machine independent approach to complexity was new for us, and it aroused keen interest in our seminar. But, when later, at a meeting with Tseitin, I began telling him about Blum's work, he interrupted me almost at once and proceeded to set forth many basic definitions and theorems. As it turned out, he had realized it for some time already, but had never discussed the subject in public!

My "gap" theorem [T67] was stimulated by Blum's theory. It illustrated a set of pathological time-bounding functions which need to be avoided in developing complexity theory. Meyer and McCreight's "Honesty Theorem" [McM69] showed how this can be done through the use of appropriate "honest" functions. (Note that often, instead of [T67], the wrong reference is given to my paper [T64a], which deals with gaps in the context of crossing-sequences.)

In 1967, I published a set of lecture notes [T67] for a course "Complexity of Algorithms and Computations" that I had given in Novosibirsk. The notes contained an exposition of results of Blum and Hartmanis-Stearns, based on their published papers, as well as results of our Novosibirsk group: the "gap" theorem, the crossing sequences techniques developped by Barzdin and myself, and other results reported on our seminar.

I sent a copy of these notes to M. Blum (by then at Berkeley). Further I am quoting Albert Meyer [M84]: "Blum passed on a copy of the Trakhtenbrot notes to me around 1970 when I was at MIT since I knew of a graduate student who was interested in translating them. His work was not very satisfactory, but then Filloti came to MIT to work as a Post-Doc with me and did a respectable job. By this time the notes began to seem outdated (about five years old in 1972!) and I decided that they needed to be revised and updated. This youthful misjudgment doomed the project since I was too impatient and perfectionist to complete the revision myself, and the final editing of the translation was never completed".

In the academic year 1970-71, V. N. Agafonov continued my 1967-course, and published the lecture notes [Ag75] as Part 2 of "Complexity of Algorithms and Computations". But, unlike Part 1, which focused on complexity of computations (measured by functions) Part 2 was dedicated to descriptive complexity of algorithms (measured by numbers). It contained a valuable exposition of the literature around bounded Kolmogorov complexity and pseudo-randomness, including contributions of Barzdin and of Valery himself.

Towards applications

In the SU it was fully in the tradition of the theory of algorithms to handle applications of two kinds: (i) proving or disproving decidability for concrete problems, (ii) algorithmic interpretation of mathematical concepts (for example, along the line of constructive analysis in the Markov School). So, it seemed natural to look for similar applications in the complexity setting.

The attitude of the "classical" cybernetics people, (notably of Yablonski) to the introduction of the theory of algorithms into complexity affairs was quite negative. The main argument they used was that the theory of algorithms is essentially a theory of diagonalization, and is therefore alien to the complexity area that requires combinatorial constructive solutions. And indeed, except some simple lower bounds supported by techniques of crossing sequences, all our early results rested on the same kind of "diagonalization" with priorities, as in classical computability theory.

But whereas in Algebra and Logic there were already known natural examples for undecidability phenomena which were earlier analyzed in the classical theory, no natural examples of provable complexity phenomena were known. This asymmetry was echoed by those who scoffed at the emptiness of the diagonal techniques with respect to applications of complexity theory. In particular, they distrusted the potential role of algorithm based complexity in the explanation of perebor phenomena, and insisted on this view even after Kolmogorov's new approach to complexity of finite objects.

In the summer of 1963, during a visit by A. N. Kolmogorov to the Novosibirsk University, I learned more about his new approach to complexity and the development of the concepts of information and randomness by means of the theory of algorithms. In the early cybernetics period it was already clear that the essence of problems of minimization of boolean functions was not in the particular models of switching circuits under consideration. Any other natural class of 'schemes', and ultimately any natural coding of finite objects (say, finite texts) could be expected to exhibit similar phenomena, and, in particular, those related to perebor. But, unlike former pure combinatorial approaches, the discovery by Kolmogorov (1965), and independently by Solomonoff (1964) and Chaitin (1966), of optimal coding for finite objects occurred in the framework of algorithm and recursive function theory. (Note that another related approach was developed by A. A. Markov (1964) and V. Kuzmin (1965).)

Algorithms and randomness

I became interested in the correlation between these two paradigms back in the fifties, when P. S. Novikov called my attention to algorithmic simulation of randomness in the spirit of von Mises-Church strategies. Ever since, I have returned to this topic at different times and for various reasons, including the controversies around perebor. Since many algorithmic problems encounter essential difficulties (non existence of algorithms or non existence of feasible ones), the natural tendency is to use devices that may produce errors in certain cases. The only requirements are that the probability or frequency of the errors does not exceed some acceptable level and that the procedures are feasible. In the framework of this general idea, two approaches seemed to deserve attention: probabilistic algorithms and frequential algorithms.

In the academic year 1969-70 I gave a course "Algorithms and Randomness" which covered these two approaches, as well as algorithmic modelling of Mises-Church randomness.

The essential features of a frequential algorithm M are generally as follows:

1. M is deterministic, but each time it is applied, it inputs a whole suitable sequence of inputs instead of an individual one, and then produces the corresponding sequence of outputs.
2. The frequency of the correct outputs must exceed a given level.

The idea of frequency computations is easily generalized to frequency enumerations, frequency reductions, etc.

I learned about a particular such model from a survey by Mc. Naughton (1961), and soon realized that as in the probabilistic case, it is impossible to compute functions that are not computable in the usual sense.

Hence, the following questions:

1. Is it possible to compute some functions by means of probabilistic or frequential algorithms with less computational complexity than that of deterministic algorithms?
2. What reasonable sorts of problems (not necessarily computation of functions) can be solved more efficiently by probabilistic or frequential algorithms than by deterministic ones?
3. Do problems exist that are solvable by probabilistic or frequential algorithms but not by deterministic algorithms?

These problems were investigated in deep by Barzdin, Freivalds and their students.

Relativized complexity

Computations with oracles are a well established topic in the Theory of Algorithms, especially since Post's classical results and the solution of his famous problem by Muchnik and Friedberg. So it seemed to me quite natural to look how such issues might be carried to the complexity setting. At this point I should mention that Meyer's confession, about the translation of my lecture notes, points only on a transient episode in our long-time contacts. Let me quote again Albert [M84]: "Repeatedly and independently our choices of scientific subareas, even particular problems, and in one instance even the solution to a problem, were the same. The similarity of our tastes and techniques was so striking that it seemed at times there was a clairvoyant connection between us. Our relationship first came about through informal channels – communications and drafts circulated among researchers, lecture notes, etc. These various links compensated for the language barrier and the scarcity of Soviet representation at international conferences. Through these means there developed the unusual experience of discovering an intellectual counterpart, tackling identical research topics, despite residing on the opposite side of the globe ... Today ... we find ourselves collaborating firsthand in an entirely different area of Theoretical Computer Science than complexity theory to which we were led by independent decisions reflecting our shared theoretical tastes". (End of quotation)

As to computations with oracles, we both were attracted by the question: to what extent can be simplified a computation by bringing in an oracle, and how accurately can the reduction of complexity be controlled depending on the choice of the oracle? This was the start point for a series of works of our students (mainly M. Degtyar, M. Valiev in Novosibirsk and N. Lynch at MIT) with similar results of two types: about oracles which do help (including the estimation of the help) and oracles which cannot help. The further development of the subject by A. Meyer and M. Fischer ended with a genuine complexity-theoretic analog to the famous Friedberg-Muchnik theorem. It reflects the intuitive idea that problems might take the same long time to solve but for different reasons! Namely: *There exist nontrivial pairs of (decidable!) sets, such that neither member of a pair helps the other be computed more quickly.*

Independently of Meyer and Fischer, and using actually the same techniques, I obtained an improvement of this theorem. That happened in the frame of my efforts to use relative algorithms and complexity in order to formalize intuitions about mutual independence of tasks and about perebor.

Formalizing intuitions

Autoreducibility. When handling relativized computations it is sometimes reasonable to analyze the effect of restricted access to the oracle. In particular, this is the case with the algorithmic definition of "collectives", i. e. of random sequences in the sense of von Mises-Church. This definition relies on the use of "selection strategies", which are relative algorithms with restricted access to oracles. A similar situation arises with the intuition about mutual independence of individual instances which make up a general problem [T70a]. Consider, for example, a first order theory T. It may well happen that there is no algorithm, which, for an arbitrary given formula $\mathcal{A}$, decides whether $\mathcal{A}$ is provable or not in T. However, there is a trivial procedure W which reduces the question about $\mathcal{A}$ to similar questions for other formulas; W just inquires about the status of the formula $(\neg(\neg\mathcal{A}))$. The procedure W is an example of what may be called *autoreduction.* Now, assume that the problem is decidable for the theory T, and hence the correct answers can be computed directly (without autoreduction). It still might happen that one cannot manage without very complex computations, whereas the autoreduction above is simple.

A guess strategy is a machine M with oracle, satisfying the condition: for every oracle G and natural number n, the machine M, having been started with n as input, never addresses the oracle with the question "$n \in G$?" (although it may put any question "$\nu \in G$?" for $\nu \neq n$). A set G is called *autoreducible* if it possesses an autoreduction, i. e. a guess strategy which, having been supplied with the oracle G, computes the value $G(n)$ for every n. Otherwise G is *nonautoreducible*, which should indicate that the individual queries "$n \in G$?" are mutual independent.

It turned out that:

1. The class of nonautoreducible sequences is essentially broader than the class of random sequences.
2. There are effectively solvable mass problems M of arbitrary complexity with the following property: autoreductions of M are not essentially less complex than their unconditional computations.

Understanding perebor. Disputes about perebor, stirred by Yablonski's paper [Y59], had a certain influence on the development, and developers of complexity theory in the SU. By and large, reflections on perebor activated my interest in computational complexity and influenced my choice of special topics, concerning the role of sparse sets, immunity, oracles, frequency algorithms, probabilistic algorithms, etc. I told this story in details in [T84]; below I will reproduce a small fragment from [T84].

The development of computational complexity created a favorable background for alternative approaches to the perebor topics: the inevitability of perebor should mean the nonexistence of algorithms that are essentially more efficient. My first attempt was to explain the plausibility of perebor phenomena related to the "frequential Yablonski-effect"; it was based on space complexity considerations. Already at this stage it became clear that space complexity was too rough and that time complexity was to be used. Meanwhile I began to feel that another interpretation of perebor was worth considering, namely, that the essence of perebor seemed to be in the complexity of interaction with a "checking mechanism", as opposed to the checking itself. This could be formalized in terms of oracle machines or reduction algorithms as follows. Given a total function f that maps binary strings into binary strings, consider Turing machines, to compute f, that are equipped with the oracle G that delivers (at no cost!)

the correct answers to queries "$f(x) = y$?" (x, y may vary, but f is always the same function). Among them is a suitable machine $M_{perebor}$ that computes $f(x)$ by subsequently addressing the oracle with the queries

$$f(x) = B(0)?, f(x) = B(1)?, \ldots, f(x) = B(i)? \ldots$$

where $B(i)$ is the ith binary string in lexicographical order. Hence, in the computation of the string $f(x)$ the number of steps spent by $M_{perebor}$ is that represented by the string $f(x)$. I conjectured in 1966 that for a broad spectrum of functions f, no oracle machine M can perform the computation essentially faster. As for the "graph predicates" $G(x, y)$ =def $f(x) = y$, it was conjectured that they would not be too difficult to compute. By this viewpoint, the inevitability of perebor could be explained in terms of the computational complexity of the reduction process. The conjecture was proved by M. I. Degtyar in his Master's thesis [1969] for different versions of what "essentially faster" should mean. Using modern terminology, one can say that Degtyar's construction implicitly provides the proof of the relativized version of the NP $\neq$ P conjecture. For the first time, this version was explicitly announced by Baker, Gill and Solovay (1975) together with the relativized version of the NP = P conjecture. Their intention was to give some evidence to the possibility that neither NP = P nor NP $\neq$ P is provable in common formalized systems. As to my conjecture it had nothing to do with the ambitious hopes to prove the independence of the NP = P conjecture. As a matter of fact, I then believed (and to some extent do so even now) that the essence of perebor can be explained through the complexity of relative computations based on searching through the sequence of all binary strings. Hence, being confident that the true problem was being considered (and not its relativization!), I had no stimulus to look for models in which perebor could be eliminated.

To the perebor account [T84] it is worth adding the following quotations from my correspondence with Mike Sipser (Feb. 1992).

S. You write that Yablonski was aware of perebor in the early 50's, and that he even conjectured that perebor is inevitable for some problems in 1953-54. But the earliest published work of Yablonski that you cite is 1959. Is there a written publication which documents Yablonski's awareness of these issues at the earlier time? This seems to be an important issue, at least from the point of establishing who was the first to consider the problem of eliminating brute force search. Right now the earliest document I have is Gödel's 1956 letter to Von-Neumann.

T. I cannot remember about any publication before 1959 which documents Yablonski's awareness of these issues but I strongly testify and confirm that (a quotation follows from my paper [T65]): *"Already in 1954 Yablonski conjectured that the solution of this problem is in essence impossible without complex algorithms of the kind of perebor searching through all the versions ..."*

He persistently advocated this conjecture on public meetings (seminars and symposia).

S. Second, is it even clear that Yablonski really understands what we presently mean by eliminating brute force search? He claimed to have proven that it could not be eliminated in some cases back in 1959. So there must be some confusion.

T. That is indeed the main point I am discussing in Section 1 of my *perebor* paper [T84]. The conclusion there is that there is no direct connection between Yablonski's result and what we presently mean by eliminating *perebor*. Hence the long year controversy with Yablonski.

S. I'd appreciate your thoughts on how to handle Yablonski's contribution to the subject.

T. I would mention three circumstances:

1. In Yablonski's conjecture the notion of *perebor* was a bit vague and did not anticipate any specific formalization of the idea of *complexity*. Nevertheless (and may be just due to this fact) it stimulated the investigation of *different approaches* to such a formalization, at least in the USSR.
2. Yablonski pointed from the very beginning on very attractive candidates for the status of problems which need essentially *perebor*. See Section 1 of [T84], where synthesis of circuits is considered in this context.
3. Finally, he made the point that for his candidates the disaster caused by *perebor* might be avoided through the use of probabilistic methods.

...let me mention that as an alternative to Yablonski's approach I advocated the idea of complexity of computations with oracles. In this terms I formulated a conjecture which presently could be interpreted as the relativised version of P not equal NP. This conjecture was proved by my student M. Degtyar [D69]. (End of quotations)

Turning points

The controversies around perebor were exacerbated by the emergence of the new approach to complexity of algorithms and computations. And it was precisely this approach which was relevant for the genuine advance in the investigation of perebor in the seminal works of Leonid Levin in the SU and the Americans, Steven Cook and Richard Karp.

The discovery of NP-complete problems gave evidence to the importance of the Theory of Computational Complexity. Soon another prominent result strengthened this perception. In 1972 A. Meyer and Stockmeyer (see [M73]) found the first genuine natural examples of inherently complex computable problems. This discovery was particularly important for me because the example came from the area of automata theory and logic in which I had been involved for a long time. Clearly, for the adherents of the algorithmic approach to complexity, including myself, these developments confirmed the correctness of their views on the subject and the worthwhileness of their own efforts in the past. However the time had also come for new research decisions, inspired by the developments in semantics, verification, lambda calculus and schematology. But that is another story!

Epilogue

For a long time I was not actively involved in automata and computational complexity, being absorbed in other topics. During that period both areas underwent impressive development, which is beyond the subject of this account.

My entry into the field happened at an early stage, when formation of concepts and asking the right questions had high priority, at least as solving well established problems. This is also reflected in my above exposition in which the emphasis was rather on the conceptual framework in the area. Some of those concepts and models occurred in very specific contexts, or were driven by curiosity rather than by visible applications. Do they make sense beyond their first motivation? I would like to conclude with some remarks about this.

The first is connected to timed automata and hybrid systems (HS). Nowadays the area is still dominated by an explosion of models, concepts and ad hoc notation, a reminder of the situation in automata theory in the fifties. It seems that the "old" conceptual framework can still help to elucidate the underlying computational intuition and to avoid the reinvention of existing ideas.

In order to properly adapt that conceptual framework and also to be aware about its limitations, it may be convenient to start with two separate and orthogonal extensions of the basic model of a finite automaton M. The first one is by interconnecting M with an oracle N, which is also an automaton, but, in general, with an infinite set of states. Whatever M can do while using N is called its relativization with respect to this oracle. The other extension is with continuous time (instead of discrete, as in the classical case), but without oracles.

For each of these extensions apart, it becomes easier to retrace the impact of classical automata theory and logic (see [RT97]). An appropriate combination of the two extensions might facilitate the formalization of hybrid systems and the adaptation of the classical heritage, whenever it makes sense.

The next remark is about a resurgence of interest in autoreducibility and frequency computations.

It was instructive to learn that the idea of restricting access to oracles, now underlies several concepts, which are in fact randomized and/or time bounded versions of autoreducibility: coherence, checkability, selfreducibility, etc. Most of these concepts were identified independently from (though later than) my original autoreducibility, and have occupied a special place in connection with program checking and secure protocols (see [BF92] for details and references).

On the other hand, the idea of frequency computation was extended to bounded query computations and parallel learning. Also interesting relationships were discovered between autoreducibility, frequency computations and various other concepts.

A final remark about the continuous conceptual succession since my youthful exercises in descriptive set theory, which I tried to emphasize in my previous exposition. In particular, it is quite evident that computational complexity is inspired by computability. But the succession can be traced back even to descriptive set theory; just keep in mind the ideas which lead from the classification of sets and functions to the classification of what is computable, and ultimately to hierarchies within computational complexity.

Acknowledgement. The help of Mrs. Diana Yellin in editing and formatting the text is gratefully appreciated.

References

[Ag75] Agafonov V. N. Complexity of algorithms and computations (part 2), Lecture Notes, Novosibirsk State University, 1975, 146 p.

[AS56] Automata Studies, Princeton, 1956, Edited by J. McCarthy and Claude Shannon.

[BF92] Beigel R. , Feigenbaum J. On being incoherent without being very hard, Computational Complexity, 2, 1-17 (1992).

[Bu62] Büchi R. , On a decision method in restricted second order arithmetic, Proc. of the 1960 Intern. Congr. on Logic, Philosophy and Methodology of Sciences, pp. 1-11, Stanford Univ. Press, 1962.

[BW53] Burks, A. , Wright J. , Theory of logical nets, Proc. IRE, 41(4) (1953).

[D69] Dekht'ar M. I. , The impossibility of eliminating complete search in computing functions from their graphs, DAN SSSR 189, 748-751 (1969).

[F79] Freivalds R. V. , Fast probabilistic algorithms, LNCS 74, 57-69 (1979).

[G62] Glushkov V. M. , Synthesis of digital automata, Fizmatgiz, Moscow, 1962.

[His70] History of Mathematics, Kiev 1970, edited by I. Z. Shtokalo (Russian).

[KT62] Kobrinski N. E. , Trakhtenbrot B. A. , Introduction to the Theory of Finite Automata, Fizmatgis, Moscow 1-404 (1962), English translation in "Studies in Logic and the Foundations of Mathematics" in North-Holland (1965).

[KS93] Kummer M. , Stephan F. , Recursion theoretic properties of frequency computations and bounded queries (1993), 3rd K. Gödel Colloquium, 1993, LNCS.

[L65] Lupanov O. B. , An approach to Systems Synthesis – A Local Coding Principle, Problems of Cybernetics, 14, 31-110 (1965) (Russian).

[M73] Meyer A. R. , Weak monadic second order theory of successor is not Elementary recursive, Proj. MAC, MIT, (1973).

[M84] Meyer A. R. , Unpublished Memo.

[McM69] McCreight E. M. , Meyer A. R. , Classes of computable functions defined by bounds on computation, 1st STOC, 79-88 (1969).

[RS97] Rozenberg G. and Salomaa A. (eds), Handbook of Formal Languages, I-III, Springer-Verlag, Berlin (1997).

[S67] Scott D. , Some definitional suggestions in automata theory, J. of Computer and Syst. Sci. , 187-212 (1967).

[T50] Trakhtenbrot B. A. , The Impossibility of an algorithm for the decidability problem on finite classes, Doklady AN SSR 70, No. 4, 569-572 (1950).

[T57] Trakhtenbrot B. A. , On operators, realizable by logical nets, Doklady AN SSR 112, No. 6, 1005-1006 (1957).

[T57a] Trakhtenbrot B. A. , Algorithms and computing machines, Gostechizdat (1957), second edition by Fizmatgiz (1960), English translation in the series "Topics in Mathematics", D. C. Heath and Company, Boston 1-101 (1963).

[T58] Trakhtenbrot B. A. , The synthesis of logical nets whose operators are described in terms of monadic predicates, Doklady AN SSR 118, No. 4, 646-649 (1958).

[T59] Trakhtenbrot B. A. , The asymptotic estimate of the logical nets with memory, Doklady AN SSR 127, No. 2, 281-284 (1959).

[T61] Trakhtenbrot B. A. , Some constructions in the monadic predicate calculus, Doklady AN SSR 138, No. 2, 320-321 (1961).

[T61a] Trakhtenbrot B. A. , Finite automata and the monadic predicate calculus, Doklady AN SSR 140, No. 2, 326-329 (1961).

[T62] Trakhtenbrot B. A. , Finite automata and the monadic predicate calculus, Siberian Mathem. Journal 3, No. 1, 103-131 (1962).

[T63] Trakhtenbrot B. A. , On the frequency computation of recursive functions, Algebra i Logika, Novosibirsk 1, No. 1, 25-32 (1963).

[T64] Trakhtenbrot B. A. , On the complexity of schemas that realize many-parametric families of operators, Problemy Kibernetiki, Vol. 12, 99-112 (1964).

[T64a] Trakhtenbrot B. A. , Turing computations with logarithmic delay, Algebra i Logika, Novosibirsk 3, No. 4, 33-48 (1964).

[T65] Trakhtenbrot B. A. , Optimal computations and the frequency phenomena of Yablonski, Algebra i Logika, Novosibirsk 4, No. 5, 79-93 (1965).

[T66] Trakhtenbrot B. A. , On normalized signalizing functions for Turing computations, Algebra i Logika, Novosibirsk 5, No. 6, 61-70 (1966).

[T67] Trakhtenbrot B. A. , The complexity of algorithms and computations, Lecture Notes, ed. by Novosibirsk University 1-258 (1967).

[TB70] Trakhtenbrot B. A. , Barzdin Ja. M. , Finite Automata (Behavior and Synthesis), Nauka, Moscow 1-400 (1970), English translation in "Fundamental Studies in Computer Science 1", North-Holland (1973).

[T70a] Trakhtenbrot B. A. , On autoreducibility, Doklady AN SSR 192, No. 6, 1224-1227 (1970).

[T73c] Trakhtenbrot B. A. , Formalization of some notions in terms of computation complexity, Proceedings of the 1st International Congress for Logic, Methodology and Philosophy of Science, Studies in Logic and Foundations of Mathematics, Vol. 74, 205-214 (1973).

[T74a] Trakhtenbrot B. A. , Notes on the complexity of probabilistic machine computations, In "Theory of Algorithms and Mathematical Logic", ed. by the Computing Center of the Academy of Sciences 159-176 (1974).

[T84] Trakhtenbrot B. A. , A survey of Russian approaches to perebor (Brute-Force Search) Algorithms, Annals of the History of Computing 6(4), 384-400 (1984).

[T85] Trakhtenbrot B. A. , Selected Developments in Soviet Mathematical Cybernetics, Monograph Series, sponsored by Delphic Associated, Washington, XIV + 122 pages, 1985.

[T] rakhtenbrot B. A. , In memory of S. A. Yanovskaya (1896-1966) on the centenary of her birth, In "Research in History of Mathematics", Second Series, vol. 2 (37), pp. 109-127, Moscow, Russian Academy of Sciences. (English translation available as Tech. Report of the Computer Science Dept. , Tel-Aviv Univ. , 1997)

[Y59] Yablonski S. V. , Algorithmic difficulties in the synthesis of minimal contact networks, Problems of Cybernetics, vol. 2, Moscow, 1959, (Russian).

[Ya59] Yanovskaya, S. , Mathematical logic and fundamentals of mathematics, in Mathematics in the USSR for 40 years, Moscow, Fizmatgiz, 1959, pp. 13-120 (Russian).

Springer Series in

Discrete Mathematics and Theoretical Computer Science

Springer-Verlag Singapore's series in *Discrete Mathematics and Theoretical Computer Science* is produced in cooperation with the Centre for Discrete Mathematics and Theoretical Computer Science of the University of Auckland, New Zealand. This series brings to the research community information about the latest developments on the interface between mathematics and computing, especially in the areas of artificial intelligence, combinatorial optimization, computability and complexity, and theoretical computer vision. It focuses on research monographs and proceedings of workshops and conferences aimed at graduate students and professional researchers, and on textbooks primarily for the advanced undergraduate or lower graduate level.

For details of forthcoming titles, please contact the publisher at:

Springer-Verlag Singapore Pte. Ltd.
#04-01 Cencon I
1 Tannery Road
Singapore 347719
Tel: (65) 842 0112
Fax: (65) 842 0107
e-mail: rebec@cyberway.com.sg
http://www. springer.com.sg

Springer Series in
Discrete Mathematics and Theoretical Computer Science

D.S. Bridges, C.S. Calude, J. Gibbons, S. Reeves, I.H. Witten (Eds.), *Combinatorics, Complexity and Logic*. Proceedings, 1996. viii, 422 pages.

L. Groves, S. Reeves (Eds.), *Formal Methods Pacific '97*. Proceedings, 1997. Viii, 320 pages.

G.J. Chaitin, *The Limits of Mathematics: A Course on Information Theory and the Limits of Formal Reasoning*. 1998. xii, 148 pages.

C.S. Calude, J. Casti, M.J. Dinneen (Eds.), *Unconventional Models of Computation*. Proceedings, 1998. viii, 426 pages.

K. Svozil, *Quantum Logic*. 1998. xx, 214 pages.

J. Grundy, M. Schwenke, T. Vickers (Eds.), *International Refinement Workshop and Formal Methods Pacific '98*. Proceedings, 1998. viii, 381 pages

G. Pāun (Ed.), *Computing with Bio-Molecules: Theory and Experiments*. 1998. x, 352 pages

C.S. Calude (Ed.), *People and Ideas in Theoretical Computer Science*. 1998.vii, 341 pages